Innovative Applications of Nanowires for Circuit Design

Balwinder Raj
National Institute of Technical Teachers Training and Research, Chandigarh, India

A volume in the Advances in Computer and Electrical Engineering (ACEE) Book Series

IGI Global
PUBLISHER of TIMELY KNOWLEDGE

Published in the United States of America by
IGI Global
Engineering Science Reference (an imprint of IGI Global)
701 E. Chocolate Avenue
Hershey PA, USA 17033
Tel: 717-533-8845
Fax: 717-533-8661
E-mail: cust@igi-global.com
Web site: http://www.igi-global.com

Library of Congress Cataloging-in-Publication Data

Names: Raj, Balwinder, 1980- editor.
Title: Innovative applications of nanowires for circuit design / Balwinder Raj, editor.
Description: Hershey, PA : Engineering Science Reference, an imprint of IGI Global, [2021] | Includes bibliographical references and index. | Summary: "The goal of the book is to provide a platform for domain experts, researchers, practitioners, and students to exchange ideas on technologies and practical experiences related to innovations in current research on Nanowire and its applications"-- Provided by publisher.
Identifiers: LCCN 2020023239 (print) | LCCN 2020023240 (ebook) | ISBN 9781799864677 (hardcover) | ISBN 9781799864684 (paperback) | ISBN 9781799864691 (ebook)
Subjects: LCSH: Nanowires. | Electronic circuit design.
Classification: LCC TK7874.85 .I46 2021 (print) | LCC TK7874.85 (ebook) | DDC 621.3815--dc23
LC record available at https://lccn.loc.gov/2020023239
LC ebook record available at https://lccn.loc.gov/2020023240

This book is published in the IGI Global book series Advances in Computer and Electrical Engineering (ACEE) (ISSN: 2327-039X; eISSN: 2327-0403)

British Cataloguing in Publication Data
A Cataloguing in Publication record for this book is available from the British Library.

All work contributed to this book is new, previously-unpublished material.
The views expressed in this book are those of the authors, but not necessarily of the publisher.

For electronic access to this publication, please contact: eresources@igi-global.com.

Advances in Computer and Electrical Engineering (ACEE) Book Series

ISSN:2327-039X
EISSN:2327-0403

Editor-in-Chief: Srikanta Patnaik, SOA University, India

MISSION

The fields of computer engineering and electrical engineering encompass a broad range of interdisciplinary topics allowing for expansive research developments across multiple fields. Research in these areas continues to develop and become increasingly important as computer and electrical systems have become an integral part of everyday life.

The **Advances in Computer and Electrical Engineering (ACEE) Book Series** aims to publish research on diverse topics pertaining to computer engineering and electrical engineering. **ACEE** encourages scholarly discourse on the latest applications, tools, and methodologies being implemented in the field for the design and development of computer and electrical systems.

COVERAGE

- Power Electronics
- VLSI Design
- Electrical Power Conversion
- Computer Science
- Circuit Analysis
- Digital Electronics
- Computer Hardware
- Chip Design
- Sensor Technologies
- Analog Electronics

IGI Global is currently accepting manuscripts for publication within this series. To submit a proposal for a volume in this series, please contact our Acquisition Editors at Acquisitions@igi-global.com or visit: http://www.igi-global.com/publish/.

The Advances in Computer and Electrical Engineering (ACEE) Book Series (ISSN 2327-039X) is published by IGI Global, 701 E. Chocolate Avenue, Hershey, PA 17033-1240, USA, www.igi-global.com. This series is composed of titles available for purchase individually; each title is edited to be contextually exclusive from any other title within the series. For pricing and ordering information please visit http://www.igi-global.com/book-series/advances-computer-electrical-engineering/73675. Postmaster: Send all address changes to above address.

Titles in this Series

For a list of additional titles in this series, please visit:
http://www.igi-global.com/book-series/advances-computer-electrical-engineering/73675

Examining Quantum Algorithms for Quantum Image Processing
HaiSheng Li (Guangxi Normal University, China)
Engineering Science Reference • ©2021 • 336pp • H/C (ISBN: 9781799837992) • US $195.00

Design and Investment of High Voltage NanoDielectrics
Ahmed Thabet Mohamed (Aswan University, Egypt & Qassim University, Saudi Arabia)
Engineering Science Reference • ©2021 • 363pp • H/C (ISBN: 9781799838296) • US $195.00

Research Advancements in Smart Technology, Optimization, and Renewable Energy
Pandian Vasant (University of Technology Petronas, Malaysia) Gerhard Weber (Poznan University of Technology, Poland) and Wonsiri Punurai (Mahidol University, Thailand)
Engineering Science Reference • ©2021 • 407pp • H/C (ISBN: 9781799839705) • US $225.00

HCI Solutions for Achieving Sustainable Development Goals
Fariza Hanis Abdul Razak (Universiti Teknologi MARA (UiTM), Malaysia) Masitah Ghazali (Universiti Teknologi Malaysia (UTM), Malaysia) Murni Mahmud (International Islamic University Malaysia, Malaysia) Chui Yin Wong (Multimedia University, Malaysia) and Muhammad Haziq Lim Abdullah (Universiti Teknikal Malaysia, Melaka, Malaysia)
Engineering Science Reference • ©2021 • 300pp • H/C (ISBN: 9781799849360) • US $195.00

Innovations in the Industrial Internet of Things (IIoT) and Smart Factory
Sam Goundar (The University of the South Pacific, Fiji) J. Avanija (Sree Vidyanikethan Engineering College, India) Gurram Sunitha (Sree Vidyanikethan Engineering College, India) K Reddy Madhavi (Sree Vidyanikethan Engineering College, India) and S. Bharath Bhushan (Sree Vidyanikethan Engineering College, India)
Engineering Science Reference • ©2020 • 300pp • H/C (ISBN: 9781799833758) • US $225.00

For an entire list of titles in this series, please visit:
http://www.igi-global.com/book-series/advances-computer-electrical-engineering/73675

IGI Global
PUBLISHER of TIMELY KNOWLEDGE

701 East Chocolate Avenue, Hershey, PA 17033, USA
Tel: 717-533-8845 x100 • Fax: 717-533-8661
E-Mail: cust@igi-global.com • www.igi-global.com

Editorial Advisory Board

Table of Contents

Detailed Table of Contents

Despite various methods available theoretically and experimentally for material characterization, the ultrasonic characterization method has proven as one of important method because of its non-destructive evaluation (NDE) technique. This chapter explores the ultrasonic characterization of wide band gap semiconducting material SiC at nanoscale. In the present chapter, diameter dependent nonlinear elastic properties (second and third order elastic constants), thermophysical properties, and ultrasonic properties of single crystalline SiC-nanowires have been computed using Lennard-Jones potentials in the high temperature regime. The size dependent mechanical and ultrasonic properties of the SiC-nanowires and their mechanical stability has also been estimated using the higher order elastic constants of the material. The correlation between size dependent ultrasonic attenuation and thermophysical properties have been established for the SiC-nanowires. Obtained results are discussed in correlation with available theoretical/experimental results for application in nanodevices.

This chapter represents some essential aspects of nanowires and their transport properties. Scaling of MOSFETs becomes a huge problem for industries due to short channel effects (SCEs) and sub-threshold leakage current. So, nanowires become a good solution to SCEs due to their structure. This chapter is divided into three parts. The first part gives a brief introduction of nanowire and different materials that can replace Si (channel material) and SiO2 (oxide material) due to their superior performance over Si. In the second part, the device structure and device structural measurement is discussed. In the third part, transport properties are discussed. This chapter shows the behavior of nanowire on changing different device materials and device dimensions. Electrical characteristics of Si and III-V based nanowires FETs will be analyzed and compared. High-k dielectric as oxide material also helps in improving device performance. HfO2 shows improvement in device characteristics over SiO2 taken as an oxide material. Junctionless nanowire MOSFET has also been designed and analyzed.

Section 3
Nanowire-Based Circuits Design

This chapter presents the design and simulation analysis nanowire-based FET (NWFET) for best possible Ig-Vgs characteristics. A NWFET is a device in which channel is wire-like structure with diameter or lateral dimension in nanometer (10-9 m) range. Performance analysis has been done for various design and process parameters variation to propose optimized parameter for best performance. Although a lot of focus has been put on homogenous Si based NWFETs, there has been a rising interest in III-V NWFETs. This is mainly due to the excellent carrier transport properties are provided by these materials. NWFETs have ability to suppress SCEs and are also good in suppressing OFF-current (IOFF), because of gate all around (GAA) configuration. Secondly, NWFETs have large ON-current (ION) due to quasi one-dimensional (1D) conduction of NWs, and as a result of low carrier scattering, conduction of NWs-based devices is very large.

This chapter encompasses the gradual requirements, basic working principle, inbuilt physics, structural and functional characteristics, and applications of nanowires, especially that of semiconductor nanowires in depth. Today, research and development in material science and electronics going hand in hand have opened up numerous directions for the exploration and utilization of several unique semiconducting materials in the design of novel field-effect-transistors (FETs) in the nano-scale architecture. The performance results of the basic NWFETs structures and hetero-structures along with methods to organize nanowires in the form of arrays to fulfill the requirement of integration of devices and circuits are described in detail. This chapter would be beneficial for students of undergraduate and postgraduate, researchers, and the industrial peoples as well who are working in the regime of the advancement of semiconductor technology because every aspect of nanowire and NWFETs is discussed here deeply in a single platform.

Gaurav Kaushal, Atal Bihari Vajpayee Indian Institute of Information Technology and Management, Gwalior, India
Balamurugan Murgan, Qualcomm, India
Manisha Pattanaik, Atal Bihari Vajpayee Indian Institute of Information Technology and Management, Gwalior, India
Chinnapurapu Naga Raghuram, Department of ECE, National Institute of Technology, Patna, India
Surendra Singh Rathod, Sardar Patel Institute of Technology, India

Radiation environment generates high soft error rates in conventional SRAM. To overcome this issue, several radiation hardened by design SRAM circuits (12TRHBD, 13TRHBD, DICE, etc.) have been developed. Although many of the radiation hardened SRAM cells are there, all the circuits mainly concern a single node upset only. In this chapter, 16T radiation hardened static random-access memory bit cell is designed and verified for a single node and multi-node upset. RHBD 16T bit cell is designed with SAED-PDK 32nm technology and compared with recently reported RHBD 12T and has a 99% improvement in recovery rate. Simulation results show that RHBD 16T is more resilient to a single node and multi-node upset. This shows that the proposed RHBD16T cell is highly tolerant against radiation strikes.

Rohit Rastogi, Dayalbagh Educational Institute, India & ABES Engineering College, Ghaziabad, India
Puru Jain, ABES Engineering College, Ghaziabad, India
Rishabh Jain, ABES Engineering College, Ghaziabad, India

In current conditions, robotization has changed into the fundamental piece of our lives. Everybody is completely subject to mechanization whether it is an extraordinary bundling or home robotization. So as to bring home automation into thought, everybody now needs a heterogeneous state security, and in our task on residential robotization, such high security highlights are completely on the best possible consumption for this reason. In light of the structure of the interruption zone, there are some fundamental interests in it. Piezoelectric sensors are compelling for sharpening appropriated wellbeing checking and structures. An intrusion detection system (IDS) is a structure that screen for suspicious movement and issues alarms when such advancement is found. While impossible to miss worthiness and presentation is, some obstruction divulgence structures are fit to take practice when poisonous improvement or peculiar action is perceived.

Preface

The electronic systems have taken rapid improvements in speed, size and cost since the invention of integrated circuits (ICs). For today's ICs, switching time is is on the order of nanoseconds, minimum feature size is on the order of nano-meters, transistor count is on the order of billions and cost is on the order of a few dollars. Scaling of conventional MOSFETs towards smaller sizes is near to end because of limited mobility, large Short Channel Effects (SCEs), high leakage currents and power dissipation. To mitigate these challenges of conventional MOSFET devices and to scale the semiconductor device further into nanoscale domain, alternative materials and device structures are required.

Nanowire is an important sector whose applications in VLSI design have huge impact for bringing revolutionary advancements in nanoscale devices, circuits and systems due to improved electronic properties of the nanowires. Nanowires are the potential devices for VLSI circuits and system applications and are highly preferred in novel nanoscale devices due to their high mobility and high driving capacity. Although currently the knowledge and resources for the fabrication of nanowires are limited but it is predicted by International Technology Road Map for Semiconductors (ITRS) that, with advancement of technology the conventional fabrication flow can be used for the nanoscale devices specifically nanowires.

There many books in this area and the question arises that is there any justification of adding one more book to the already large stock of books on the subject of VLSI. Perhaps there is. This is the only book in which subject matter is dealt from namowire materials to the advance level of devices and circuits applications of nanowire in a unique manner which will certainly fascinate the readers. The book that covers basics to advanced modeling concept of nanowires along with various applications.

Acknowledgment

I wish to express my gratitude to my loving family members for their support and blessings throughout my life. I would like extend my sincerest thanks to Director, NITTTR Chandigarh for providing me this opportunity and helping environment for initial start of this book project. I also that to Dr. Sandeep Singh Gill, Professor and head, Department of Electronics and Communication Engineering, National Institute of Technical Teachers Training and Research, Chandigarh for guidance and support. I have been fortunate in receiving support, assistance and encouragement from our colleagues and scholars in the Department of Electronics and Communication Engineering, National Institute of Technical Teachers Training and Research, Chandigarh, India and NIT Jalandhar. I acknowledge the contribution of authors of book chapters, which have been written very well with modifications number of times for improvement of quality of chapters and their timely response.

A special thanks to the reviewers for their valuable points, comments and suggestions, which have been greatly, helped us in giving final shape to the book. We gratefully acknowledge the contribution of each of them, in particular, Dr. S Dasgupta, IIT Roorkee, Dr. Arun Kumar, PEC Chandigarh, Dr. Rajeevan Chandel, NIT Hamirpur, Dr. AshwaniRana, NIT Hamirpur, Dr. Sunny Anand, Amity University, Dr. Jeetendra Singh, NIT Sikkim, Mrs. Shashi Bala, Chandigarh Engineering College, Dr. Balwinder Singh, CDAC Mohali, Dr. Gurmohan Singh, CDAC Mohali, Dr. Karmjit Singh, Thapar University Patiala, Dr. Balwant Raj, SSG Panjab University RC Hoshiarpur, Dr. Sanjeev Kumar Sharma, Dr. Intekhab Amin, Jamia Millia Islamia, University Delhi, Mr. Inderjit Singh, DAV University Jalandhar, Mr. GirishWadhwa, Mr. Shailendra Singh, Ms. Tulika Chawla, research scholars, Dept. of ECE NIT Jalandhar, India.

Thanks are also due to publisher, IGI Global USA, its editorial and production teams for providing us with necessary support for overseeing various aspects of production of this book.

Balwinder Raj

Section 1
Introduction to Nanowires

Chapter 1
Introduction to Nanowires:
Types, Proprieties, and Application of Nanowires

Sanjeev Kumar Sharma
Dr. B. R.Ambedkar National Institute of Technology, Jalandhar, India

Parveen Kumar
Dr. B. R. Ambedkar National Institute of Technology, Jalandhar, India

Balwinder Raj
National Institute of Technical Teachers Training and Research, Chandigarh, India

ABSTRACT

This chapter describes nanowires (NWs), their types, characteristic features, and their use in sensor and transistor applications as well. Introductory part in general explains briefly about nanowires (NWs), their characteristics features, types, as well as their most significant properties. The types of nanowires (NWs) can be of conducting materials such as Ni, Pt, Au; semiconducting materials like Si, InP, GaN; as well as insulating materials like that of SiO2, TiO2, etc. The property of nanowires includes mechanical, electrical, chemical, optical, and thermal properties. Also, a small description of nanowires (NWs) and sensors are explained with their performance parameters. Furthermore, nanowires (NWs)-based transistors are discussed in addition with their characteristics and applications. Finally, this chapter concludes with the significance of nanowires (NWs) in contemporary era.

DOI: 10.4018/978-1-7998-6467-7.ch001

INTRODUCTION

Nanowires are those nanostructures which have diameters (Radial dimension) in the range of nanoscalethat is 10^{-9} (tens of nanometers) and lateral dimension upto range of micro meter (µm) as shown in Figure.1.

Figure 1. Represents typical Nanowire, which is equivalent to size of biological DNA

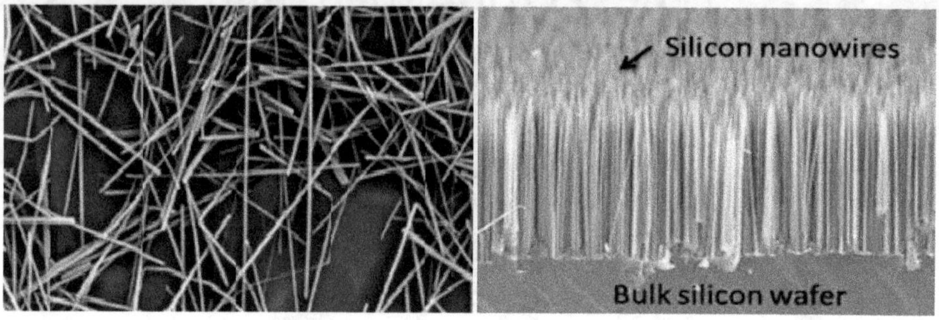

Nanowires (NW), can be of conducting materials such as (Al, Pt, Au), while semiconducting materials such as (Si, Ge, GaN) as well of insulating materials such as (SiO_2, HfO_2 TiO_2) etc (Wu et al., 2008). Nanowires can be fabricated by using two approaches such as top-down approach and bottom-up approach but bottom-up is most preferably used because following parameters of Nanowires (Lu et al., 2008) like length, radius, chemical composition and direction of growth can be precisely controlled (Sohn et al., 2009; Yu & Meyyappan, 2006). Nanowires (NW) can used to design various types of sensorsas reported by various investigations (Najam, 2008). Further, because of large Young's moduli, nanowires (NW) is used to design nano electromechanical systems (NEMS) (Lu et al., 2008). Nanowires (NWs) by virtue of excellent optical, electrical, chemical as well as mechanical properties are considered as popular nanomaterials (Muskins, 2008). The conductivity of Nanowires (NW) can be managed by field-effect mechanism, due to this action it can be used as future generation Transistors, Radom Access Memory (RAM) and advanced sensors (Hochbaum et al., 2008).

As per International Technology Roadmap of Semiconductors (ITRS) future generation transistor size can be scaled down to 18nm and gate length will scaled down to 7nm. At this dimensation conventional MOSFETs proves to be ineffective. Nanowires based MOSFETs are potential candidates at such device size. Moreover, for the diagnosis of diseases the future sensor must have fast real-time response, low cost and high resolution. The amazing properties of Nanowires give born to various types of biosensors, optosensors, gas sensor as well as electrochemical sensors.

Properties of Nanowire

Nanowires (NWs) possess extremely extraordinary properties which makes them useful for Nanoelectronics applications. The various properties are explained as follows:

Electrical Properties

The mechanism such as charge trapping and scattering electrical transport can be controlled easily with respect to both minority and majority carrier in Nanowire (NW). When light falls on Nanowires due to dielectric constant reduces the difference for light absorption by order of magnitude from ~250μm making Nanowires suitable for photovoltaics applications (Boukai et al., 2008).

Thermal Properties

Thermal properties depend upon the scattering of phonon within the Nanowires as thermal transport is due to phonons mostly. Due to small diameter the scattering of phonon reduces leads to reduction in heat transport. Thus making Nanowires suitable for thermoelectric applications (Chan et al., 2008; Cui et al., 2003).

Chemical Properties

Chemical properties depend upon large effective areas and surfaces. As nanowirespossess large area/volume ratio. Thus making Nanowires a potential candidate for Chemical and Sensing applications (Zheng et al., 2004).

Magnetic Property

Because of high energy of anisotropy with a super magnetic in Nanowires making them suitable for electromagnetic applications.

Catalytic Property

The large surface area of Nanowires formed due to transition oxide materials shows inspiring catalytic properties and can be further improved by using gold and platinum.

Types of Nanowires

Nanowires are often of two types because of its structural compositions.

Homogenous Structures

Homogeneous structure based nanowires with only one types of material composition such as Silicon (Si), germanium (Ge), etc.

Silicon Nanowires (SiNWs)

Silicon Nanowires (SiNWs) are mostly used material for making Nanodevices. The reason behind this silicon is abundance in semiconductors industry. Moreover, their structure and doping can easily be controlled during growth of silicon Nanowires (Sze, 1981), (Wang et al., 2003).

Germanium Nanowires (GeNWs)

As compared to Silicon nanowires (SiNWs), Germanium Nanowires (GeNWs) have higher electron and hole mobility (Bryllert et al., 2006a). Also GeNW devices observe smaller contact effects due to smaller Ge bandgap which leads to lower Schottky barrier at the interface of metal/NW.Reported studies of p-type GeNW devices with Pd S/D contactsgave thehole mobility of 600 $cm^2/V.s$(Wan et al., 2006).

Indium Arsenide Nanowires (InAsNWs)

Due to small effective electron mass (0.023 m_0), high electron mobility formation of electron gas layer, because of pinning of fermi level at the conduction band, small band gap (0.35 eV) and relatively transparent contacts. InAs becomes an attractive material for formation of nanodevices (Wan et al., 2007).

One of fascinating III-V family compound is InAs which have band gap in the range of ~0.354 eV. These materials can be used for various applications such as photodiodes, phototransistors, biosensors, lasers and amplifiers. Particularly, InAs is used for high mobility FETs because of its high carrier mobility which can be further used for designing of various devices as well as circuit applications. Metal organic chemical vapor deposition (MOCVD) method is suitable approach for growing InAs NWs particularly in nanoscale regime.

In comparison to quantum dots as well as thin films structures, one dimensional (1D) nanostructures generally demonstrate better physical and chemical properties. This is due to its 1D structure that helps in maintaining electron transport within the structure. It will be highly recommended to design various types of detectors and sensors by taking InAs NWs as channel material. Due to small bandgap, wonderful electrical conductivity and comparatively large electron mean free path (λ), InAs NWs are potential candidate for making different types of sensors.

Metal Oxide Nanowires (MoNWs)

The most interesting class of Nanowires that can be fabricated from metal oxides such as Zinc Oxide (ZnO), SnO_2 and In_2O_3 etc. are mostly used nowadays. Among them Lightly doped SnO_2 nanowires and moderately doped SnO_2 Nanowiresare sensitive to gas and UV sensors respectively. Also these Nanowires can be used to design devices heaving mobility up to 550 $cm^2/V{\cdot}s$ (Park & Yi, 2004). While degenerately doped SnO_2 and In_2O_3 Nanowires represents metallic properties with resistivity $10^{-4}\Omega{\cdot}cm$ (Wan et al., 2004). These types of nanowires used for various applications such as optoelectronics devices (Xiang et al., 2006), field-effect transistors, ultrasensitive nanoscale gas sensors (Sharma et al., 2017), etc.

Heterogeneous Structure Nanowires

Performance of the nanowire-based devices can be enhanced with the formation of heterogeneous structure nanowire by changing band structure engineering due to their low contact resistance with lightly doped or intrinsic channel. The schematics of SiGe heterogeneous structure Nanowire is given in Figure. 2.

Figure 2. Schematics of Si/Ge core/shell Nanowire structure (Sharma et al., 2016a).

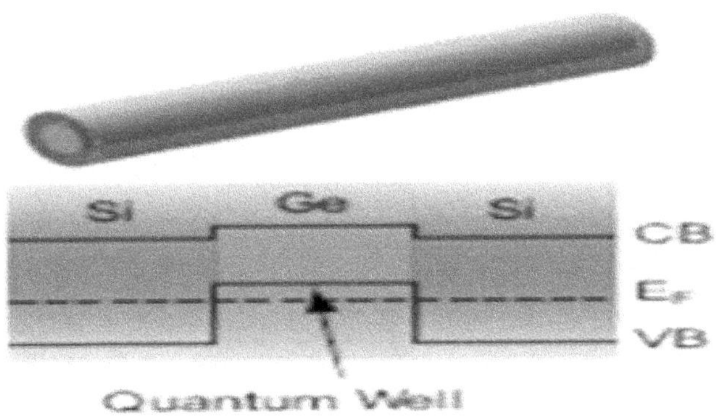

NANOWIRE APPLICATIONS

Wrap Gate MOSFET

Nanowire transistors next generation devices among different types of MOS transistors, because of predicted properties like that of good conductance, short channel effect immunity, low leakage current and so on, as we scale down the devices (Bryllert et al., 2006b; Raj, 2007; Sharma et al., 2016b; Sharma, Khosla, & Raj, 2016; Sharma et al., 2014a; Sharma et al., 2014b; Suk, 2005; Vishvakarma et al., 2007). However, most Nanowire transistors using bottom-up process have not been well estimated their performance due to process limitation and reproducibility. The effects of bottom parasitic transistor, nanowire and temperature dependency of subthreshold swing (SS) improves these limitations.

Field-effect transistor (FET) having three contacts source and drain as well as gate. It is observed that vertically fabricated Nanowire have excellent electrostatic control over the channel because of better control of gate surrounding the channel during the fabrication. However, it is also possible to make Nanowire in lateral dimension, but is difficult to fabricate these types of Nanowires as compared to vertical shaped Nanowires (Gong et al., 2009). The evolution of Field-effect transistor (FET) is shown in (a) (b) (a) (b) (c) of Figure.3. (a) planner Metal-oxide-semiconductor FET (MOSFET (b) FINFET: the gate surrounds the channel from all but one side. (c) Nanowire FINFET: complete, but nonuniform wrap-around gate. (d) Vertical nanowire FET: the gate has a complete, uniform wrap-around gate. (e) Vertical array of nanowires after wrap-gate formation (Liu & Lin, 2010).

Figure 3. Evolution of the field-effect transistor (Liu & Lin, 2010)

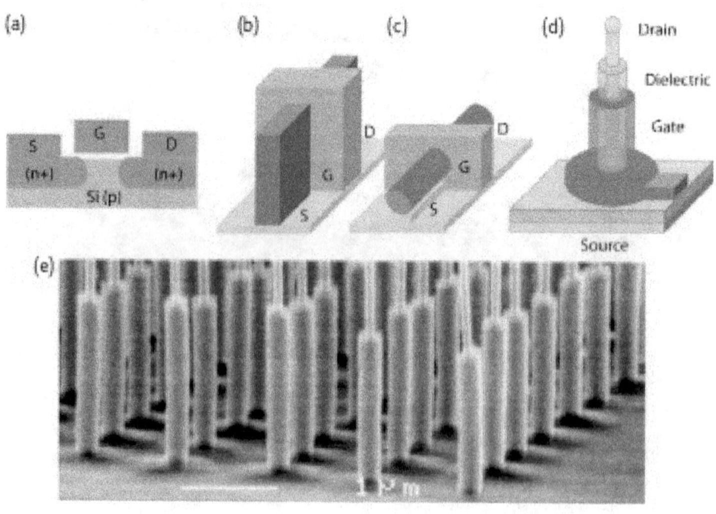

Nanowire Based Photosensors

One of potential application of Nanowire is low cost and highly sensitive Photosensor. These types of sensor have been used for different applications such as biomedicine (Lin et al., 2005), interconnects (Sarasqueta et al., 2011), optical storage (Sharma et al., 2018) and circuit compatible Field- Effect Transistors (FETs). Due to exceptionally large sensitivity, quite miniaturization, large-production and low power leakage features of MOSFET based Photodetectors. These types of detectors have been studied by various research groups (ATLAS User's Manual 3-D Device Simulator, ; Sharma, Jain, & Raj, 2016). This chapterdescribes about the designing of MOSFET based sensor and also comparison has been made between double gate (DG) MOSFET as well as cylindrical gate (CG) MOSFET for performance parameters.

2D-cross structure of CG MOSFET and simulated structure of 3D CG MOSFET in presence of incident radiation has been shown in Figure 4 as follows. In this structures (R) radius of channel, (L) channel length, length of source (L_S), length of drain (L_D), direction along channel (z), and thickness of oxide material (t_{ox}).

Figure 4. (a) 2D-cross structure of CG MOSFET (b) Simulated structure of 3D CG MOSFET (Sze, 1981).

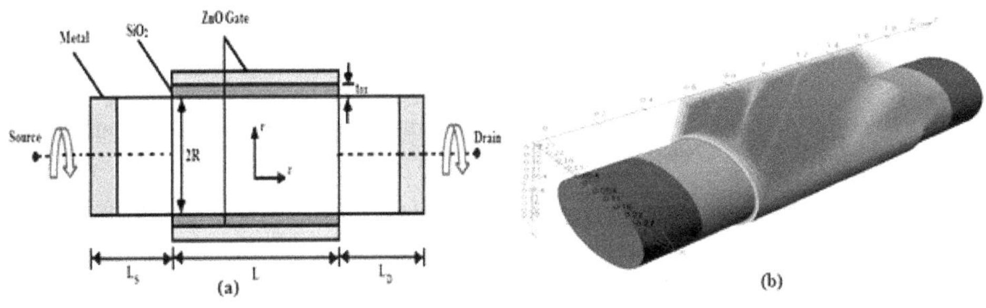

Table 1 represent the numerical value taken for designing and simulation of device structures.

Under Dark condition, to design the device and obtaining the characteristics SILVACO ATLAS-TCAD simulator (Sze, 1981) has been used and to obtain the device characteristics under incident radiation LUMINOUS-3D optical simulator has been used which incorporate Ray Trace method is used for evaluating photogeneration rate and with the help of BEAM keyword which is present in LUMINOUS-3D module of SILVACO ATLAS the most of the optical parameters of radiation which is incident on the device such as intensity of radiation, wavelength of radiation and location of incident radiation is set. Field Dependent Mobility model (FDM) and Shockley

Table 1. Parameters for designing sensor

Value	Symbol	Parameter
L	Length of channel	1.0μm
t_{si}	Silicon film thickness	0.5μm
R	Channel radius (t_{si}=2R)	0.25μm
L_S	Source length	0.5μm
L_D	Drain length	0.5μm
t_{ox}	Oxide thickness	10nm
N_A	Channel doping	$1\times10^{16}cm^3$
N_D	Source/Drain doping	$1\times10^{20}cm^3$

Read Hall model (SRH) used as models for the simulation. The dimension of oxide layer such as SiO_2 and gate material like ZnO which acts a transparent material for incident light is taken in such a way that maximum light will reach to the channel material used for device design.

The equation for reflection coefficients is given as follows (SOPRA Infobase, n.d.):

$$R_c = \frac{(n_c - n_{c+1})^2 + k_{c+1}{}^2}{(n_c + n_{c+1})^2 + k_{c+1}{}^2}$$

Here R_1, R_2 and R_3 are interfaces coefficients such as air to gate (ZnO), gate to oxide (SiO_2) and oxide to semiconductor. The real part of refractive index is represented by n_1, n_2, n_3, n_4 and their imaginary part are represented by k_c's for

Table 2. The value of refractive index for different material ZnO, SiO2, and Si (Gautam et al., 2013)

(μm)	ZnO			SiO_2			Si			
	n_1	k	R_1	n_2	K	R_2	n_3	k	R_3	$\alpha\ (cm^{-1})\times10^6$
0.25	1.92	0.383	0.115	1.600	0.008	0	1.58	3.63	0.56	1.827
0.30	1.99	0.404	0.126	1.578	0.013	0	5.00	4.168	0.48	1.746
0.35	2.14	0.457	0.150	1.565	0.024	0	5.43	2.989	0.41	1.073
0.40	2.27	4×10^{-3}	0.151	1.557	0.035	0	5.57	0.387	0.31	0.121
0.45	2.11	4×10^{-18}	0.127	1.552	0.023	0	4.67	0.139	0.25	0.03886
0.50	2.05	0	0.119	1.548	0.019	0	4.29	0.069	0.22	0.01746
0.55	2.02	0	0.114	1.545	0.018	0	4.08	0.043	0.20	0.00985

different material as air, gate, oxide and semiconductor respectively. To extract the real and imaginary values of different materials, SOPRA database (Sze, 1981) is taken as reference.

The equation for generation rate is given by (Gong et al., 2009).

$$G_r = a\phi d$$

Where ϕd the photon flux density and its equation is given by ().

$$\phi_d = I_o \frac{h}{c\lambda}(1-R_1)(1-R_2)(1-R_3)$$

Where I_o intensity of light, c speed of light, and h is the Planck's constant.

RESULTS AND DISCUSSION

The result obtain after simulations is shown in Figure 5 i.e, VI characteristics for varying light intensities (I_0). It is observed from the Figure 5 that conductance of the channel increases as the intensity of light increases from darkness towards the higher values results into larger available photo current. The variation of photo current as

Figure 5. VI characteristics of CG MOSFET for varying light intensities (Sze, 1981)

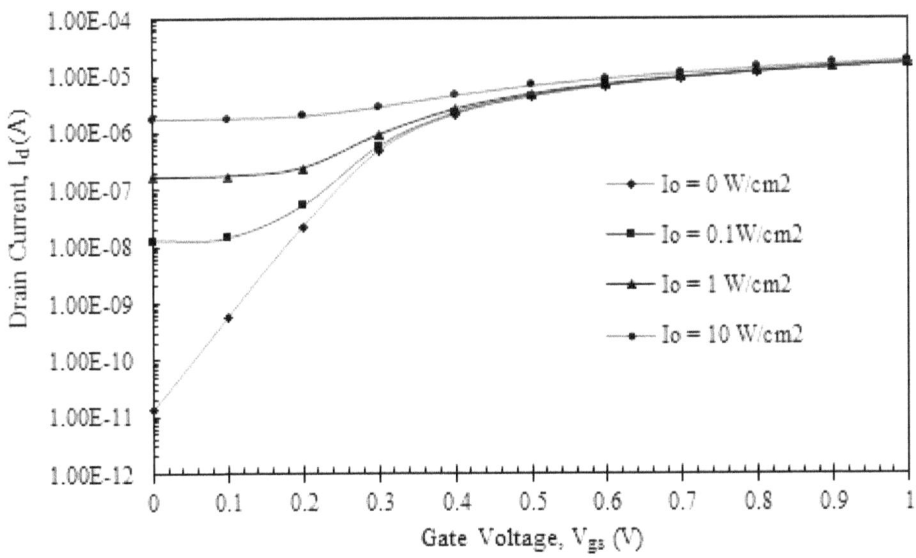

observed from Figure 5 is more towards the subthreshold region in comparision to linear or saturation region. Hence variation in subthreshold region is key parameter to design UV-visible Photosensor in this work.

Figure 6. VI characteristics of CG MOSFET for varying wavelength of light (Sze, 1981)

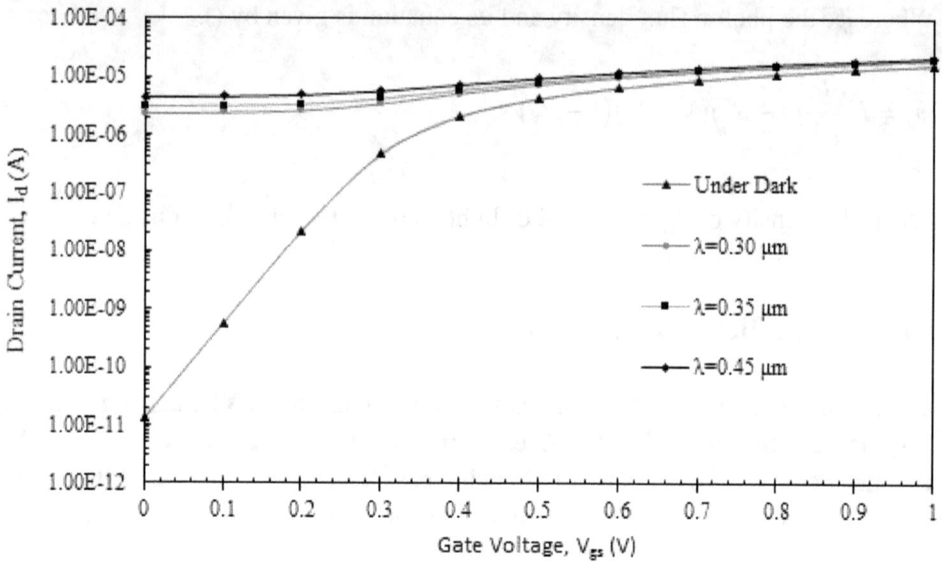

Figure 6 shows VI characteristics for different wavelength of light. It is observed that higher the wavelength of light more will be illumination current and the maximum value of I_{on}/I_{off} ratio obtained at radius R=0.25µm at λ=0.45µm thus thissensor is used in thisregion of spectrum.

Table 3 represents the comparison of DG MOSFET and CG MOSFET for different parameters such as threshold voltage (V_{th}) and I_{on}/I_{off} ratio with the intensity of

Table 3. Performance Comparison (Sze, 1981)

I_o W/cm²	DG MOSFET (V_{th})	CG MOSFET (V_{th})	DG MOSFET (I_{on}/I_{off})	CG MOSFET (I_{on}/I_{off})
0	0.35	0.310	1.75×10^5	12.6×10^5
0.1	0.34	0.302	0.925×10^3	1.30×10^3
1	0.30	0.272	0.716×10^2	1.03×10^2
10	0.19	0.212	7.19	10.9

incident radiation at λ=0.25µm. from comparison it is observed that CG MOSFET is better device for low power photosensor.

CONCLUSION

Nanowires are the most important materials for the designing of nanoscales devices. As per international Road map semiconductors Technology (ITRS) futures nanoscales are based on Nanowires as they have 1D conduction, low scattering high mobility, better gate control making Nanowires suitable candidates for nano range applications. In this chapter we have describes the various properties of Nanowires such as Electrical, Mechanical, Catalytic optical and soon which makes Nanowires an exceptionally suitable for low power applications. Also we have described various types of Nanowires and their properties which will helpful to studies various types of devices and circuits in electronics. Further, the performance of Nanowires can also be improved by making various types of heterostructure like we have explained SiGe heterostructure nanowires for the sake of understanding for the readers. Moreover, few applications which are based on Nanowire are also explained in this chapter such as MOSFET based on Nanowires in the form of Evolutionary trends and also in this chapter we have explained the designing and performances of the Nanowires based photosensors. These types of sensors detect from ultraviolet to visible radiations and can be used for applications such as biomedical, electronics appliances, interconnects as well as space technology. For the future sensing application devices based on Nanowire as sensitive materials is highly recommended as their performance parameters are extra ordinary in accordance with sensing application. Also their performance parameters can be imroved by using different device engineering techniques.

REFERENCES

ATLAS User's Manual 3-D Device Simulator. (2010). SILVACO International, Version 5.14.0.R.

Boukai, A. I., Bunimovich, Y., Tahir-Kheli, J., Yu, J.-K., Goddard, W. A. III, & Heath, J. R. (2008). Silicon nanowires as efficient thermoelectric materials. *Nature*, *451*(7175), 168–171. doi:10.1038/nature06458 PMID:18185583

Bryllert, T., Wernersson, L. E., Froberg, L. E., & Samuelson, L. (2006a, May). Vertical high-mobility wrap-gated InAs nanowire transistor. *IEEE Electron Device Letters*, *27*(5), 323–325. doi:10.1109/LED.2006.873371

Bryllert, T., Wernersson, L.-E., Löwgren, T., & Samuelson, L. (2006b). Vertical wrap-gated nanowire transistors. *Nanotechnology, 17*(11), S227–S230. doi:10.1088/0957-4484/17/11/S01

Chan, C. K., Peng, H., Liu, G., McIlwrath, K., Zhang, X. F., Huggins, R. A., & Cui, Y. (2008). High-performance lithium battery anodes using silicon nanowires. *Nature Nanotechnology, 3*(1), 31–35. doi:10.1038/nnano.2007.411 PMID:18654447

Cui, Y., Zhong, Z. H., Wang, D. L., Wang, W. U., & Lieber, C. M. (2003, January). High performance silicon nanowire field effect transistors. *Nano Letters, 3*(2), 149–152. doi:10.1021/nl0258751

Fagas, G., & Greer, J. C. (2009). Ballistic conductance in oxidized Si. *Nanoletters, 9*(5), 1856-1860. Available at: http://apps.isiknowledge.com

Gautam, Saxena, Gupta, & Gupta. (2013). Analytical Model of Double Gate MOSFET for High Sensitivity Low Power Photosensor. *Journal of Semiconductor Technology and Science, 13*(5).

Gong, X., Tong, M. H., Xia, Y. J., Cai, W. Z., Moon, J. S., Cao, Y., Yu, G., Shieh, C. L., Nilsson, B., & Heeger, A. J. (2009). High-Detectivity Polymer Photodetectors with Spectral Response from 300 nm to 1450 nm. *Science, 325*(5948), 1665–1667. doi:10.1126cience.1176706 PMID:19679770

Hochbaum, A. I., Chen, R., Delgado, R. D., Liang, W., Garnett, E. C., Najarian, M., Majumdar, A., & Yang, P. (2008). Enhanced thermoelectric performance of rough silicon nanowires. *Nature, 451*(7175), 163–167. doi:10.1038/nature06381 PMID:18185582

Lin, H. W., Ku, S. Y., Su, H. C., Huang, C. W., Lin, Y. T., Wong, K. T., & Wu, C. C. (2005). Highly Efficient Visible-Blind Organic Ultraviolet Photodetectors. *Advanced Materials, 17*(20), 2489–2493. doi:10.1002/adma.200401622

Liu, C. W., & Lin, C. H. (2010). Metal-Insulator-Semiconductor Photodetectors. *Sensors (Basel), 10*(10), 8797–8826. doi:10.3390101008797 PMID:22163382

Lu, W., Xie, P., & Lieber, C. M. (2008). Nanowire Transistor Performance Limits and Applications. *IEEE Transactions on Electron Devices, 55*(11), 2859-2876. Available at: http://apps.isiknowledge.com

Muskins, O. L. (2008). Article. *Nano Letters, 8*, 2638. doi:10.1021/nl0808076 PMID:18700806

Najam, S. F. (2008). *Silicon Nanowire: Synthesis and Applications*. Available at: http://swot.ecs.soton.ac.uk/137/3/IRP_edited.pdf

Park, W. I., & Yi, G. C. (2004, January). Electroluminescence in n-ZnOnanorod arrays vertically grown on p-GaN. *Advanced Materials, 16*(1), 87–90. doi:10.1002/adma.200305729

Raj, S. K. (2007). Analytical Modeling of Nanoscale Double Gate FinFET Device. *International Journal of Intelligent Electronics System, 1*, 633–637.

Sarasqueta, G., Choudhury, K. R., Subbiah, J., & So, F. (2011). Organic and Inorganic Blocking Layers for Solution-Processed Colloidal PbSe Nanocrystal Infrared Photodetectors. *Advanced Functional Materials, 21*(1), 167–171. doi:10.1002/adfm.201001328

Sharma, Raj, & Khosla. (2016a). Comparative Analysis of MOSFET, CNTFET and NWFET for High Performance VLSI Circuit Design: A Review. *Journal of VLSI Design Tools & Technology, 6*(2), 19-32.

Sharma, Raj, & Khosla. (2016b). A Gaussian Approach for Analytical Subthreshold Current Model of Cylindrical Nanowire FET with Quantum Mechanical Effects. *Microelectronics Journal, 53*, 65-72.

Sharma, Raj, & Khosla. (2017). Subthreshold Performance of In1-xGaxAs based Dual Metal with Gate Stack Cylindrical/Surrounding Gate Nanowire MOSFET for Low Power Analog Applications. *Journal of Nanoelectronics and Optoelectronics, 12*, 171–176.

Sharma, S., Jain, A., & Raj, B. (2018, May). Analysis of Triple Metal Surrounding Gate (TM-SG) III-V Nanowire MOSFET for Photosensing Application. *Opto-electronics Journal, Elsevier, 26*(2), 141–148. doi:10.1016/j.opelre.2018.03.001

Sharma, S. K., Jain, A., & Raj, B. (2016). *Design and Analysis of High Sensitivity Photosensor Using Cylindrical Surrounding Gate MOSFET. In Engineering Science and Technology, an International Journal*. Elsevier.

Sharma, S. K., Khosla, M., & Raj, B. (2016). Performance Enhancement of Junctionless Nanowire FET with Laterally Graded Channel Doping and High-κ Spacers. *IEEE Global Conference on Consumer Electronics.*

Sharma, V. K., Pattanaik, M., & Raj, B. (2014a). PVT variations aware low leakage INDEP approach for nanoscale CMOS Circuits. *Microelectronics Reliability, Elsevier, 54*(1), 90–99. doi:10.1016/j.microrel.2013.09.018

Sharma, V. K., Pattanaik, M., & Raj, B. (2014b). ONOFIC approach: Low power high speed nanoscale VLSI circuits design. *International Journal of Electronics, Taylor & Francis, 101*(1), 61–73. doi:10.1080/00207217.2013.769186

Sohn, Y. S., Park, J., Yoon, G., Song, J., Jee, S. W., Lee, J. H., Na, S., Kwon, T., & Eom, K. (2009). Mechanical Properties of Silicon Nanowires. *Nanoscale Research Letters*, 1-6. Available at: www.scopus.com

SOPRA Infobase. (n.d.). http://refractiveindex.info

Suk, S. D. (2005). *Tech.* Dig. IEDM.

Sze, S. M. (1981). *Physics of Semiconductor Devices* (2nd ed.). John Wiley and Sons.

Vishvakarma, Agarwal, Raj, Saxena, & Dasgupta. (2007). Two Dimensional Analytical Potential Modeling of Nanoscale Symmetric Double Gate (SDG) MOSFET with Ultra Thin Body (UTB). *Journal of Computational and Theoretical Nanoscience, 4*(5), 1144–1148.

Wan, Dattoli, & Lu. (2007). Transparent metallic Sb-doped SnO2 nanowires. *Appl. Phys. Lett., 90*(22).

Wan, Q., Dattoli, E. N., Fung, W. Y., Guo, W., Chen, Y., Pan, X., & Lu, W. (2006, December). High-performance transparent conducting oxide nanowires. *Nano Letters*, *6*(12), 2909–2915. doi:10.1021/nl062213d PMID:17163729

Wan, Q., Li, Q. H., Chen, Y. J., Wang, T. H., He, X. L., Li, J. P., & Lin, C. L. (2004, May). Fabrication and ethanol sensing characteristics of ZnO nanowire gas sensors. *Applied Physics Letters*, *84*(18), 3654–3656. doi:10.1063/1.1738932

Wang, D. W., Wang, Q., Javey, A., Tu, R., Dai, H. J., Kim, H., McIntyre, P. C., Krishnamohan, T., & Saraswat, K. C. (2003, September). Germanium nanowire fieldeffect transistors with SiO2 and high-κ HfO2 gate dielectrics. *Applied Physics Letters*, *83*(12), 2432–2434. doi:10.1063/1.1611644

Wikipedia. (2009). *Nanowire*. Available at: https://en.wikipedia.org/wiki/Nanowire

Wu, X., Kulkarni, J. S., Collins, G., Petkov, N., Almecija, D., Boland, J. J., Erts, D., & Holmes, J. D. (2008). Synthesis and Electrical and Mechanical Properties of Silicon and Germanium Nanowires. *Progress in Chemistry, 20*(19), 5954-5967. Available at: http://apps.isiknowledge.com

Xiang, J., Vidan, A., Tinkham, M., Westervelt, R. M., & Lieber, C. M. (2006, December). Ge/Si nanowire mesoscopic Josephson junctions. *Nature Nanotechnology*, *1*(3), 208–213. doi:10.1038/nnano.2006.140 PMID:18654188

Yu, B., & Meyyappan, M. (2006). Nanotechnology-Role in emerging nanoelectronics. *Solid-State Electronics, 50*(4), 536-544. Available at: http://apps.isiknowledge.com

Zheng, G. F., Lu, W., Jin, S., & Lieber, C. M. (2004, September). Synthesis and fabrication of high-performance n-type silicon nanowire transistors. *Advanced Materials*, *16*(21), 1890–1893. doi:10.1002/adma.200400472

Section 2

Applications of Nanowires and Modeling

Chapter 2
Various Applications of Nanowires

Zahied Azam
National Institute of Technology, Srinagar, India

Amandeep Singh
https://orcid.org/0000-0002-2373-3368
National Institute of Technology, Srinagar, India

ABSTRACT

In this era of nanotechnology, nanowires (NWs) propose potential impact on electronics, computing, memory, data storage, communications, manufacturing, health, medicine, national security, and other economic sectors as well. NWs offer excellent surface-to-volume ratio (interface phenomena), low defect density, high optical output, and controllable n-type conductivity, making them more appropriate and sensitive for sensing applications. NWs with their ultra-sensitive and real-time detection capabilities lend their applications in nanobiosensors, chemical sensors, gas sensors, and electrochemical sensors. NWs have been used in improving the optical absorption as well as for the collection efficacy in photovoltaic devices. NWs having small size, low weight, low cost for mass production, and are also compatible with commercial planar processes for large-scale circuitry. In the chapter, the authors are focusing to summarize the recent advances in NWs with their potential applications in various fields such as research, health, security, education, entertainment, and power generation.

DOI: 10.4018/978-1-7998-6467-7.ch002

INTRODUCTION

Nanowires (NWs) are actually the nanoscale material that has applications in next generation devices to enrich their functionality, superior performance and also allowing high integrity and low cost. In the last two decades, a number of metallic, compound semiconducting and dielectric material have been used grown in two dimensional thin films via the cutting-edge techniques like Chemical vapor deposition (CVD), Metal organic chemical vapor deposition (MOCVD) and Molecular beam epitaxy (MBE). The wave length coverage of this material is from ultraviolet to infrared and its band gap ranges from 0.4 to 4.0 eV. High dimensional control has

Table 1. Various types of nanowires with their diverse applications

S. No	Types of Nanowires	Applications of Nanowires	References
1.	Silicon	Electronics, biosensors, Electroluminescence	(Huo et al., 2004), (Kurkina et al., 2012),
2.	Germanium	Electronics, IR detectors	(Das et al., 2010), (Qi et al., 2007)
3.	GaAs	High speed electronics, Photoconductive roles	(Joyce, 2011), (Wang, 2013)
4.	InAs	Photodetector, Solar cell, NWFET, Tunnel diode, TFET, SET	(Pettersson et al. 2006), (Bessire et al. 2011), (Wei et al., 2009)
5.	InP	NWFET, Single-photodetectors, Single-photon detector, Tunnel diode	(Borgstrom et al. 2011), (Kobayashi et al.,2012), (Bulgarini et al., 2012), (Wallentin et al., 2010)
6.	GaP	Single-photonsource, Photodetector, Solar cell	(Borgstrom et al. 2005), (Gutsche et al. 2012), (Krogstrup et al., 2013)
7.	Ag NWs	Wearable devices	(Wang et al., 2010), (Hong et al., 2015)
8.	CuO NWs	Electronic-conductor	(Yang et al., 2014)
9.	Zn NWs	Optoelectronics, light emitters and lasers	(Kar et al., 2006)
10.	DNWs	Radiation particle detectors, UV-light detectors and emitters, high-speed and high-power field effect transistors, field emission sources, position-sensitive biochemical substrates, and room temperature-stabilized high-efficiency single-photon emitters.	(Monroy,2003), (Koizumi, 2001), (Ito, 2000), (Isberg, 2002), (Okano, 1996), (Yang, 2002), (Babinec, 2010)
11.	Cadmium telluride	IR detectors for various wavelengths	(Rakovich et al.,2007)

been gained as multiple quanta layering in III–V material as GaAs/AlGaAs with a layering thickness of 1 nm or fewer. The capacity of its growth as a thin epitaxial layer also in a controlled way that leads towards a number of its commercial applications in logic as well as in microelectromechanical systems, sensors, lasers, detectors, photovoltaics, memory, logic devices and others. NWs surfaces may be tailored fromsensitive to chemical or biological species (Seker, 2000), (Taylor and Schultz, 1996). These interesting properties with others of nanowires make them better candidates for their extensive range of applications as seen in Table 1 and discussed in details as follows. Although there are a number of the applications of the various nanowires to be discussed, but here in this chapter few applications are explained in details with examples and their uses in various fields.

APPLICATIONS OF SEMICONDUCTING NANOWIRES

Semiconducting nanowireshave potential in optoelectronics, sensors and electronics applications. In semiconductor nanowires, Silicon nanowires (SiNWs) are getting the most attention due to its ultimate role in the integrated circuits (IC) industry. Its constant performance is also appreciable and necessary as an alternative to thin film based planar CMOS transistors in the shape of 3D structures, pioneering architectures and amalgamation with other useful components (Hutchby et al., 2002). So, these nanowires may prove in providing the new avenues in this direction. As, silicon is a well-known direct band gap semiconductor in the nanoscale dimensions (Yu et al., 1999), so it has thus fascinating optoelectronics applications as well. This interest in optical interconnects in copper era especially in the IC development with a wish to cover all silicon-based procedure in place of the usage of III–V materials for optoelectronics etc possibly will profit from mature nanowires technology if and when it became accessible. Lately, SiNWs are very much needed in the production and manufacturing of DNA sensors (Li et al., 2003). Like SiNWs, Other nanowires like Ge is also of great interest as providing the competitive avenue, receiving the consideration in nanowires community. NWs of III–V semiconductors likeInP, GaAS, $Al_xGa_{1-x}As$, InAs and wide band gap III–V nitridesalso gained too much importance. In general, thesemiconducting nanowires as nanomaterials are proved as best gears in the biomedical fields as part of applications. As these nanowires has the ability to directly identify a range of species electrically without labels (Sang, 2015).

Silicon Nanowires

The primary inspiration in the SiNWs is the electronics. The two major and common things in the silicon epitaxy involve $SiCl_4/H_2$ and SiH_4/H_2, and are also used in its growth. Electroluminescence of the room temperature is also achievable with these NWs (Huo et al., 2004). Also, the detection and the characterization of various biological and chemical species for the purpose of disease diagnosis and discovery is an important task in the biomedical purposes. Nanomaterials like the carbon nanotubes, nanowires and nanoparticle are important tools for this primary biomedical usage, as described by (Kurkina & Balasubramanian, 2012) and shown in Figure 1. All the above given nanomaterials provide diverse magnetic, electrical andoptical properties for imaging as well as sensing. As shown by colloidal gold and semiconductor nanocrystals, which are used as the detection tags for the purpose of disease indicators and otherbiotic species, these materials are also used as contrast agents for their use in optical imaging and magnetic resonance too. Figure. 1C explains a general idea of the schematics and the various uses of nanostructures which are gaining importance in biological and medical applications. In this way semiconducting nanowires directly help in the detection of various species electrically without labels.

Figure 1. Nanostructures as sensors for biomolecules

(A) Showing various profitsas high surface to volume ratios indetecting a specified type of aimed particles, viewing that most of the areasmay be used as detecting areas.(B) Size of several biomolecules and nanostructures is almost similar, this assuresthat nanostructures are the good meansof treating biomolecules. (C) Nanostructures shown which are frequently usefulfor the purpose of biological and therapeutic applications.Those with blue backgrounds functions to stop receptors or labels and with yellow backgrounds are used as labels, and others with green backgrounds works as both labels as well as supports. The insets shown especially in the white backgroundsdisplay the key bioconjugation procedurewhich is applied to immobilize receptors on the conforming nanomaterial. The green antibody present in the insets characterizes a basic biomolecule(Adopted from Kurkina et al., 2012).

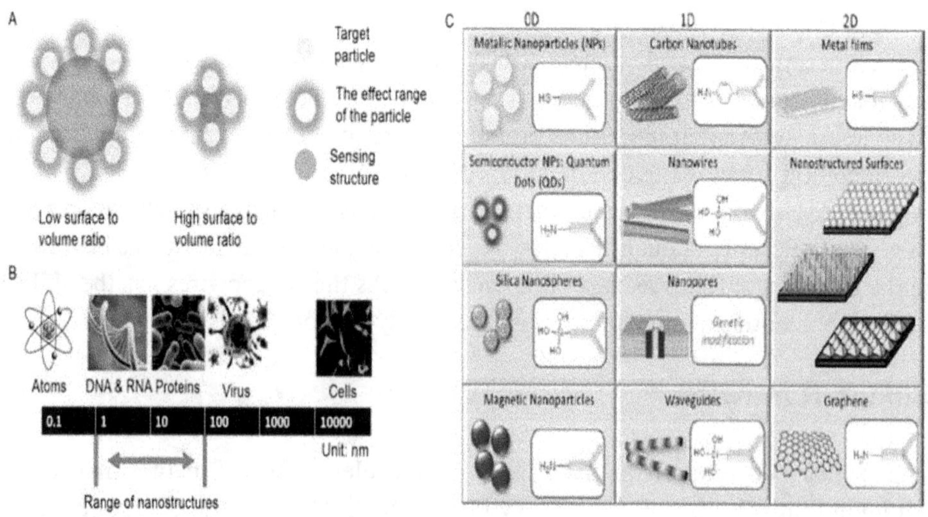

Figure 2. Detection of virus using nanowire biosensors: (A) Showing the method of virus detection from Exhaled breath condensate (EBC) samples by means of SiNWs sensing device, (B) Explaining the consequences for the exposure of the H3N2 virus deprived of the nanoparticle treatment, (C) Showing outcomes for the uncovering of H3N2 virus especially with the nanoparticle treatment
Source: (Adopted from Shen et al., 2012)

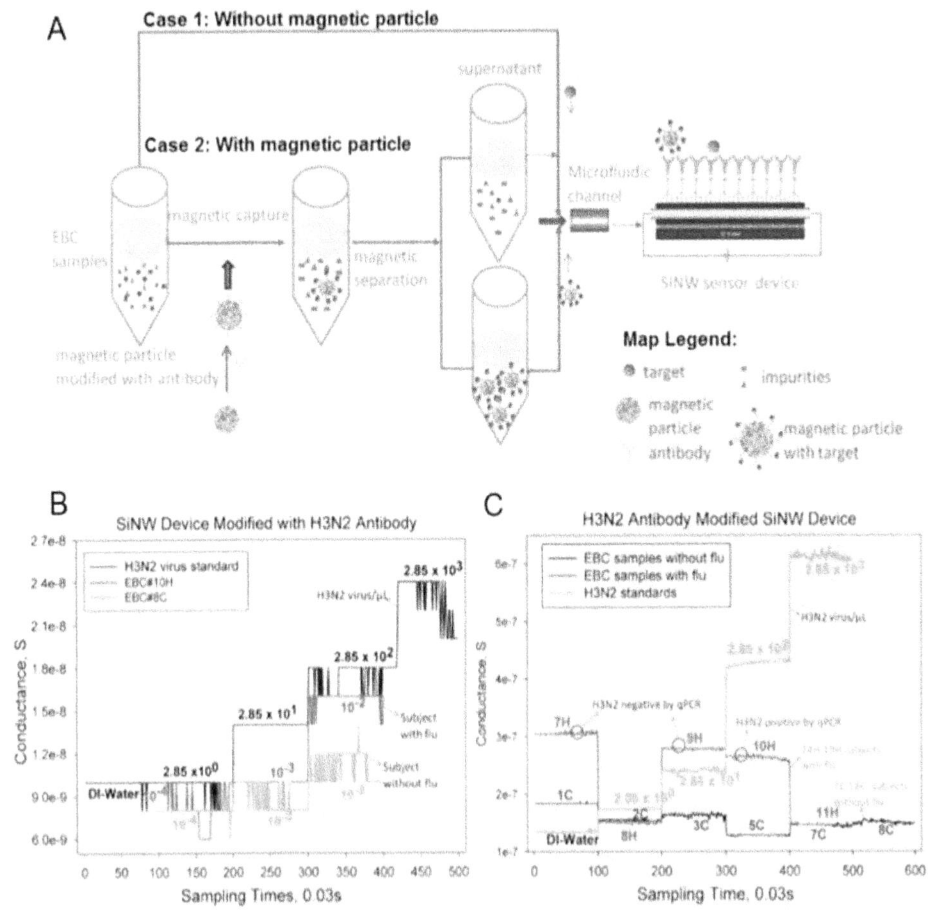

Viruses are for the most part widespread reason of human disease (Stadler, 2003), and no doubt there is growing unease concerning their ability to be used as a biological weapon (Atlas, 2003). Till date, Si nanowire sensors have fruitfully detected a lot of hazardous viruses, i.e. Influenza A H3N2 (Shen et al., 2012), and shown in Figure 2. Dengue (Zhang, 2010), HIV (Inci, 2013) and H1N1 (Kao, 2011). Shen et al. group also successfully developed Si nanowires-basedbiosensor especially in quickly diagnosing the flu.Si nanowires-based biosensor is capable of detecting

Figure 3. Nanowire sensor based on breath volatolome: (A) Detection process flow of nanowire sensor based on breath volatolomeStep 1: Fabricating standard nanowire sensor (Provided boxes point out the location of nanowires), Step 2: Nanowire surface modifications, Step 3: Biosensor exposure to VOCs connected with gastric cancer situations, Step 4: Biosensor exposure to real breath samples, (B) Typical source-drain current (IDS) vs. back-gate voltage (VGS) curves of the sensor upon vacuum exposure to N2 and upon exposure to elevated concentrations of VOC2 and (C) VOC5

Source: (Adopted from Shehada et al., 2015).

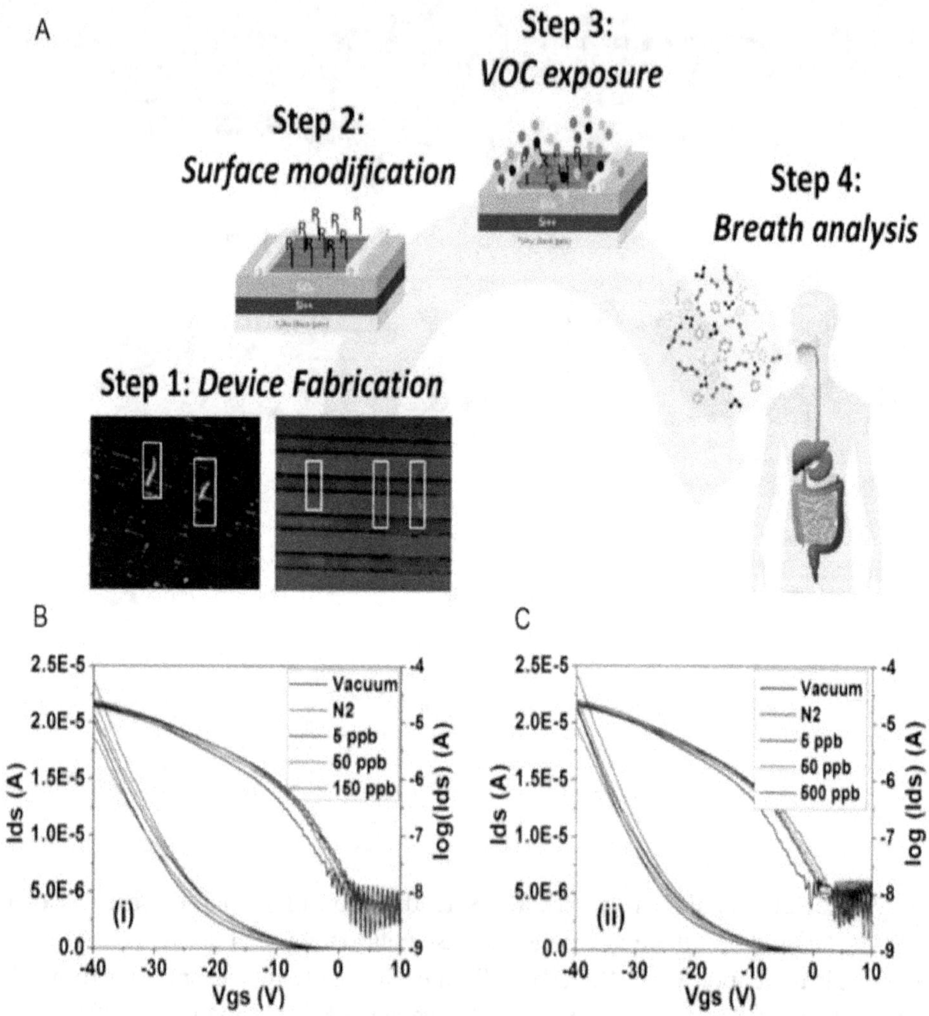

nearly 29 viruses/μL at the same time from the respired breath condensed sample. This device needs nanomaterials to attain the high sensitivity and also providing fast, little cost besides a reliable diagnostic technique replacing the typical qPCR method. (Shen et al., 2012).

Figure 4. Biosensors from nanowires for the investigation of the connections amongst protein and DNA: (A) Showing its working principle, (B) Interaction specificity of ERα by means of three different EREs i.e. mutant, wild-type and scrambled ERE, (C) Showing response of wild type-ERE functionalized biosensor to different concentrations of specific ERα, (D) Response of the wild type-ERE functionalized biosensor to the nuclear extracts from MCF-7 and MDA MB231 breast cancer cells
Source: (Adopted from Zhang et al., 2011).

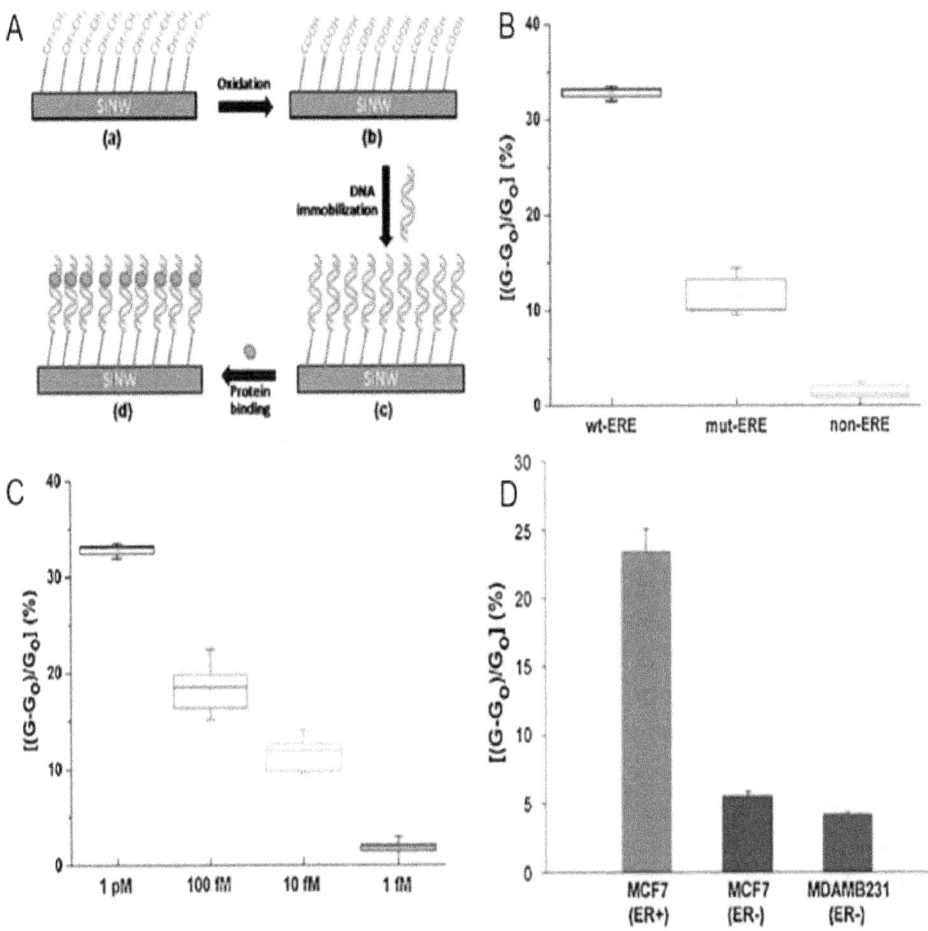

Taking into consideration breath volatolome, Shehada et al. group successfully manufactured cancer diagnosis system, which is totallynanowire-based (Shehada, 2015). This sensor by applying a nanowire is capable of sensing volatile organic compounds (VOCs), with its modified surface that can be additionally linked with gastric cancer conditions. Also, it is worth noted that the sensor is 85% accurate in sensing these VOCs as an agent for cancer. So, it has potential as a simple, transportable, painless as well as a speedy method for the initial diagnosis of gastro-intestinal-tract diseases like cancer etc (Shehada et al. 2015) as shown in Figure 3.

SiNWs sensor also has the capability for detecting the DNA ranging from 10 femtomolar level (Hahm, 2004), that is proved much better than the already existing quartz-crystal microbalance (Hook, 2001), nanoparticle-enhanced surface plasmon resonance (SPR) (He, 2000) and SPR (Jensen, 1997) in the recognition of DNA. A scientist, Dorvel et al. also stated that an ultrasensitive SiNWs-based biosensor having high-k oxide dielectrics helps detect microRNA (Dorvel, 2012). Zhang et al group explained the interactions between protein and DNA and also defined the Estrogen receptor alpha (ERα) and showed the interaction of ERα and three different EREs i.e. mutant, wild-type and scrambled ERE (Zhang et al., 2011), as shown in Figure 4 below.

Nanowire biosensors are reportedly applied in measuring the electrophysiological signal generation as well as in the chemotherapy of Doxorubicin (DOX), i.e. as a drug delivery agent (Bhattacharrya et al., 2000). Peng et al group explained the technique utilized to overcome the Multidrug resistance (MDR) consequence of free DOX, Si nanowire DOX complexes as well as cell viability and MCF-7/ADR with different concentrations for the treatment (Peng et al., 2014), Figure 5. Nanowire biosensors also has applications in the intracellular and extracellular action potentials generated by beating cardiomyocytes (Qing, 2014; Duan, 2011) and from brain tissues and cells (Cohen-Karni, 2012), (Qing, 2010). Various types of thiols i.e. Cysteine, Homocysteine and Glutathione are generally distributed in tissues and cells. These are equally important in many of the physiological processes like maintaining the redox, gene regulation and detoxification purposes. So, its abnormal level creates too much complexities and health related issues, So, for the same concern scientist Huang, et al., group reported that silicon quantum dots and 5,5-dithiobis-(2-nitrobenzoic acid) (SiQDs-DTNB) is a selective and stable detection method for the detection of thiols in serum samples and living cells. Their assay really provides a biocompatible nano-platform for thiols detection (Huang, B. H, et al., 2020). This simulation also provides the foundation stone for the biomedical applications of silicon sensor in the near future too. A surface protein namely S-layer protein (SLP) present in many of the microorganisms like bacteria etc usually show self-assembly ability for too many applications. Zhuo et al group has successfully developed its detection method, which is quite simple and reliable. This group developed a label-free as well as highly selective and sensitive assay, by using TiO2-coated porous silicon (PSi)-based microfluidic biosensor. In this assay the monoclonal antibodies related to SLP was prepared, which was then immobilized on Psi. The developed microfluidic biosensor has a broad detection range i.e. 0.001to 10 μg/mL and the lower one is 3.2±1.7×10-5 μg/mL (0.70±0.37 pM) for SLP.

Figure 5. Techniques utilized to overcome Multidrug resistance (MDR) effect of free Doxorubicin (DOX) (A) Si nanowire DOX complexes (B) Cell viability of (C) MCF-7 cells and (D) MCF-7/ADR while its treatment with various concentrations of DOX and nanowire-DOX

Source: (Adopted from Peng et al., 2014).

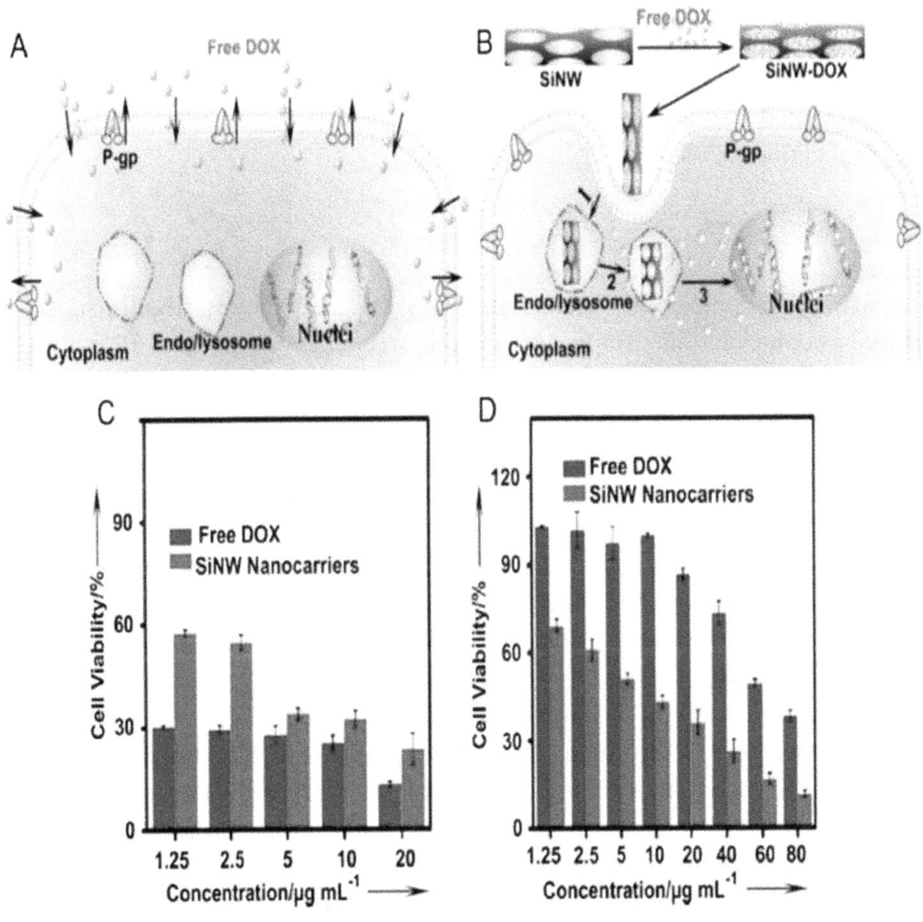

GERMANIUM NANOWIRES

Germanium nanowire (GeNW) has its own importance as a pure element or as an alloy with the silicon i.e. (Si_xGe_{1-x}) for logic as well as other devices too. As the direct and indirect band gap of germanium is 0.8 eV and 0.6 eV respectively it behaves as a good semiconductor. It also offers high internal mobilities as compared to silicon and thus enables faster switching and also the high frequency devices. It has too many additional advantages which include higher intrinsic carrier concentration i.e. 2.4 × 1013 cm^{-3}versus 1.45 × 1010 $cm^{-3}$with greater bulk excitonic Bohr radius, 24.3 nm versus 4.7 nm for the purpose of more prominent quantum confinement effects too. It also possesses compatibility with high-k dielectrics which enable its integration to the semiconductor processing technology. Germanium is also well-suited with III–V materials as in case of high-quality lattice matching with GaAs. On the other hand, its oxides also possess exciting optical properties suitable for attractive integrated optoelectronic circuits. All the above-mentioned properties have prompted its widespread investigations in GeNW growth, its characterization and applications (Das et al., 2010; Qi et al., 2007).

III–V Nanowires

III–V complex semiconductor materials is really of great interest in both electronics as well as optoelectronics for its use in higher frequency transistors, LEDs, Lasers, Resonant tunneling diodes (RTDs), solar cells, photodetectors as well as others. Their importance in all these applications was well established and finely reported especially in the last two decades. Their quantum wells and thin films in InxGa1-xAs, GaP, InP, AlxGa1-xAs and GaAs etc grown using some specialized techniques like MOCVD and MBE. In recent times, identical material is being grown as one-dimensional nanowire because quantum curbed low-dimensional structure encompass the potential to offer improved performance in all of the above-mentioned applications.

GaAs Nanowires

Gallium arsenide (GaAs)is also a direct band semiconductor having band gap of 1.424 eV and electron saturation velocity of 10^7 cm/s with its carrier mobility 8500 and 450 cm2/V s for electrons and holes too respectively. GaAs is grown in its nanowire form by MBE (Ihn et al., 2007), laser-assisted catalytic growth (Duan et al., 2000), template-guided approach (Noborisaka et al., 2005) and MOCVD (Parkinson et al., 2007; Paladugu et al., 2007). Like III–V complex semiconductor materials GaAs nanowire remarkably has high mobility as well as a very short lifetime. GaAs nanowire is also usable as ideal material for its valuable uses in high speed electronics (Joyce, 2011)and less poweredextremely sensitive Triple Metal Surrounding Gate (TM-SG) Nanowire MOSFETphotosensorcan be made. It shows that Core-Shell structures may also offer a better choice for photodetection invisible region (Sharma et al., 2018).Including the single GaAs NWs,photoconductive roles also has been widely considered and reported (Wang, 2013). It has been also well established that III-V compound especially(GaAs) structures has the capability of noticeably lessening the material thickness necessary for absorbing 90% of power (optical thickness) to submicron region in comparison with Si (Kayes, 2005).

InAs Nanowires

InAs nanowire possess small energy band gap i.e. (0.35 eV) with high electron mobility (33000 cm^2/V s) and large quantum confinement energy. There has been a number of information on the preparation of InAs nanowires (Mandl et al., 2006; Martensson et al., 2007). As Mandl et al. group reportedly presented InAs growth by MOCVD with a thin layer of SiO serving as catalyst on InP (111), InAs (001), and Si (001) substrates (Mandl et al., 2006). The application of InAs is well documented in the Tunnel diode (Bessire et al. 2011), Solar cell (Wei et al. 2009) andPhotodetector (Pettersson et al. 2006).

InP Nanowires

InP is also a direct band gap semiconductor having energy gap of 1.34 eV, which has been extensively documented and applied for electronics and optoelectronics applications. Its growth in the form of nanowire includes the technique of laser-assisted growth (Duan et al., 2001), by applying the sublimation of InP powder MBE (Poole, 2003), (Tang et al., 2003) and MOCVD (Mattila et al., 2007). InP nanowire has applications in NWFET (Borgstrom et al. 2011), Single-photodetectors (Kobayashi et al., 2012), Single-photon detector (Bulgarini et al., 2012) and Tunnel diode (Wallentin et al., 2010).

GaP Nanowires

GaP as anextensive band gap semiconductor possessing an indirect gap of 2.26 eV, with its electron mobility bulk as GaP as 160 $cm^2/V.s$. GaP is favored for the high-temperature electronic applications and having light emission especially in visible range. Doping with nitrogen make it appear with bright orange color in place of typical greenish yellow color undoped wires allowing it for the production of yellow LEDs. GaP nanowire has application in Single-photon source (Borgstrom et al. 2005), Photodetector (Gutsche et al. 2012) and Solar cell (Krogstrup et al., 2013).

APPLICATIONS OF SUPERCONDUCTING NANOWIRE

Like semiconductors the superconducting Nanowire (SNWs) are also gaining much importance after the manufacturing and use of Superconducting Nanowire Single-Photon Detectors (SNSPD), which came into use in 2001 with some attractive properties. These materials propose its potential as single-photon detector from X-rays to terahertz wavelengths. This novelkind of single-photon detector which emerged in 2001 is niobium nitride (NbN) nanowire (Skocpol et al., 1974). This SNSPD functions at ~4 K with a temperature achievable either through liquid helium or modern closed-cycle cooling technology. It also shows its potential in infrared single-photon sensitivity as well as with picosecond timing resolution and specifically in little dark counts. In other words, we can say that SNSPD has insignificant dark counts, broadband sensitivity and picosecond timing resolution. SNSPDs also have its capability to outperform the already existing photon counting technologies based on semiconductors too, provided only if the difficulty of less temperature operation could be achieved. For doing the same a multichannel SNSPD system based on a closed-cycle refrigeratorhas been manufactured and tested. In this way an efficient optical coupling has been obtained by cautiously aligning optical fibre. Coupling of the fiber of SNSPDs was founded, which was based on the NbN on MgO substrates and NbTiN on oxidized Si substrates, in this way it has been seen that NbTiN on oxidized Si substrates was with improved performance at telecom wavelengths, thus manipulating the reflection from the Si/SiO2 interface. The basic operational principle of the SNSPD is shown in Figure 6 and explained herein as follows. Schematic illustration of the detection cycle(i) (a) SNWs were maintained well less than the critical temperature is direct current (DC) biased just lower than the critical current (ii)Resistive hotspot created after absorbing the photon by a nanowire (iii) As the size of the NbN nanowires are finethus local current density near this hotspot increase, resulting inincreased superconducting critical current density because supercurrent here is enforcedfor flowing along the periphery of the hotspot. (iv) So, a resistive fence is automatically designed across its width (Gol'Tsman et al., 2001). (v) Here Joule heating also is responsible for the growth of fenced region along the periphery of the nanowire (Yang et al., 2007) unless this current flow is stopped as well as the bias current is totally stopped through the external circuit. (vi) This process permitsits resistive region to diminishthus the wire again uses to behave as a superconductor. This biased currentwith the help of nanowire proceeds to the original value (i)(b)The simple electrical equivalent circuit of a SNSPD is shown here, Rn and Lk is the hotspot resistance and kinetic inductance of the SNSPD respectively. The SNSPD is shown as the current biased at Ibias.Thus, closing and opening of the switch initiates the absorption of a single photon and in this way an output pulse is counted across the load resistor Z0 (Hadfield et al., 2005;Kerman et al., 2006). (i)

(c) Showing the simulation of output voltage pulse related to SNSPD, values of Lk =500 nH and Rn =500Ω were implemented for the same simulation (For ease, Rn is presumedas fixed, while a more detailed treatment (Yang et al., 2007) demonstrates Rn(t)). Here solid blue line is the mostimportantpart of the SNSPD output pulse, and other one i.e. dotted red one represents the trailing edge. The time constants pertaining to the phases of the detection cycle related tothese SNSPDs thus have significant application areas like space-to-ground communications, optical quantum computing, quantum key distribution,integrated circuit testing, characterization of quantum emitters, fibre sensing and time-of-flight ranging (Natarajan et al., 2012).

Figure 6. Demonstration of the basic operational principle for superconducting nanowire single-photon detector (SNSPD)
Source: (Adopted from Natarajan et al., 2012).

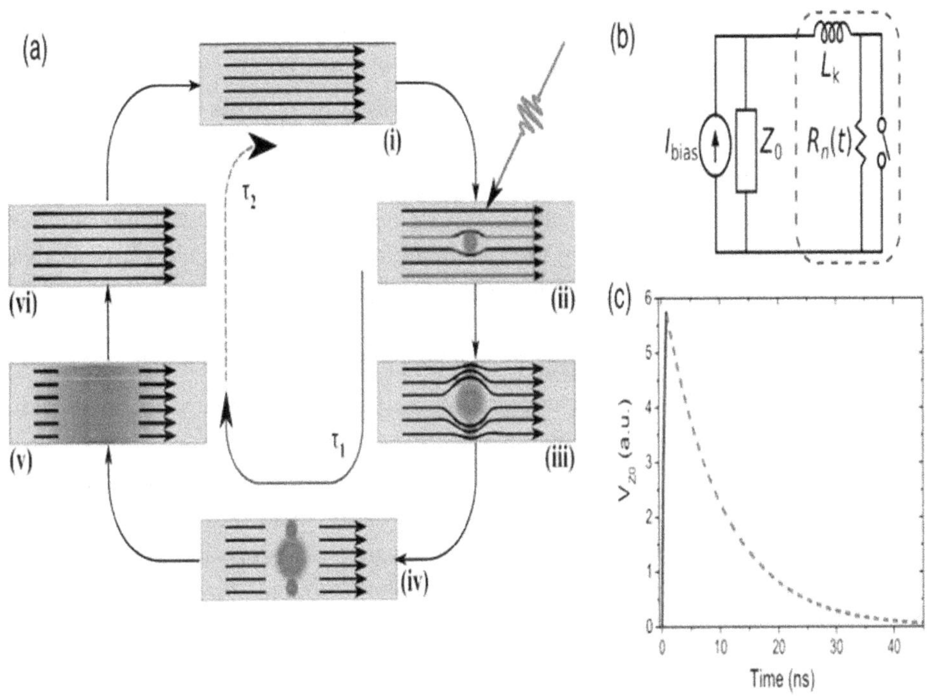

APPLICATIONS OF METALLIC NANOWIRES

Due to the rapid manufacture of flexible and stretchable devices like Google-glass, iWatch, Fitbit force products and Smarty ring etc there is an increasing demand of the new material too. So, in this way to meet the new demands the stretchability as well as the flexibility of the devices, thus their conductive parts which are used within these devices preferably have ideally pliable and deformable characters. As indium tin oxide (ITO) having low sheet resistivity with high optical transparency seems an essential conductive material which is extensively functional in such type of devices, but its inadequate supply with intrinsic brittleness largely restricted its applications too. This ultimately resulted in its urgent need for the development of novel conductive material. More over recently carbon nanotubes (CNTs) (Wang et al., 2011), metal grids and graphene (You, 2015), were made to substitute ITO which can be used in ultra-thin LEDs, bendable touchscreens, in the flexible solar cells and wearable devices due to their good conductivity as well as flexibility. Butconductivity and flexibility of the CNT and graphene mainly depends on its superiority with fabrication process. The emergence of silver and copper nanowires i.e. the two important forms of metallic nanowires (MNWs) could just meet these necessities and thus enthralled the global interest because of its excellent conductivity, flexibility, transparency and simplicity in its operation too (De et al., 2009). MNWs possessing less sheet resistance with greater optical transmittance as well as flexibility in nanowire-based electrodes also attracted substantial interests in applicationsviz. optoelectronic devices i.e. touch display panels, solar cells, LEDs,chemical sensors, heaters, surface enhanced Raman scattering, nanosolder and substrates.

On the other hand, all the above properties of metallic nanowires are also providing the amazing chance in the manufacturing of free bendable solar cells to specifically flexible organic solar cells (FOSCs) leading to its considerable attention as an alternate and potent next-generation cost-efficient as well as energy-harvesting device pertaining to its light weight, low fabrication costs, and superior flexibility (Li, 2012), (Brabec., 2008). MNWs no doubt also offers an attractive model system for studying the nanostructural effects such as localization effects, quantized conductance and other physical phenomena too. MNWs has also application in optoelectronics, touchscreens, heaters, LEDs, magnetics, solar cells, electronics and nano joints really opening new scientific opportunities because of its good conductivity, and many physical properties like electrical, excellent optical, thermal conductivity and magnetic properties. Nanowires of bismuth (Tian et al., 2006), gold (Tan et al., 2007), platinum (Lu et al., 2007), copper (Molares et al., 2001), iron (Lin et al., 2010), cobalt (Yuan et al., 2003), palladium (Atashbar et al., 2004), indium (Oh et al., 2008), silver (Ding et al., 2006), molybdenum (Zach, 2000), tungsten (Vaddiraju et al., 2003), nickel (Tian et al., 2007) and zinc (Kar et al., 2006) have been grown effectively. The most appreciable approach used for the growth of these nanowires is by using some type of template in a chemical or electrochemical synthesis. But the anodized alumina (AA) template used is the trendy one, others templates included are polycarbonate-etched ion-track membranes (Molares et al., 2001), polymer templates, carbon nanotubes (Lu et al., 2007), DNA (Monson et al., 2003), molecular sieves, zeolites (Ding et al., 2006), and block copolymers. One of the examples of the wearable and transparent nanowire devices is the Ag nanowire (STAN) heater which is shown in Figure 7 and discussed under the head Ag nanowires as follows.

Silver Nanowires

AgNWs also have also gained too much consideration because of their high thermal and electrical conductivities.Conductive metal-oxide i.e. Indium tin oxide (ITO), is the most frequently used conducting material for conventional transparent heaters, although its applications in stretchable material is very limited owing to its brittle ceramic nature (Cairns et al., 2000) and ruthless processing conditions (Liu et al., 2009). So, in the current situation i.e. the latest development of next generation future electronic devices (Jung et al., 2014) both in industry as well as the academics realize that recent electrical heater should be preferably flexible, transparent as well as highly stretchable, which would be finally applicable for wearable electronics. It has been reported now that Ag NWs are continuously making their path in the applications like antibacterial agents, photography, conductive inks, photonics, electronics and catalysis. One such example in which Ag nanowire is highly appreciated is the mostStretchable and Transparent Ag Nanowire (STAN) heater meant for wearableelectronic applications developed for the first time is based on moderately embedded Ag NW percolation network on polydimethylsiloxane (PDMS) film. This operates under strain over 60% with other real-time deformations mainlyrelated with human motionsi.e. bending and twisting. Its schematic structure is illustrated and shown in Figure7. a, that is made up of uniformly spread Ag NW at casual directions with some sort of percolation network specifically on a PDMS film. So, for developing the Ag NW percolation network, vacuum filtration transfer (Hu et al., 2010) is directly applied on the elastic PDMS film. Then, after vacuum filtration PDMS film transfer of Ag NW is trimmed and adjustedin between two glass substrates for the purpose of its actual use working as a stretchable heater. Thus, a stable DC biased voltage is carried at both the terminals of Ag NW network electrode specifically to encourage resistive Joule heating which is basically electrically driven. The conductance material Ag NW network and the underlying substrate i.e. PDMS both may be stretched also while keeping the constant electrical conductivity as well as the optical transmittance of the heater. Figure 7. b, c demonstrates well the projected transparent and stretchable heater, which is being captured by infrared (IR) camera for the heating purpose in stretched positions both i.e. 0% to 60%. In view of the fact that temperature alteration may be verified instantaneously via IR camera.So, the real functional voltage can be controlled in situ to reach 60 °C in all the cases. In this way it can be verified that this heater works well in the above-

Figure 7. Diagrammatic representation of a stretchable and transparent heater. a) Schematic illustration showing the stretchable and transparent heater made of Ag NW percolation network specifically on PDMS film. b, c) Presentation of the Pseudo color image gained at room temperature is shown in left and infrared (IR) camera thermal image is shown in right at 60 °C with (b) no strain condition and (c) at 60% strain condition.(Adopted from Hong et al., 2015).

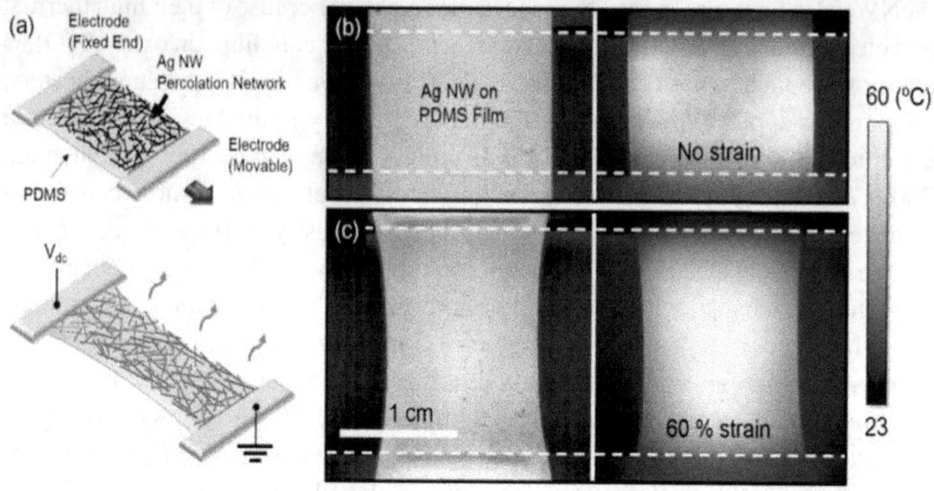

mentioned situations with stability that is satisfactory for general purpose wearable devices (Wang et al., 2010; Hong et al., 2015).

Ismuth Nanowires

Bismuth as a semimetal possesses low electron effective mass, long carrier i.e. free path with high anisotropic Fermi surface (Zhu et al., 2006). The energy overlap connecting both i.e. L and T-point as conduction and valence band respectively is little, about 38 meV at 77 K, it can ultimately lead to semimetal-semiconductor transition ultimately with a decreasing diameter of nearly 60 nm at 77 K (Li et al., 2003). Potential of the bismuth nanowires in thermoelectric devices has encouraged the study of bismuth and its alloy with antimony much attractive. Anodized alumina membrane (AAM) can be efficiently used to grow bismuth and Bi_xSb_{1-x} nanowires.

Copper Nanowires

Cu NWs is currently gaining importance as an interconnect material in the silicon CMOS technology because of its higher electrical conductivity. In view of the fact that, copper NWs have a high surface-to-volume ratio, these are also more susceptible to oxidation. For the same cause, Cu NWs are of utmost interest in examining their potential use in 32 nm generation CMOS and beyond. The oxidation behavior of the Cu nanorods at room temperature (Liu et al., 2003) indicates that its oxidation resistance is very much low as compared to the bulk copper. Actually, the oxide layer made on the rods and wires is porous and unstable. In comparison, when copper nanowires are within 230 nm deep SiO2 trenches the oxidation does not appear to be a question and the resultant resistivity is only a factor of 2.6 high (40 nm wire) than its bulk value (Steinhogl et al., 2002). Another chief asset of interest is the thermal expansion coefficient. For copper nanowires, this coefficient is an order of magnitude smaller than for bulk copper (1.2 × 10−6K−1 versus 148 × 10−6K−1). As a result, the lower thermal expansion coefficient of copper nanowires is desirable for cermet preparation.CuO nanorods also acts as electronic-conductor (Yang et al., 2014).

Zinc Nanowires

Zn NWs exhibiting superconducting also with their luminescence properties make them striking in optoelectronics applications such as light emitters and lasers. Thermal evaporation of ZnS powder at 1225°C under an argon flow in a quartz tube reactor appears to be a simple approach to produce these wires (Kar et al., 2006).

APPLICATIONS OF OXIDE NANOWIRES

Metal oxides as an important division of materials find their valuable applications in a wide range of fields. ZnO is also a directwide band gap semiconductor, it also has large exciton binding energy (~60 meV), found its important applications in ultraviolet (UV) and near UV photonics. Many other metal oxides too find their applications either as catalyst supports or vigorous catalyst resources and conducting channels in gas sensors. Several others find their way in electrochemical energy devices i.e. electrode materials, lithium ion batteries and solar cells etc. No doubt the metal oxides are also eligible as structural materials and composites. Metal oxides with their well-defined crystallinity and surface reactivity, electronic properties and high surface area have a number of applications in too many devices and electronics. These nanowires with diameters less than few nanometers are usable in the understanding of quantum confinement effects. Also, the nanowires with radius larger than the corresponding Bohr radius also never show their distinct properties compared to bulk due to the reasons i.e. low thresholds for percolation, high surface area, fast charge transport and their well-defined crystal surfaces. The low and high melting point metal oxides are Ga_2O_3, In_2O_3, Al_2O_3, MgO, SnO_2, ZnO and WO_3, Ta_2O_5, NiO, MoO_3respectively. Many of the studies are generally focused on the synthesis of Zinc oxide (ZnO) NWs due to interest in ZnO as an optoelectronic material. Excitonic solar cells (SCs) together with organic and dye-sensitized solar cells (DSSCs), have significant potential as a low-cost alternative to conventional inorganic photovoltaic (PV) devices. Till date, a number of metal oxide nanostructures like TiO_2, ZnO, Fe_2O_3, ZrO_2, Nb_2O_5, Al_2O_3 and CeO_2 have been effectively engaged as photoelectrodes in SCs. From all of the above defined metal oxide nanowires, TiO_2 and ZnO are the extensively studied one as these are suggested promising candidates as photoelectrode employed in SCs. By making comparison between TiO_2 and ZnO, ZnO are having high electron mobility also similar bandgap with conduction band energies. This property confirms that it the best candidate pertaining to stable room temperature luminescent and lasing devices. Thus, it has been noticed that, ZnO nanowires is a substitute candidate for highlycompetent SCs (Qiang Peng and Yuancheng Qin, 2011).

APPLICATIONS OF NITRIDE NANOWIRES

III-nitride materials i.e. aluminum nitride and indium nitride have a broad array of applications in optoelectronics, LEDs, electronics, lasers and photo electrochemical systems. Aluminum nitride with its very wide band gap of 6.4 eV, high dielectric constant, and thermal conductivity possess impending applications in as a deep UV emitter, optical storage media and as a substrate for high-temperature devices. In the same manner gallium nitride with a 3.4 eV band gap in the UV region with high thermal voltage as well as chemical stability and has applications in high-power, high-speed, high-temperature electronics, blue lasers and photo electrochemical hydrogen production. Indium nitride with its narrow band gap of 0.7 eV has budding applications in near infrared lasers, solar cells LEDs, coating solar cells and microcavities (Ishikawa, F., 2017). III-nitrides also show their less sensitivity to ionizing radiation, allowing them for their use in solar cells in outer space. To one side from their applications of binary species, the most important attribute of III-nitrides as a system is that their alloys can exist as solid solutions, which translates to a composition-dependent control over the properties. Thus, making it possible in tuning the properties of III-nitrides all the way from InN to AlN by varying the alloy composition. The band gap bowing property is also worth working means the band gap does not vary linearly with composition. These alloys can encompass properties additional than the range of binary III-nitrides, which opens up new applications for ternary and quaternary III-nitride alloys.

APPLICATIONS OF CARBON NANOWIRES

Diamond Nanowires

Diamond nanowires (DNWs) are well-known material of extremes, as they have outstanding properties by means ofhardness, thermal conductivity,chemical inertness, band gap, resistance to particle bombardment and electron affinity. These have potential benefits in radiation particle detectors (Monroy, 2003), ultraviolet (UV) light detectors and emitters (Koizumi, 2001, Ito, 2000), high-speed and high-power field effect transistors (Isberg, 2002), field emission sources (Okano, 1996), position-sensitive biochemical substrates (Yang, 2002) and room temperature-stabilized high-efficiency single-photon emitters (Babinec, 2010). Following are the main utilities that have been demonstrated robustly. These applications include; the field emission applications i.e. DNWs with their negative electron affinity are used for the field emission studies. Recent research suggested that,Electron field emission (EFE) properties of CVD-developed ultracrystalline. DNWs are also used for the mass analysis related to small molecules i.e. Coffinier et al. group demonstrated well the matrix-free laser desorption/ionization (D/I) mass spectrometric utilization of boron-doped DNWs in the direction of small molecular analysis (Coffinier, 2012), It is also worth mentioning here that S/N ratios of Uranium doped DNWs are very much low as compared to the boron doped DNWs thus signifying, that the capability of BDD NWs in mass analysis of little molecules is proved. Other than this DNWs are also used in nanoelectromechanical switches in the operation of single-crystalline DNWs as (Nanoelectromechanical) NEM appraised by Liao and coworkers (Liao, 2010), as these switches show low leakage current (<0.1 pA) with a high ON/OFF ratio, overall competing the Si-NEMS structures. DNWs also acts as electrochemical sensors i.e. DNWs were efficientlyused in many other electrochemical sensory studies too, these were also suggested by Yang and Nebel, as they utilized the vertically aligned DNWs for the purpose of DNA detection using electrochemical approach (Yang, 2009). Above all DNWs has also contributed a lot in the ultrasensitive force microscopy as Tao et al. defined the usefulness of DNWs as tips for ultrasensitive force microscopy observations (Tao, 2015).

APPLICATIONS OF OTHER NANOWIRES

Besides the elemental i.e. Ge, Si and compound (III–V) semiconductors, oxides, nitrides and metals a diverse range of other material is also grown in the form of nanowires. Their general inspiration is due to the lack of lattice-matching conditions with the substrate also with an additional option of higher absorption and higher quantum efficiencies because of their quantum confinement; also, their band gap is allowing operation over the wider wavelengths too. Their applications include the optoelectronics and optics in the ultraviolet (UV), visible and IR regimes i.e. in lasers, optical switches, detectors, and in some other areas such as electronics, lasers, interconnect, sensors, thermoelectric devices and field emitters. Other nanowires developed are covered under the heads antimonides (Vaddiraju et al., 2007),

tellurides (Trahey et al., 2007), selenides (Zhu et al., 2007), sulfides (Bao et al., 2007), silicides (Kim et al., 2006). Boron with boron and others i.e. silicon carbide, aluminum borate, lithium fluoride, boron carbide and magnesium boride etc.

Antimonides

GaSb and InSb have a room temperature band gap of 0.72 and 0.17 eV, respectively. Ternary alloy nanowires, $Ga_xIn_{1-x}Sb$ can offer a constant disparity in band gap between 0.17 and 0.72 eV. Thus, the electron effective mass in both GaSb and InSb is low i.e. 0.042mo and 0.013,5 mo respectively. It thus explains that altering the diameter i.e. just in the 40 nm range may lead to the variation in band gap. As it has been observed that the diameter variation from 65 nm down to 10 nm may vary the band gap of InSb from 0.17 to 0.4 eV. So, the significance of these materials has long been documented for mid- and long-wave IR detection and lasing applications. Adding together, it can be concluded that InSb is valuable for the fabrication of thermoelectric devices and magneto-resistive sensors.

Zinc Selenide

Like others ZnSe also has a direct energy band gap of 2.7 eV at room temperature. In both thin as well as bulk film forms II–VI substance has been broadly considered for its valuable applications in mid-IR laser source for remote sensing, blue-green laser diode, modulated waveguides, optical switching, q-switches, photodetectors and optoelectronics applications. The interest in these nanowire forms has been on apex for additional miniaturization of these devices with improved performance.

Other Selenides

CdSe is also very interesting II–VI material showing its growth as nanowires with applications in LEDs and photodetectors etc. For growing these nanowires an AAM template with a metal i.e. Ag substrate as a cathode and a counter electrode can be used to carry forward the electrodeposition to manufacture these nanowires. This electrolyte solution has 0.05 M/L of CdCl2 and saturated elemental selenium in dimethyl sulfoxide. A direct current density of 0.85 mA/cm^2 for 30 to 60 minutes also needed for filling the pores thus manufacturing the stoichiometric nanowires with diameters matching the pore size. This method is highly appreciated for the growth of CdSe nanowires with a stoichiometry very close to 1:1. Other selenides which are also useful for the same purpose are GaSe and PbSe, these can be easily grown by Au-catalyzed VLS using the respective powders at 800° and 700°C, respectively (Zhu et al., 2007).

Bismuth Telluride

Bi2Te3 with its related substances are extensively investigated for their potential role in the thermoelectric refrigeration and power generation. Bulk $Bi2Te^3$ is thought to be having a relatively high thermoelectric figure of merit, ZT (~1.0). Trahey et al., 2007 et al group became interested in the growth of these nanowires because of its increasedZT due to higher density of states and increased phonon scattering (Trahey et al., 2007). AAM template-based growth is a general idea to manufacture $Bi2Te^3$ nanowires. An array of nanowires can be shaped to create a meaningful level of current transport in contrast to a single nanowire (Prieto et al., 2001). Also, a classic electrolyte has 0.035 M $Bi(NO3)^3.5H_2O$ and 0.05 M $HTeO^{2+}$; the latter can be reached at reacting Te powder with 5 M HNO3. Also, the pH needs to be adjusted to 1 by adding 1 M HNO3. In addition, a graphite plate as well as a gold layer at the bottom will be serving as the cathode and counter electrode will comprise the electrochemical cell. Also, a current density of 2.5 mA/cm^2 for 2 h produce $Bi2Te^3$ nanowires by filling the pores too. These nanowires are dense, vertical and help gaining the pore opening size this process yields a stoichiometry close to 2:3.

Cadmium Telluride

CdTe is a key material in photovoltaics and CdTe/CdS junction cells have reached the commercial market to meet the current insatiable demand for solar cells. In addition to this, alloying CdTe with HgTe allows construction of IR detectors for various wavelengths. In spite of such potential, there is very little work on CdTe nanowires (Rakovich et al., 2007).The emergence of CdTe nanowires through the VLS technique appears to be rather difficult. A successful process starts with the help of sapphire substrate that is coated by polyvinyl alcohol (PVA), followed by annealing at 70°C to smooth out the alcohol layer (Neretina et al., 2007). Then a thin layer (2 nm) of bismuth is applied as catalyst by pulsed laser deposition. Growth of CdTe nanowires is carried out using a CdTe target at 365°C and 400 mTorr under a helium flow.

Other Tellurides

ZnTe is an II–VI semiconductor with a direct band gap of 2.26 eV at room temperature. Having its applications in green LEDs, electro-optic detectors, thermoelectric devices, solar cells and even the nanowire form has been of interest (Huo et al., 2006). The exciton Bohr radius is 6.2 nm and even for nanowires less than 40 nm in diameter, a relatively strong quantum confinement effect is seen. These nanowires can be grown by VLS technique using ZnTe powder and gold catalyst film. The synthesis is carried out at 1000°C, which yields long (10 mm) nanowires 30–80 nm in diameter is about 30 min (Huo et al., 2006). The XRD analysis of the sample indicates formation of zinc blende ZnTe nanowires. NWs are single crystals having their cubic structures in the orientation of (111) as established by HRTEM. ThePL spectra of these ZnTe nanowires at room temperature show BE emission at 2.26 eV (548 nm), which is indicative of high optical quality. Post growth copper doping can be achieved by immersing in Cu(NO3)2 solution for p-type doping ZnTe nanowires. A ZnTe FET with p-type wires shows a channel mobility of 1 cm2/V s (Huo et al., 2006). Other telluride nanowires include HgTe for detectors and Sb2Te3 (Jin et al., 2005) for thermoelectric devices.

Zinc Sulfide

ZnS is a semiconductor with its direct band gap 3.7 eV, possessing a huge excitation binding energy of nearly 40 meV. It has been reportedly previously as having potential in IR windows, electroluminescent devices, displays and in other optoelectronics applications. So, there is an intensive interest in the growth of ZnS nanowires through VLS and AAM-template approaches. Especially in the VLS approach, the submission of ZnS powder at 900°C creates source vapor, a thin gold film catalyst helps in the growth of ZnS nanowires (Wang et al., 2002) thus the material grown appears as a white sponge consisting of a nanowire with its range varying between of 30-60 nm in diameter.

Other Sulfides

In addition to the ZnS nanowires, other sulphide nanowires are also gaining importance which include lead sulfide (Ge et al., 2005), cadmium sulfide (Kang et al., 2007) and bismuth sulfide (Bao et al., 2007) etc. Thin film of CdS is very popular for the construction of CdTe/CdS heterojunction solar cells. At room temperature its band gap is 2.54 eV. The electrolyte needed for the construction of the CdS nanowires by AAM templating approach comprise of the equimolar (0.01 M) Na2S and CdCl2 (Mondal et al., 2007). Same is the case with the growth of Lead sulfide (PbS) nanowire, as its growth needs the mixing of sulfur, carbon and PbC12 at a ratio of 4:1:1. PbS nanowire is reported with narrow band gap of (0.41 eV) also with an exciton Bohr radius of 9 nm. Also, a carbothermal process too can be used for the production of the PbS nanowires, this process yields black sintered product of PbS nanowires at a temperature of 600°C (Zhang et al., 2006). Bismuth sulfide is also a direct band gap material i.e. 1.3 eV and is used in the in photodiodes and thermoelectric devices. For the growth of the Bismuth sulfide in an AAM-templated approach, the electrolyte needed for the deposition of Bi2S3 requires 0.055 m/L of BiC13 and 0.19m/L ofelemental Sulphur dioxide with DMSO. In this process the nanowire yielded is of 40 nm in diameter with an adsorption band gap of 1.56 eV (Peng et al., 2001).

Silicides

Silicides of iron (Liang et al., 2006), cobalt (Okino et al., 2005), nickel (Kim et al., 2007), titanium and tantalum (Xiang et al., 2005), manganese (Zou et al., 2007) and platinum (Liu et al., 2007) are grown in the form of nanowires. As low resistance ohmic contacts in gate and interconnect formation are critical for high-performance devices (Kim et al., 2006). Owing to its low resistivity and low formation temperature Nickel silicide has been extensively investigated for this purpose. Iron silicide exhibit a number of appealing phases that include metallic, semiconducting at 0.87 eV with applications in 1.5 mm wavelength optical fiber communications, n and s phases, which are metastable (Liang et al., 2006). Similarly alloys of cobalt and iron silicides are ferromagnetic semiconductors useful in spintronics applications. A particularsourceorganometallic precursor of Fe(SiC13)2(co)4seems to be valuable to grow FeSi nanowires in a CVD setup without the aid of any catalyst metal. Titanium silicide is also a low-resistivity material with its potential to meet interconnect needs in microelectronics. It can be prepared using an extremely simple physical vapor deposition method (Xiang et al., 2005).

CONCLUSION

During the last decade a significant progress has been noticed specifically in the field of nanowires and their valuable application in the fields of research, health, security, education, entertainment, power generation, electronics and optoelectronics etc. To the large extent work has been focused towards the fabrication of the various types of nanowires with their multiple uses in various fields with their integration in devices to demonstrate biosensing functions. Undoubtedly, there is immense scientific as well as technological potential hidden in these nanowires, which need to be explored. Thus, scientific society is hopeful that these new composites will have a profound impact on new applications. We expect to see further research aimed at overcoming the drawbacks pertaining to the already existing nanowires for their best implementation in various cutting-edge technologies.

REFERENCES

Atashbar, M. Z., Banerji, D., Singamaneni, S., & Bliznyuk, V. (2004). Deposition of parallel arrays of palladium nanowires and electrical characterization using microelectrode contacts. *Nanotechnology*, *15*(3), 374–378. doi:10.1088/0957-4484/15/3/025

Atlas, R. M. (2003). Bioterrorism and biodefence research: Changing the focus of microbiology. *Nature Reviews. Microbiology*, *1*(1), 70–74. doi:10.1038/nrmicro728

Babinec, T. M., Hausmann, B. J., Khan, M., Zhang, Y., Maze, J. R., Hemmer, P. R., & Lončar, M. (2010). A diamond nanowire single-photon source. *Nature Nanotechnology*, *5*(3), 195–199. doi:10.1038/nnano.2010.6

Bao, H., Cui, X., Li, C. M., Gan, Y., Zhang, J., & Guo, J. (2007). Photoswitchable semiconductor bismuth sulfide (Bi2S3) nanowires and their self-supported nanowire arrays. *The Journal of Physical Chemistry C*, *111*(33), 12279–12283. doi:10.1021/jp073504t

Bhattacharrya, S., Saha, S. K., & Chakravorty, D. (2000). Nanowire formation in a polymeric film. *Applied Physics Letters*, *76*(26), 3896–3898. doi:10.1063/1.126813

Borgström, M. T., Wallentin, J., Heurlin, M., Fält, S., Wickert, P., Leene, J., Magnusson, M. H., Deppert, K., & Samuelson, L. (2010). Nanowires with promise for photovoltaics. *IEEE Journal of Selected Topics in Quantum Electronics*, *17*(4), 1050–1061. doi:10.1109/JSTQE.2010.2073681

Borgström, M. T., Zwiller, V., Müller, E., & Imamoglu, A. (2005). Optically bright quantum dots in single nanowires. *Nano Letters*, *5*(7), 1439–1443. doi:10.1021/nl050802y

Brabec, C. J., & Durrant, J. R. (2008). Solution-processed organic solar cells. *MRS Bulletin*, *33*(7), 670–675. doi:10.1557/mrs2008.138

Bulgarini, G., Reimer, M. E., Hocevar, M., Bakkers, E. P., Kouwenhoven, L. P., & Zwiller, V. (2012). Avalanche amplification of a single exciton in a semiconductor nanowire. *Nature Photonics*, *6*(7), 455–458. doi:10.1038/nphoton.2012.110

Cairns, D. R., Witte, R. P. II, Sparacin, D. K., Sachsman, S. M., Paine, D. C., Crawford, G. P., & Newton, R. R. (2000). Strain-dependent electrical resistance of tin-doped indium oxide on polymer substrates. *Applied Physics Letters*, *76*(11), 1425–1427. doi:10.1063/1.126052

Chueh, Y. L., Ko, M. T., Chou, L. J., Chen, L. J., Wu, C. S., & Chen, C. D. (2006). TaSi2 nanowires: A potential field emitter and interconnect. *Nano Letters*, *6*(8), 1637–1644. doi:10.1021/nl060614n

Coffinier, Y., Szunerits, S., Drobecq, H., Melnyk, O., & Boukherroub, R. (2012). Diamond nanowires for highly sensitive matrix-free mass spectrometry analysis of small molecules. *Nanoscale*, *4*(1), 231–238. doi:10.1039/C1NR11274K

Cohen-Karni, T., Casanova, D., Cahoon, J. F., Qing, Q., Bell, D. C., & Lieber, C. M. (2012). Synthetically encoded ultrashort-channel nanowire transistors for fast, pointlike cellular signal detection. *Nano Letters*, *12*(5), 2639–2644. doi:10.1021/nl3011337

Das, K., Chakraborty, A. K., NandaGoswami, M. L., Shingha, R. K., Dhar, A., Coleman, K. S., & Ray, S. K. (2007). Temperature dependent shape transformation of Ge nanostructures by the vapor-liquid-solid method. *Journal of Applied Physics*, *101*(7), 074307. doi:10.1063/1.2718282

De, S., Higgins, T. M., Lyons, P. E., Doherty, E. M., Nirmalraj, P. N., Blau, W. J., Boland, J. J., & Coleman, J. N. (2009). Silver nanowire networks as flexible, transparent, conducting films: Extremely high DC to optical conductivity ratios. *ACS Nano*, *3*(7), 1767–1774. doi:10.1021/nn900348c

Ding, X., Briggs, G., Zhou, W., Chen, Q., & Peng, L. M. (2006). In situ growth and characterization of Ag and Cu nanowires. *Nanotechnology*, *17*(11), S376–S380. doi:10.1088/0957-4484/17/11/S24

Dorvel, B. R., Reddy Jr, B., Go, J., Duarte Guevara, C., Salm, E., Alam, M. A., & Bashir, R. (2012). Silicon nanowires with high-k hafnium oxide dielectrics for sensitive detection of small nucleic acid oligomers. *ACS Nano*, *6*(7), 6150–6164. doi:10.1021/nn301495k

Duan, X., Gao, R., Xie, P., Cohen-Karni, T., Qing, Q., Choe, H. S., Tian, B., Jiang, X., & Lieber, C. M. (2012). Intracellular recordings of action potentials by an extracellular nanoscale field-effect transistor. *Nature Nanotechnology*, *7*(3), 174–179. doi:10.1038/nnano.2011.223

Duan, X., Huang, Y., Cui, Y., Wang, J., & Lieber, C. M. (2001). Indium phosphide nanowires as building blocks for nanoscale electronic and optoelectronic devices. *Nature*, *409*(6816), 66–69. doi:10.1038/35051047

Duan, X., Wang, J., & Lieber, C. M. (2000). Synthesis and optical properties of gallium arsenide nanowires. *Applied Physics Letters*, *76*(9), 1116–1118. doi:10.1063/1.125956

Ge, J. P., Wang, J., Zhang, H. X., Wang, X., Peng, Q., & Li, Y. D. (2005). Orthogonal PbS nanowire arrays and networks and their Raman scattering behavior. *Chemistry (Weinheim an der Bergstrasse, Germany)*, *11*(6), 1889–1894. doi:10.1002/chem.200400633

Gol'tsman, G. N., Okunev, O., Chulkova, G., Lipatov, A., Semenov, A., Smirnov, K., Voronov, B., Dzardanov, A., Williams, C., & Sobolewski, R. (2001). Picosecond superconducting single-photon optical detector. *Applied Physics Letters*, *79*(6), 705–707. doi:10.1063/1.1388868

Gutsche, C., Lysov, A., Braam, D., Regolin, I., Keller, G., Li, Z. A., Geller, M., Spasova, M., Prost, W., & Tegude, F. J. (2012). n-GaAs/InGaP/p-GaAs Core-Multishell Nanowire Diodes for Efficient Light-to-Current Conversion. *Advanced Functional Materials*, *22*(5), 929–936. doi:10.1002/adfm.201101759

Hadfield, R. H., Miller, A. J., Nam, S. W., Kautz, R. L., & Schwall, R. E. (2005). Low-frequency phase locking in high-inductance superconducting nanowires. *Applied Physics (Berlin)*.

Hahm, J. I., & Lieber, C. M. (2004). Direct ultrasensitive electrical detection of DNA and DNA sequence variations using nanowire nanosensors. *Nano Letters*, *4*(1), 51–54. doi:10.1021/nl034853b

Hashim, A. A. (Ed.). (2011). *Nanowires: Implementations and Applications*. BoD–Books on Demand. doi:10.5772/1025

He, L., Musick, M. D., Nicewarner, S. R., Salinas, F. G., Benkovic, S. J., Natan, M. J., & Keating, C. D. (2000). Colloidal Au-enhanced surface plasmon resonance for ultrasensitive detection of DNA hybridization. *Journal of the American Chemical Society*, *122*(38), 9071–9077. doi:10.1021/ja001215b

Hong, S., Lee, H., Lee, J., Kwon, J., Han, S., Suh, Y. D., Cho, H., Shin, J., Yeo, J., & Ko, S. H. (2015). Highly stretchable and transparent metal nanowire heater for wearable electronics applications. *Advanced Materials*, *27*(32), 4744–4751. doi:10.1002/adma.201500917

Höök, F., Ray, A., Nordén, B., & Kasemo, B. (2001). Characterization of PNA and DNA immobilization and subsequent hybridization with DNA using acoustic-shear-wave attenuation measurements. *Langmuir*, *17*(26), 8305–8312. doi:10.1021/la0107704

Hu, L., Kim, H. S., Lee, J. Y., Peumans, P., & Cui, Y. (2012). ACS Nano 4, 2955 (2010). *P. Lee, J. Lee, H. Lee, J. Yeo, S. Hong, KH Nam, D. Lee, SS Lee and SH Ko. Advanced Materials*, *24*, 3326.

Huang, B. H., Shen, S. S., Wei, N., Guo, X. F., & Wang, H. (2020). Fluorescence biosensor based on silicon quantum dots and 5, 5′-dithiobis-(2-nitrobenzoic acid) for thiols in living cells. *Spectrochimica Acta. Part A: Molecular and Biomolecular Spectroscopy*, *229*, 117972. doi:10.1016/j.saa.2019.117972

Huo, H. B., Dai, L., Liu, C., You, L. P., Yang, W. Q., Ma, R. M., Ran, G. Z., & Qin, G. G. (2006). Electrical properties of Cu doped p-ZnTe nanowires. *Nanotechnology*, *17*(24), 5912–5915. doi:10.1088/0957-4484/17/24/002

Huo, H. B., Dai, L., Xia, D. Y., Ran, G. Z., You, L. P., Zhang, B. R., & Qin, G. G. (2006). Synthesis and optical properties of ZnTe single-crystalline nanowires. *Journal of Nanoscience and Nanotechnology*, *6*(4), 1182–1184. doi:10.1166/jnn.2006.179

Huo, J., Solanki, R., Freeouf, J. L., & Carruthers, J. R. (2004). Electroluminescence from silicon nanowires. *Nanotechnology*, *15*(12), 1848–1850. doi:10.1088/0957-4484/15/12/027

Hutchby, J. A., Bourianoff, G. I., Zhirnov, V. V., & Brewer, J. E. (2002). Extending the road beyond CMOS. *IEEE Circuits and Devices Magazine*, *18*(2), 28–41. doi:10.1109/101.994856

Ihn, S. G., Song, J. I., Kim, T. W., Leem, D. S., Lee, T., Lee, S. G., Koh, E. K., & Song, K. (2007). Morphology-and orientation-controlled gallium arsenide nanowires on silicon substrates. *Nano Letters*, *7*(1), 39–44. doi:10.1021/nl0618795

Inci, F., Tokel, O., Wang, S., Gurkan, U. A., Tasoglu, S., Kuritzkes, D. R., & Demirci, U. (2013). Nanoplasmonic quantitative detection of intact viruses from unprocessed whole blood. *ACS Nano*, *7*(6), 4733–4745. doi:10.1021/nn3036232

Isberg, J., Hammersberg, J., Johansson, E., Wikström, T., Twitchen, D. J., Whitehead, A. J., ... Scarsbrook, G. A. (2002). High carrier mobility in single-crystal plasma-deposited diamond. *Science*, *297*(5587), 1670–1672. doi:10.1126cience.1074374

Ishikawa, F., & Buyanova, I. (Eds.). (2017). *Novel compound semiconductor nanowires: materials, devices, and applications*. CRC Press. doi:10.1201/9781315364407

Ito, T., Nishimura, M., Yokoyama, M., Irie, M., & Wang, C. (2000). Highly efficient electron emitting diode fabricated with single-crystalline diamond. *Diamond and Related Materials*, *9*(9-10), 1561–1568. doi:10.1016/S0925-9635(00)00293-4

Jensen, K. K., Ørum, H., Nielsen, P. E., & Nordén, B. (1997). Kinetics for hybridization of peptide nucleic acids (PNA) with DNA and RNA studied with the BIAcore technique. *Biochemistry*, *36*(16), 5072–5077. doi:10.1021/bi9627525

Jin, C., Zhang, G., Qian, T., Li, X., & Yao, Z. (2005). Large-area Sb2Te3 nanowire arrays. *The Journal of Physical Chemistry B*, *109*(4), 1430–1432. doi:10.1021/jp046100z

Joyce, H. J., Gao, Q., Tan, H. H., Jagadish, C., Kim, Y., Zou, J., ... Johnston, M. B. (2011). III–V semiconductor nanowires for optoelectronic device applications. *Progress in Quantum Electronics*, *35*(2-3), 23–75. doi:10.1016/j.pquantelec.2011.03.002

Jung, S., Lee, J., Hyeon, T., Lee, M., & Kim, D. H. (2014). Fabric-based integrated energy devices for wearable activity monitors. *Advanced Materials*, *26*(36), 6329–6334. doi:10.1002/adma.201402439

Kang, C. C., Lai, C. W., Peng, H. C., Shyue, J. J., & Chou, P. T. (2007). Surfactant- and Temperature-Controlled CdS Nanowire Formation. *Small*, *3*(11), 1882–1885. doi:10.1002mll.200700390

Kao, L. T. H., Shankar, L., Kang, T. G., Zhang, G., Tay, G. K. I., Rafei, S. R. M., & Lee, C. W. H. (2011). Multiplexed detection and differentiation of the DNA strains for influenza A (H1N1 2009) using a silicon-based microfluidic system. *Biosensors & Bioelectronics*, *26*(5), 2006–2011. doi:10.1016/j.bios.2010.08.076

Kar, S., Ghoshal, T., & Chaudhuri, S. (2006). Simple thermal evaporation route to synthesize Zn and Cd metal nanowires. *Chemical Physics Letters*, *419*(1-3), 174–178. doi:10.1016/j.cplett.2005.11.054

Kayes, B. M., Atwater, H. A., & Lewis, N. S. (2005). Comparison of the device physics principles of planar and radial p-n junction nanorod solar cells. *Journal of Applied Physics*, *97*(11), 114302. doi:10.1063/1.1901835

Kerman, A. J., Dauler, E. A., Keicher, W. E., Yang, J. K., Berggren, K. K., Gol'Tsman, G., & Voronov, B. (2006). Kinetic-inductance-limited reset time of superconducting nanowire photon counters. *Applied Physics Letters*, *88*(11), 111116. doi:10.1063/1.2183810

Kim, J., & Anderson, W. A. (2006). Direct electrical measurement of the self-assembled nickel silicide nanowire. *Nano Letters*, *6*(7), 1356–1359. doi:10.1021/nl0602894

Kim, J., Shin, D. H., Lee, E. S., Han, C. S., & Park, Y. C. (2007). Electrical characteristics of single and doubly connected Ni silicide nanowire grown by plasma-enhanced chemical vapor deposition. *Applied Physics Letters*, *90*(25), 253103. doi:10.1063/1.2749430

Koblmüller, G., Hertenberger, S., Vizbaras, K., Bichler, M., Bao, F., Zhang, J. P., & Abstreiter, G. (2010). Self-induced growth of vertical free-standing InAs nanowires on Si (111) by molecular beam epitaxy. *Nanotechnology*, *21*(36), 365602. doi:10.1088/0957-4484/21/36/365602

Koizumi, S., Watanabe, K., Hasegawa, M., & Kanda, H. (2001). Ultraviolet emission from a diamond pn junction. *Science*, *292*(5523), 1899–1901. doi:10.1126cience.1060258

Krogstrup, P., Jørgensen, H. I., Heiss, M., Demichel, O., Holm, J. V., Aagesen, M., Nygard, J., & Fontcuberta i Morral, A. (2013). Single-nanowire solar cells beyond the Shockley–Queisser limit. *Nature Photonics*, *7*(4), 306–310. doi:10.1038/nphoton.2013.32

Kurkina, T., & Balasubramanian, K. (2012). Towards in vitro molecular diagnostics using nanostructures. *Cellular and Molecular Life Sciences*, *69*(3), 373–388. doi:10.100700018-011-0855-7

Li, G., Zhu, R., & Yang, Y. (2012). Polymer Solar Cells. *Nature Photonics*, *6*(3), 153–161. doi:10.1038/nphoton.2012.11

Li, L., Zhang, Y., Li, G., & Zhang, L. (2003). A route to fabricate single crystalline bismuth nanowire arrays with different diameters. *Chemical Physics Letters*, *378*(3-4), 244–249. doi:10.1016/S0009-2614(03)01264-8

Li, Z., Chen, Y., Li, X., Kamins, T. I., Nauka, K., & Williams, R. S. (2004). Sequence-specific label-free DNA sensors based on silicon nanowires. *Nano Letters*, *4*(2), 245–247. doi:10.1021/nl034958e

Liang, S., Islam, R., Smith, D. J., Bennett, P. A., O'brien, J. R., & Taylor, B. (2006). Magnetic iron silicide nanowires on Si (110). *Applied Physics Letters*, *88*(11), 113111. doi:10.1063/1.2185610

Liao, M., Hishita, S., Watanabe, E., Koizumi, S., & Koide, Y. (2010). Suspended Single-Crystal Diamond Nanowires for High-Performance Nanoelectromechanical Switches. *Advanced Materials*, *22*(47), 5393–5397. doi:10.1002/adma.201003074

Lin, J. L., Petrovykh, D. Y., Kirakosian, A., Rauscher, H., Himpsel, F. J., & Dowben, P. A. (2001). Self-assembled Fe nanowires using organometallic chemical vapor deposition and CaF 2 masks on stepped Si (111). *Applied Physics Letters*, *78*(6), 829–831. doi:10.1063/1.1345830

Liu, B., Wang, Y., Dilts, S., Mayer, T. S., & Mohney, S. E. (2007). Silicidation of silicon nanowires by platinum. *Nano Letters*, *7*(3), 818–824. doi:10.1021/nl062393r

Liu, J., Wu, D., & Zeng, S. (2009). Da Wu, and Shengnan Zeng. "Influence of temperature and layers on the characterization of ITO films. *Journal of Materials Processing Technology*, *209*(8), 3943–3948. doi:10.1016/j.jmatprotec.2008.09.016

Liu, Z., Bando, Y., & Tang, C. (2003). Synthesis of tungsten oxide nanowires. *Chemical Physics Letters*, *372*(1-2), 179–182. doi:10.1016/S0009-2614(03)00397-X

Lu, S., Sivakumar, K., & Panchapakesan, B. (2007). Sonochemical Synthesis of Platinum Nanowires and its Application as Electro-Chemical Actuators. *Journal of Nanoscience and Nanotechnology*, *7*(7), 1. doi:10.1166/jnn.2007.427

Ma, D. D. D., Lee, S. T., Mueller, P., & Alvarado, S. F. (2006). Scanning tunneling microscope excited cathodoluminescence from ZnS nanowires. *Nano Letters*, *6*(5), 926–929. doi:10.1021/nl052507j

Mandl, B., Stangl, J., Mårtensson, T., Mikkelsen, A., Eriksson, J., Karlsson, L. S., Bauer, G., Samuelson, L., & Seifert, W. (2006). Au-free epitaxial growth of InAs nanowires. *Nano Letters*, *6*(8), 1817–1821. doi:10.1021/nl060452v

Mårtensson, T., Wagner, J. B., Hilner, E., Mikkelsen, A., Thelander, C., Stangl, J., Ohlsson, B. J., Gustafsson, A., Lundgren, E., Samuelson, L., & Seifert, W. (2007). Epitaxial growth of indium arsenide nanowires on silicon using nucleation templates formed by self-assembled organic coatings. *Advanced Materials*, *19*(14), 1801–1806. doi:10.1002/adma.200700285

Mattila, M., Hakkarainen, T., Lipsanen, H., Jiang, H., & Kauppinen, E. I. (2007). Enhanced luminescence from catalyst-free grown InP nanowires. *Applied Physics Letters*, *90*(3), 033101. doi:10.1063/1.2431711

Molares, M. T., Buschmann, V., Dobrev, D., Neumann, R., Scholz, R., Schuchert, I. U., & Vetter, J. (2001). Single-crystalline copper nanowires produced by electrochemical deposition in polymeric ion track membranes. *Advanced Materials*, *13*(1), 62–65. doi:10.1002/1521-4095(200101)13:1<62::AID-ADMA62>3.0.CO;2-7

Mondal, S. P., Das, K., Dhar, A., & Ray, S. K. (2007). Characteristics of CdS nanowires grown in a porous alumina template using a two-cell method. *Nanotechnology*, *18*(9), 095606. doi:10.1088/0957-4484/18/9/095606

Monroy, E., Omnès, F., & Calle, F. (2003). Wide-bandgap semiconductor ultraviolet photodetectors. *Semiconductor Science and Technology*, *18*(4), R33–R51. doi:10.1088/0268-1242/18/4/201

Monson, C. F., & Woolley, A. T. (2003). DNA-templated construction of copper nanowires. *Nano Letters*, *3*(3), 359–363. doi:10.1021/nl034016+

Natarajan, C. M., Tanner, M. G., & Hadfield, R. H. (2012). Superconducting nanowire single-photon detectors: Physics and applications. *Superconductor Science and Technology*, *25*(6), 063001. doi:10.1088/0953-2048/25/6/063001

Neretina, S., Hughes, R. A., Britten, J. F., Sochinskii, N. V., Preston, J. S., & Mascher, P. (2007). Vertically aligned wurtzite CdTe nanowires derived from a catalytically driven growth mode. *Nanotechnology*, *18*(27), 275301. doi:10.1088/0957-4484/18/27/275301

Neubrech, F., Kolb, T., Lovrincic, R., Fahsold, G., Pucci, A., Aizpurua, J., Cornelius, T. W., Toimil-Molares, M. E., Neumann, R., & Karim, S. (2006). Resonances of individual metal nanowires in the infrared. *Applied Physics Letters*, *89*(25), 253104. doi:10.1063/1.2405873

Noborisaka, J., Motohisa, J., & Fukui, T. (2005). Catalyst-free growth of GaAs nanowires by selective-area metalorganic vapor-phase epitaxy. *Applied Physics Letters*, *86*(21), 213102. doi:10.1063/1.1935038

Oh, S. S., Kim, D. H., Moon, M. W., Vaziri, A., Kim, M., Yoon, E., Oh, K. H., & Hutchinson, J. W. (2008). Indium nanowires synthesized at an ultrafast rate. *Advanced Materials*, *20*(6), 1093–1098. doi:10.1002/adma.200702134

Okano, K., Koizumi, S., Silva, S. R. P., & Amaratunga, G. A. (1996). Low-threshold cold cathodes made of nitrogen-doped chemical-vapour-deposited diamond. *Nature*, *381*(6578), 140–141. doi:10.1038/381140a0

Okino, H., Matsuda, I., Hobara, R., Hosomura, Y., Hasegawa, S., & Bennett, P. A. (2005). In situ resistance measurements of epitaxial cobalt silicide nanowires on Si (110). *Applied Physics Letters*, *86*(23), 233108. doi:10.1063/1.1948519

Paladugu, M., Zou, J., Guo, Y. N., Auchterlonie, G. J., Joyce, H. J., Gao, Q., Hoe Tan, H., Jagadish, C., & Kim, Y. (2007). Novel growth phenomena observed in axial InAs/GaAs nanowire heterostructures. *Small*, *3*(11), 1873–1877. doi:10.1002mll.200700222

Parkinson, P., Lloyd-Hughes, J., Gao, Q., Tan, H. H., Jagadish, C., Johnston, M. B., & Herz, L. M. (2007). Transient terahertz conductivity of GaAs nanowires. *Nano Letters*, *7*(7), 2162–2165. doi:10.1021/nl071162x

Peng, F., Su, Y., Ji, X., Zhong, Y., Wei, X., & He, Y. (2014). Doxorubicin-loaded silicon nanowires for the treatment of drug-resistant cancer cells. *Biomaterials*, *35*(19), 5188–5195. doi:10.1016/j.biomaterials.2014.03.032

Peng, F., Su, Y., Ji, X., Zhong, Y., Wei, X., & He, Y. (2014). Doxorubicin-loaded silicon nanowires for the treatment of drug-resistant cancer cells. *Biomaterials*, *35*(19), 5188–5195. doi:10.1016/j.biomaterials.2014.03.032

Peng, X. S., Meng, G. W., Zhang, J., Zhao, L. X., Wang, X. F., Wang, Y. W., & Zhang, L. D. (2001). Electrochemical fabrication of ordered Bi2S3 nanowire arrays. *Journal of Physics. D, Applied Physics*, *34*(22), 3224–3228. doi:10.1088/0022-3727/34/22/304

Pettersson, H., Trägårdh, J., Persson, A. I., Landin, L., Hessman, D., & Samuelson, L. (2006). Infrared photodetectors in heterostructure nanowires. *Nano Letters*, *6*(2), 229–232. doi:10.1021/nl052170l

Poole, P. J., Lefebvre, J., & Fraser, J. (2003). Spatially controlled, nanoparticle-free growth of InP nanowires. *Applied Physics Letters*, *83*(10), 2055–2057. doi:10.1063/1.1608486

Prieto, A. L., Sander, M. S., Martín-González, M. S., Gronsky, R., Sands, T., & Stacy, A. M. (2001). Electrodeposition of ordered Bi2Te3 nanowire arrays. *Journal of the American Chemical Society*, *123*(29), 7160–7161. doi:10.1021/ja015989j

Qi, C., Goncher, G., Solanki, R., & Jordan, J. (2007). SiGe nanowire growth and characterization. *Nanotechnology*, *18*(7), 075302. doi:10.1088/0957-4484/18/7/075302

Qing, Q., Jiang, Z., Xu, L., Gao, R., Mai, L., & Lieber, C. M. (2014). Free-standing kinked nanowire transistor probes for targeted intracellular recording in three dimensions. *Nature Nanotechnology*, *9*(2), 142–147. doi:10.1038/nnano.2013.273

Qing, Q., Pal, S. K., Tian, B., Duan, X., Timko, B. P., Cohen-Karni, T., Murthy, V. N., & Lieber, C. M. (2010). Nanowire transistor arrays for mapping neural circuits in acute brain slices. *Proceedings of the National Academy of Sciences of the United States of America*, *107*(5), 1882–1887. doi:10.1073/pnas.0914737107

Rakovich, Y. P., Volkov, Y., Sapra, S., Susha, A. S., Döblinger, M., Donegan, J. F., & Rogach, A. L. (2007). CdTe nanowire networks: Fast self-assembly in solution, internal structure, and optical properties. *The Journal of Physical Chemistry C*, *111*(51), 18927–18931. doi:10.1021/jp076622p

Sang, S., Wang, Y., Feng, Q., Wei, Y., Ji, J., & Zhang, W. (2016). Progress of new label-free techniques for biosensors: A review. *Critical Reviews in Biotechnology*, *36*(3), 465–481.

Seker, F., Meeker, K., Kuech, T. F., & Ellis, A. B. (2000). Surface chemistry of prototypical bulk II– VI and III– V semiconductors and implications for chemical sensing. *Chemical Reviews*, *100*(7), 2505–2536. doi:10.1021/cr980093r

Sharma, S. K., Jain, A., & Raj, B. (2018). Analysis of triple metal surrounding gate (TM-SG) III–V nanowire MOSFET for photosensing application. *Opto-Electronics Review*, *26*(2), 141–148. doi:10.1016/j.opelre.2018.03.001

Shehada, N., Brönstrup, G., Funka, K., Christiansen, S., Leja, M., & Haick, H. (2015). Ultrasensitive silicon nanowire for real-world gas sensing: Noninvasive diagnosis of cancer from breath volatolome. *Nano Letters*, *15*(2), 1288–1295. doi:10.1021/nl504482t

Shen, F., Wang, J., Xu, Z., Wu, Y., Chen, Q., Li, X., Jie, X., Li, L., Yao, M., Guo, X., & Zhu, T. (2012). Rapid flu diagnosis using silicon nanowire sensor. *Nano Letters*, *12*(7), 3722–3730. doi:10.1021/nl301516z

Skocpol, W. J., Beasley, M. R., & Tinkham, M. (1974). Self-heating hotspots in superconducting thin-film microbridges. *Journal of Applied Physics*, *45*(9), 4054–4066. doi:10.1063/1.1663912

Stadler, K., Masignani, V., Eickmann, M., Becker, S., Abrignani, S., Klenk, H. D., & Rappuoli, R. (2003). SARS—Beginning to understand a new virus. *Nature Reviews. Microbiology*, *1*(3), 209–218. doi:10.1038/nrmicro775

Steinhögl, W., Schindler, G., Steinlesberger, G., & Engelhardt, M. (2002). Size-dependent resistivity of metallic wires in the mesoscopic range. *Physical Review B: Condensed Matter and Materials Physics*, *66*(7), 075414. doi:10.1103/PhysRevB.66.075414

Tan, L. K., Chong, A. M., Tang, X. E., & Gao, H. (2007). Combining atomic layer deposition with a template-assisted approach to fabricate size-reduced nanowire arrays on substrates and their electrochemical characterization. *The Journal of Physical Chemistry C*, *111*(13), 4964–4968. doi:10.1021/jp066841v

Tang, C., Bando, Y., Liu, Z., & Golberg, D. (2003). Synthesis and structure of InP nanowires and nanotubes. *Chemical Physics Letters*, *376*(5-6), 676–682. doi:10.1016/S0009-2614(03)01077-7

Tao, Y., & Degen, C. L. (2015). Single-crystal diamond nanowire tips for ultrasensitive force microscopy. *Nano Letters*, *15*(12), 7893–7897. doi:10.1021/acs.nanolett.5b02885

Taylor, R. F., & Schultz, J. S. (1996). Handbook of Chemical and Biological Sensors. Institute of Physics Publishing.

Tian, F., Zhu, J., & Wei, D. (2007). Phase transition and magnetism of Ni nanowire arrays. *The Journal of Physical Chemistry C*, *111*(19), 6994–6997. doi:10.1021/jp070344c

Tian, M., Wang, J., Kumar, N., Han, T., Kobayashi, Y., Liu, Y., Mallouk, T. E., & Chan, M. H. (2006). Observation of superconductivity in granular Bi nanowires fabricated by electrodeposition. *Nano Letters*, *6*(12), 2773–2780. doi:10.1021/nl0618041

Trahey, L., Becker, C. R., & Stacy, A. M. (2007). Electrodeposited bismuth telluride nanowire arrays with uniform growth fronts. *Nano Letters*, *7*(8), 2535–2539. doi:10.1021/nl070711w

Vaddiraju, S., Chandrasekaran, H., & Sunkara, M. K. (2003). Vapor phase synthesis of tungsten nanowires. *Journal of the American Chemical Society*, *125*(36), 10792–10793. doi:10.1021/ja035868e

Vaddiraju, S., Sunkara, M. K., Chin, A. H., & Ning, C. Z. (2007). Dholakia GR and Meyyappan M. *The Journal of Physical Chemistry C*, *2007*, 111.

Wallentin, J., Persson, J. M., Wagner, J. B., Samuelson, L., Deppert, K., & Borgström, M. T. (2010). High-performance single nanowire tunnel diodes. *Nano Letters*, *10*(3), 974–979. doi:10.1021/nl903941b

Wang, C. A. D. C. M. M. E., Garcia, A., Ingram, D. C., Lake, M., & Kordesch, M. E. (1991). Cold field emission from CVD diamond films observed in emission electron microscopy. *Electronics Letters*, *27*(16), 1459–1461. doi:10.1049/el:19910914

Wang, F., Gao, C., Kuklane, K., & Holmér, I. (2010). A review of technology of personal heating garments. *International Journal of Occupational Safety and Ergonomics*, *16*(3), 387–404. doi:10.1080/10803548.2010.11076854

Wang, S., Zeng, Q., Yang, L., Zhang, Z., Wang, Z., Pei, T., Ding, L., Liang, X., Gao, M., Li, Y., & Peng, L. M. (2011). High-performance carbon nanotube light-emitting diodes with asymmetric contacts. *Nano Letters*, *11*(1), 23–29. doi:10.1021/nl101513z

Wang, Y., Zhang, L., Liang, C., Wang, G., & Peng, X. (2002). Catalytic growth and photoluminescence properties of semiconductor single-crystal ZnS nanowires. *Chemical Physics Letters*, *357*(3-4), 314–318. doi:10.1016/S0009-2614(02)00530-4

Wei, W., Bao, X. Y., Soci, C., Ding, Y., Wang, Z. L., & Wang, D. (2009). Direct heteroepitaxy of vertical InAs nanowires on Si substrates for broad band photovoltaics and photodetection. *Nano Letters*, *9*(8), 2926–2934. doi:10.1021/nl901270n

Xiang, B., Wang, Q. X., Wang, Z., Zhang, X. Z., Liu, L. Q., Xu, J., & Yu, D. P. (2005). Synthesis and field emission properties of TiSi 2 nanowires. *Applied Physics Letters*, *86*(24), 243103. doi:10.1063/1.1948515

Yamada, T., Hayamizu, Y., Yamamoto, Y., Yomogida, Y., Izadi-Najafabadi, A., Futaba, D. N., & Hata, K. (2011). A stretchable carbon nanotube strain sensor for human-motion detection. *Nature Nanotechnology*, *6*(5), 296–301. doi:10.1038/nnano.2011.36

Yang, J. K., Kerman, A. J., Dauler, E. A., Anant, V., Rosfjord, K. M., & Berggren, K. K. (2007). Modeling the electrical and thermal response of superconducting nanowire single-photon detectors. *IEEE Transactions on Applied Superconductivity*, *17*(2), 581–585. doi:10.1109/TASC.2007.898660

Yang, N., Uetsuka, H., Williams, O. A., Osawa, E., Tokuda, N., & Nebel, C. E. (2009). Vertically aligned diamond nanowires: fabrication, characterization, and application for DNA sensing. *Physica Status Solidi (a), 206*(9), 2048-2056.

Yang, W., Auciello, O., Butler, J. E., Cai, W., Carlisle, J. A., Gerbi, J. E., ... Smith, L. M. (2002). DNA-modified nanocrystalline diamond thin-films as stable, biologically active substrates. *Nature Materials*, *1*(4), 253–257. doi:10.1038/nmat779

Yang, Z., Bai, S., Yue, H., Li, X., Liu, D., Lin, S., Li, F., & He, D. (2014). Germanium anode with lithiated-copper-oxide nanorods as an electronic-conductor for high-performance lithium-ion batteries. *Materials Letters*, *136*, 107–110. doi:10.1016/j.matlet.2014.08.034

You, P., Liu, Z., Tai, Q., Liu, S., & Yan, F. (2015). Efficient semitransparent perovskite solar cells with graphene electrodes. *Advanced Materials*, *27*(24), 3632–3638. doi:10.1002/adma.201501145

Yu, D. P., Bai, Z. G., Wang, J. J., Zou, Y. H., Qian, W., Fu, J. S., ... Xu, J. (1999). Direct evidence of quantum confinement from the size dependence of the photoluminescence of silicon quantum wires. *Physical Review B: Condensed Matter and Materials Physics*, *59*(4), R2498–R2501. doi:10.1103/PhysRevB.59.R2498

Yuan, X. Y., Wu, G. S., Xie, T., Lin, Y., & Zhang, L. D. (2003). Self-assembly synthesis and magnetic studies of Co–P alloy nanowire arrays. *Nanotechnology*, *15*(1), 59–61. doi:10.1088/0957-4484/15/1/011

Zach, M. P. (2000). KH Ng and RM Penner. *Science*, *290*, 2120. doi:10.1126cien ce.290.5499.2120

Zhang, G. J., Huang, M. J., Liu, E. T., & Desai, K. V. (2011). Self-assembled monolayer-assisted silicon nanowire biosensor for detection of protein–DNA interactions in nuclear extracts from breast cancer cell. *Biosensors & Bioelectronics*, *26*(7), 3233–3239. doi:10.1016/j.bios.2010.12.032

Zhang, G. J., Zhang, L., Huang, M. J., Luo, Z. H. H., Tay, G. K. I., Lim, E. J. A., Kang, T. G., & Chen, Y. (2010). Silicon nanowire biosensor for highly sensitive and rapid detection of Dengue virus. *Sensors and Actuators. B, Chemical*, *146*(1), 138–144. doi:10.1016/j.snb.2010.02.021

Zhang, H., Zuo, M., Tan, S., Li, G., & Zhang, S. (2006). Carbothermal reduction/sulfidation synthesis and structural characterization of PbS nanobelts and nanowires. *Nanotechnology*, *17*(12), 2931–2936. doi:10.1088/0957-4484/17/12/018

Zhu, J., Peng, H., Chan, C. K., Jarausch, K., Zhang, X. F., & Cui, Y. (2007). Hyperbranched lead selenide nanowire networks. *Nano Letters*, *7*(4), 1095–1099. doi:10.1021/nl0700393

Zhu, Y., Dou, X., Huang, X., Li, L., & Li, G. (2006). Thermal properties of Bi nanowire arrays with different orientations and diameters. *The Journal of Physical Chemistry B*, *110*(51), 26189–26193. doi:10.1021/jp0652493

Zhuo, S., Xun, M., Li, M., Kong, X., Shao, R., Zheng, T., & Li, Q. (2020). Rapid and Label-free Optical Assay of S-layer Protein with High Sensitivity Using TiO2-coated Porous Silicon-based Microfluidic Biosensor. *Sensors and Actuators. B, Chemical*, *321*, 128524. doi:10.1016/j.snb.2020.128524

Zou, Z. Q., Wang, H., Wang, D., Wang, Q. K., Mao, J. J., & Kong, X. Y. (2007). Epitaxial growth of manganese silicide nanowires on Si (111)-7×7 surfaces. *Applied Physics Letters*, *90*(13), 133111. doi:10.1063/1.2717580

Chapter 3
Comparative Performance Analysis of Nanowire and Nanotube Field Effect Transistors

Raj Kumar
Panjab University, Chandigarh, India

Shashi Bala
Chandigarh Engineering College, Landran, India

Arvind Kumar
Panjab University, Chandigarh, India

ABSTRACT

To have enhanced drive current and diminish short channel effects, planer MOS transistors have migrated from single-gate devices to three-dimensional multi-gate MOSFETs. The gate-all-around nanowire field-effect transistor (GAA NWFET) and nanotube or double gate-all-around field-effect transistors (DGGA-NTFET) have been proposed to deal with short channel effects and performance relates issues. Nanowire and nanotube-based field-effect transistors can be considered as leading candidates for nanoscale devices due to their superior electrostatic controllability, and ballistic transport properties. In this work, the performance of GAA NWFETs and DGAA-NT FETs will be analyzed and compared. III-V semiconductor materials as a channel will also be employed due to their high mobility over silicon. Performance analysis of junctionless nanowire and nanotube FETs will also be compared and presented.

DOI: 10.4018/978-1-7998-6467-7.ch003

INTRODUCTION

In the last few decades, the IC technology industry has seen rapid growth as the feature size of the device has shifted to a nanometer from the micrometer regime to achieve the significant packaging density and remarkable processing speed. However, device size needed to reduce to meet the unceasing demand of having the maximum number of CMOS devices per unit area of an IC. Thus, numerous detrimental effects appear in the switching characteristics of the transistors. Although, some scaling rules are required to follow for decreasing the feature size of the device from one technology node to another to minimize these damaging effects. The Subthreshold characteristics of nanoscale devices deteriorate severely, limit more scaling of conventional MOS transistors. To do continue scaling of ICs along with following the International Technology Roadmap for Semiconductors (ITRS) (ITRS, 2015), (S.M. Sze, 2009), (J.P. Colinge, 2008) there is a need to develop and investigate some other non-conventional CMOS devices. The gate length of MOSFET has reached a sub-22nm regime within a few years. The advanced architectures and materials have been proposed to overcome device scaling concerns. Thereby, the performance of the device gets enhanced for high processing speed, low operating power, and standby power applications (Skotnicki et.al, 2005), (ITRS, 2010), (J.M. Larson and J.P. Snyder,2006),(Colinge et al.,2010),(Kranti et al.,2010),(Goel, et al.,2017).

There are various devices such as FinFET, ultra-thin body-based single, and multiple-gate silicon-on-insulator (SOI) FETs. Then, cylindrical nanowire wrapped in a unique circular gate produces GAA MOSFETs (Y.B. Kim, 2008), (J. P. Colinge, 2028). Although, Si-nanowire GAA MOSFET exhibits tremendous short-channel-effects immunity, some technical issues including a large area of contact and high series resistance refrain its usefulness for ULSI (H.M. Fahad and M.M. Hussain, 2012). Moreover, its drive current is also lower than other MOSFETs (J. Dura et al., 2011), (Autran et al., 2005). In contrast to that, a novel silicon-nanotube (Si-NT) FET has been proposed, fabricated, and investigated, maybe a worthy option for future ULSI (D.Tekleab, 2014), (Tekleab, et al.,2012). The Si-NT FET consists of a tube based Si-channel surrounded by a cylindrical shell (outer) and core (inner) gates. It has tremendous short-channel effects immunity attributed to the presence of two gates, responsible for volume inversion. Thereby, it possesses excellent ON and OFF state electrical characteristics (Fahad et al.,2011), (Tiwari et al.,2016), (Kumar et al.,2017),(Scarlet et al.,2017).

In this work, a performance comparison of Nanotube-NT and nanowire-NW FETs has been presented. The variation effect of the design parameter of both NT and NW FETs have been Compared. This analysis has been carried out for various gate dielectrics i.e. SiO2 and HfO2. High-k Gate oxide material and its thickness affect the performance of the device by decreasing leakage current and suppressing short channel effects (SCEs) (Baidya et al., 2016).High- k gate oxide devices can be used in future technologies for CMOS circuit applications.

Multi-Gate Transistor

Multi-Gate transistors have given the alternative to silicon-on-insulator MOSFETs to increase the drive current and the short-channel effects immunity. Double and Tri-Gate FETs, FinFETs, and Gate-All-Around nanowire FETs are some examples of Multi-Gate Transistors. In double-gate, two metals are fabricated on top and bottom of the device. Similarly, for Tri-gate MOFFET, a single gate is fabricated on three sides of the transistor. However, two different gate electrodes are biased with different voltages for MIGFET (Multiple-Independent Gate FET).

Double Gate and FinFET

Double gate MOSFET as shown in Fig 1(a), designed by inserting a fully-depleted SOI device between two gate electrodes of the same potential for controlling SCEs. The double gate device possesses better electrostatic control over the channel region compared with the planner SOI MOSFET. The electric field at drain impacts the channel region very less. Initially, depleted Lean-channel double-Gate MOSFET fabricated Iñiguez et al., 2006) FinFET is the modification of DELTA as shown in Fig 1(b), as having a similar structure to DELTA except for an oxide layer on the channel, referred as hard mask. The parasitic channels created in the inversion corners of the device can be prevented by the hard mask. The Multiple-Independent Gate FET (MIGFET) is similar to double gate FET with an exception of different biasing at gate electrodes. The threshold voltage of the MIGFET can be tuned by altering the potential at one of the gates, considered as most effective characteristics of the device (G. P. Wiederrecht,2010),(C. A. Richter et al.,2008),(Ahmad et al.,2011).

Figure 1. (a) Structure of a Double-Gate Transistor (b) Structure of a FinFET.

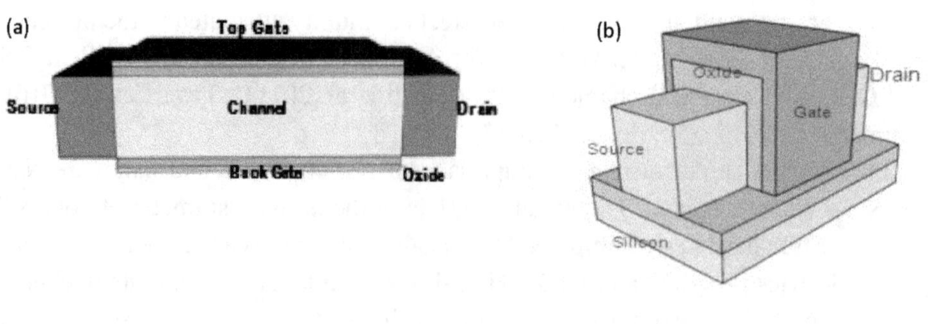

Gate-All-Around Nanowire FETs (GAA-NW FETs)

Gate-all-around (GAA) nanowire MOSFET is considered one of the most attractive multiple-gate structures to diminish the SCEs in SOI MOSFETs and suitable for scaling of the CMOS devices. As the channel thickness of the device shrinks electrostatic control improves, as shown in Fig 2. GAA MOSFETs offer better carrier transport properties than planar devices owing to high carrier mobility on sides of the device. The appropriate design parameter of GAA MOSFETs required to be selected for achieving enhanced transport properties (Young, et al., 2006). The ratio of channel width to the gate length can be increased by using GAA MOSFETs. SCEs can be suppressed up to a great extent subjected to optimize design parameters of GAA MOSFETs. GAA structure of nanowire FET may be an ultimate successor to the FinFET device as offers tremendous electrostatic control over the channel to suppress SCEs. NW GAA MOSFETs integrate a channel and high- dielectric into the gate-all-around device structure (J. J. Gu. et al., 2012). The GAA MOSFETs are more immune to radiation due to its unique structure, maybe interested candidate space applications.

Figure 2. GAA Nanowire FET a) 3D structure of NW-FET b) Circular cross-section view c) 2D cross-sectional view

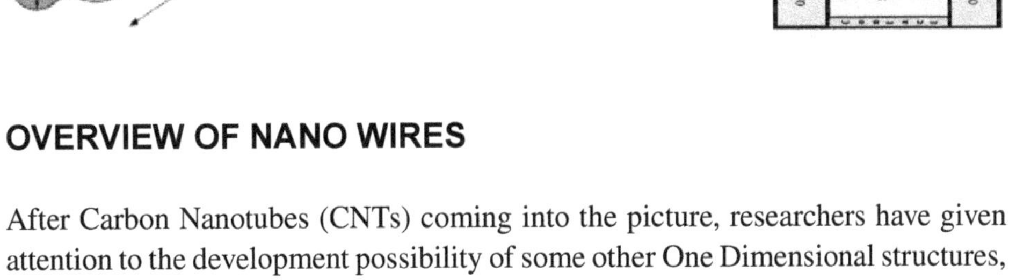

OVERVIEW OF NANO WIRES

After Carbon Nanotubes (CNTs) coming into the picture, researchers have given attention to the development possibility of some other One Dimensional structures, such as nanorods, nanoribbon, NWs and nanobelts (Wan, et al.,2009), (Rao et al.,2005) (Wolf et al.,2006). Some recent investigations have declared, some inorganic NWs as active components in transistors. Various inorganic materials based NWs have been synthesized and characterized in the last few years (Cogenda, 2017), (Aakash et al.,2018), (Sanjeev et al.,2017), (Aakash et al.,2016), (Sanjeev et al.,2016), (Sanjeev et al.,2015).

In one dimensional NW, the length to width ratio must be greater than 10 and limit the width up to a few tens of nanometers (Raj et al., 2017). Nowadays, wires of atomic and molecules dimensions are capable of providing remarkable physical properties regardless of their geometrical characteristics. NWs of the smallest dimension are capable of efficient transport of charge carriers. Thereby, these are used as channel of devices and interconnect in nanoelectronics and nano-optoelectronics (Vishvakarma et al., 2007), (Vijay et al.,2014), (Sharma et al.,2007). NWs have become very attractive for nanotechnology applications. NWs exhibit two quantum-confined directions unlike other low dimensional systems. However, one remains unconfined direction for electrical conduction (Baral et al., 2020). Nowadays, numerous types of NWs are being fabricated and characterized.

Electrical Properties of NWs

Since NWs exhibit distinct mechanisms, electrical characteristics of nanostructures and nanomaterials become quite complex with the variation of its size. These mechanisms are categorized into four types: surface scattering encompasses grain boundary scattering, Coulomb charging, quantized conduction along with ballistic conduction, tunneling, and widening of the bandgap. Moreover, the electrical conductivity of nanostructures and nanomaterials may be enhanced by the reduction of impurity, structural defects, and dislocations (H. E. Schaefer, 2010).

Electrons and holes get spatially confined with a reduction of material size less than the de-Broglie wavelength, leading to creating electric dipoles. Moreover, discrete electronic energy levels get formed in all types of materials. The energy gap between two adjacent levels increases as dimensions of material reduces. The Electron Density of States (DOS) may also depend on the dimension of nanomaterials (G. Cao, 2004). Whereas DOS is square-root dependent for bulk based systems, for Two Dimensional (2D) quantum well structures a staircase response has been found, spikes have been shown by 1D quantum wires (QWs), and discrete features can be seen in Zero Dimensional quantum dots (Zheng et al., 2004).as depicted in Figure 3.

Ballistic conduction occurs in quantum wires, as the length of the device is reduced less than the electron mean free path. In Ballistic conduction, total conductance is realized by each transverse waveguide mode or conducting channel. The energy dissipation and elastic scattering are also prohibited in ballistic conduction (G. Cao, 2004). In the end, tunneling conduction happens, when charge carriers travel through a dielectric medium, sandwiched between two conductors. Electric lines of force from gate and channel tend to overlap inside the dielectric material attributed to thin oxide (P. J. Burke, 2007).

Figure 3. DOS for 3D, 2D, 1D and 0D quantum architecture (P. J. Burke, 2007)

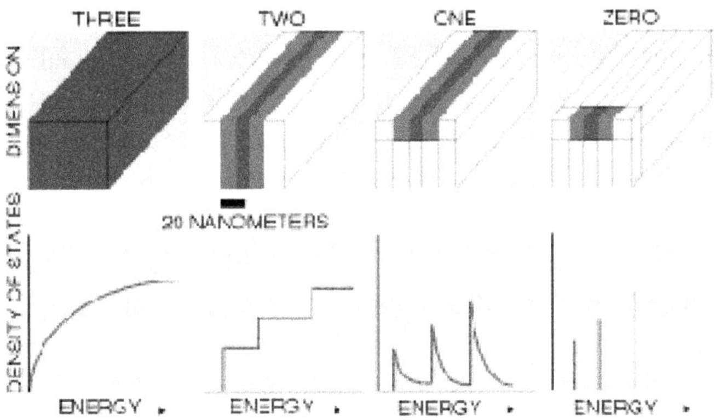

SiNWs have been declared as one of the potential candidates for future nanoscale devices such as MOSFETs, sensors, batteries, and solar cells (Wan, et al., 2009). NWs are attractive due to their inherent ability to realize nanoscale devices for many years. Vertical NWs of several materials and heterostructures are embedded within these wires to make quantum dots and resonant-tunneling diodes (P. J. Burke, 2007).

Carrier Transport in NW FETs

As the channel length of the device less than mean free path, carriers do not undergo any of the scattering mechanisms during their transport within the device and known as ballistic transport (Vilouras et al., 2017). Since, the feature size of SiNW FETs in the range of nanometer posses ballistic transport. In real devices, the charge carrier's mobility may be affected by the scattering events in immeasurable ways. This encompasses carrier to carrier scattering, surfaces, interfacial roughness scattering, acoustic phonons, plasmons, impurities, and optical phonons (J. E. Spanier, 2006) Although, recent studies claim ballistic transport in nanoscale transistors, in real devices shows some types of scattering. The performance of SiNW FETs is highly affected by scattering mechanisms (M. T. L. Peng, 2006) Apart from that, Remote-Coulomb Scattering (RCS), phonon scattering, and Surface-Roughness (SR) are analyzed. A non-perturbative approach has been used to investigate SR and RCS in devices with random realizations of rough Si/SiO2 interface and fixed-charge distributions for a high-κ/SiO2 interface. The effective mobility of SiNW FETs can be observed by seeing the effect of SR and RCS on carrier's transportation. Effective mobility is considered as an imperative performance metric even in the quasi-ballistic regime. The different scattering mechanisms have been analyzed even at 10 nm gate length and gives valuable information on mobility (P. H. Aaen et al., 2007).

STRUCTURE OF NANOTUBE TRANSISTORS

In 2011, a novel structure of Si-NT FETs was proposed, having silicon tube based channel wrapped around by inner and outer gates (Colinge et al.,2010). They have found that the Nanotube device due to its unique structure has tremendous electrical control over the channel leading to provide better SCEs resistance and volume inversion. The fabrication of NT structure was done by (Tekleab, et al., 2012)and registered patent in the year 2012. Then, NT MOSFETs described superior to GAA nanowire MOSFETs by presenting subthreshold Characteristics (D. Tekleab, 2014).

Fig 4(a) depicts the 3D structure of NT MOSFET consists of Silicon tube based channel surrounded by the inner and outer gates. The 2D cross-sectional view of device is shown in Fig. 4(b). The inner gate wrapped by the oxide layer is responsible for extra hold over the channel. Here, thickness of inner and outer gate oxide would be same, but the effective oxide thickness is different, known as Symmetrical Nanotube FETs (Tiwari et al., 2016), (Kumar et al., 2017) Fig 4(c) shows the circular cross-sectional view to present different sections of the device. Both inner and outer gates need to be interconnected to apply same gate voltage V_G.

Figure 4. GAA Nanotube FET a) 3D structure of Si- NT FET b) 2D cross-sectional view c) Circular cross-section view

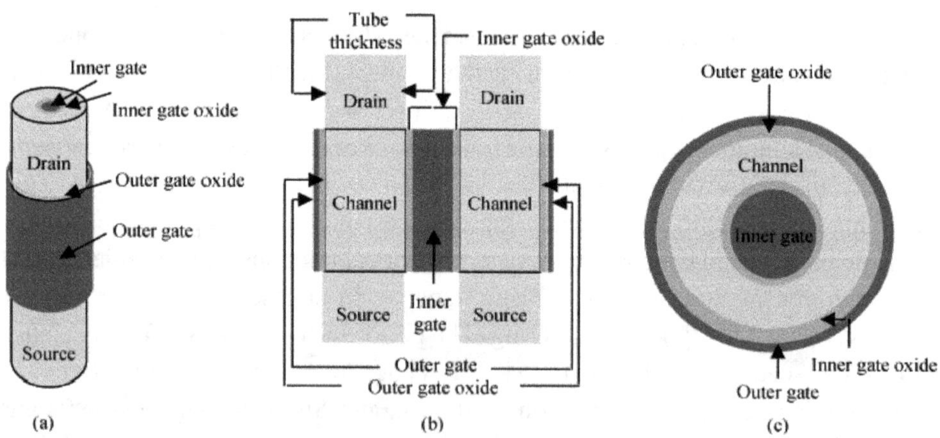

RESULTS AND DISCUSSION

The design parameters used to design the GAA-NW and DGAA-NT MOSFETs have been described in table.1. Visual TCAD, a genius 3D device simulator by Cogenda Pvt. Ltd. has been used to simulate the NT and NW FET's Structures. Lombardi surface mobility model is taken for doping dependent bulk mobility, mobility degradation due to acoustic phonon scattering, and surface roughness scattering. Fermi-carrier statistics have also been included along with impact ionization of carrier under the high lateral electric field for simulation.

In this section, I-V transfer characteristics of Si-NW and Si-NT FETs with the same and different channel diameter have been compared. Leakage current variations have also been presented on varying gate oxide thickness and channel thickness of NW and NT FETs. Band diagrams are also presented to understand the tunneling leakage in the off state of the device.

Table 1. Design parameter for simulation

Design Parameter	GAA-NW	DGAA-NT
Gate length (L_g)	20nm	20nm
Channel thickness (t_{si})	5-20nm	5-10nm
Oxide thickness (t_{ox})	1-5 nm	1-5 nm
Inner Gate core radius (t_c)	-	2.5 nm
Oxide material	Sio_2, Hfo_2	Sio_2, Hfo_2
Chanel material	Silicon	Silicon
Source/ Drain doping N_D	1×10^{20} cm^{-3}	1×10^{20} cm^{-3}
Channel doping N_A	1×10^{15} cm^{-3}	1×10^{15} cm^{-3}
Source/ Drain length	25nm	25nm

The drive current of Si-NT FET is achieved significantly larger than Si-GAA-NW FET with gate length (L_g) 20 nm and 10 nm thick channels for both, attributed to the channel being surrounded by the gates from inner as well outer sides unlike in a GAA MOSFET. The gates control the channel charges effectively give rise to the improved sub-threshold properties of SiNT FETs. The full volume inversion is responsible for high carrier mobility in the channel as oxide-interface trapped charges and surface roughness related scattering reduces. The cross-sectional area of Si-NT FET is also one of the reasons to have better ON current than GAA MOSFET. However, the OFF current is significantly increased in NT-FETs than NW FETs, as depicts in Fig. 5(a). The tunnelling width from source to drain for NT FET is less than NW FET is responsible for more leakage at off state as depicted in Fig. 5(b). It is to be noted that the off current (I_{off}) drop-down and drive current (I_{on}) reduces as well with increasing gate length L_g. it is generally seen as moves from the short gate length to the long-gate length regime. Therefore, in the case of the multiple-gate device, it is important to keep the channel thickness of nanotube less than or equal to half of the gate length (Lg/2) to gain sufficient gate control.

The drive current of Si-NT FET is remaining larger than Si-GAA-NW FET with a 20 nm gate length (L_g) and 10 nm thick (channel thickness) for NT and 20nm for NW. However, OFF-state current increases for NW FETs as the thickness of NW increases as shown in Fig 6.

Figure 5. (a) Comparison of Log I-V characteristics of Nanowire and Nanotube MOSFET (b) Energy Band diagram for NW and NT at L_g=20nm, t_{ox}=2nm, Vds=0.05V, Tsi=10nm(for both NW and NT)

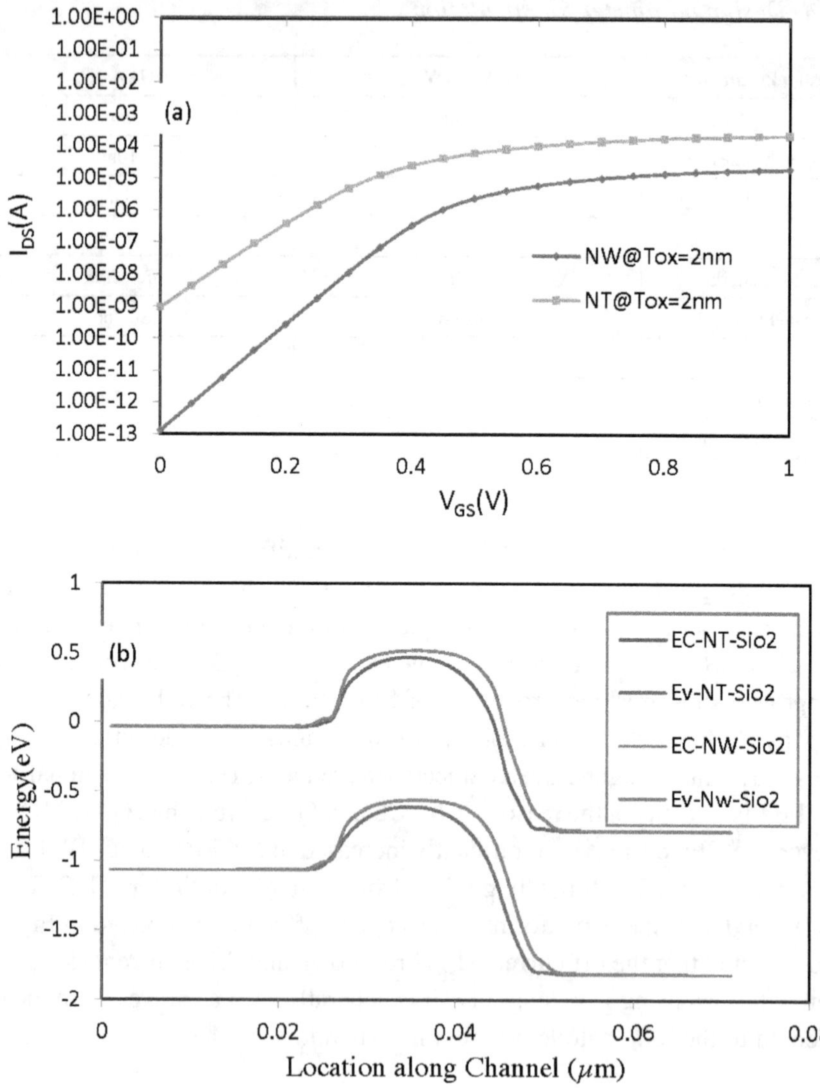

Figure 6. Log Id-Vg characteristics of Nanowire and Nanotube MOSFET at L_g=20nm, t_{ox}=2nm, Vds=0.05V for different channel thickness (NT=10nm and NW=20nm)

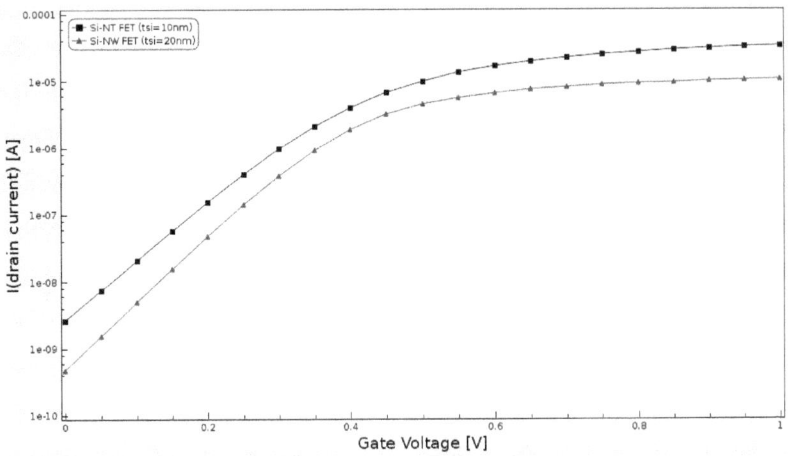

Figure 7. Comparison of Log I_{ds} vs V_{gs} characteristics of Nanowire and Nanotube MOSFET at L_g=20nm, t_{ox}=2nm, Vds=0.75V for different channel thickness

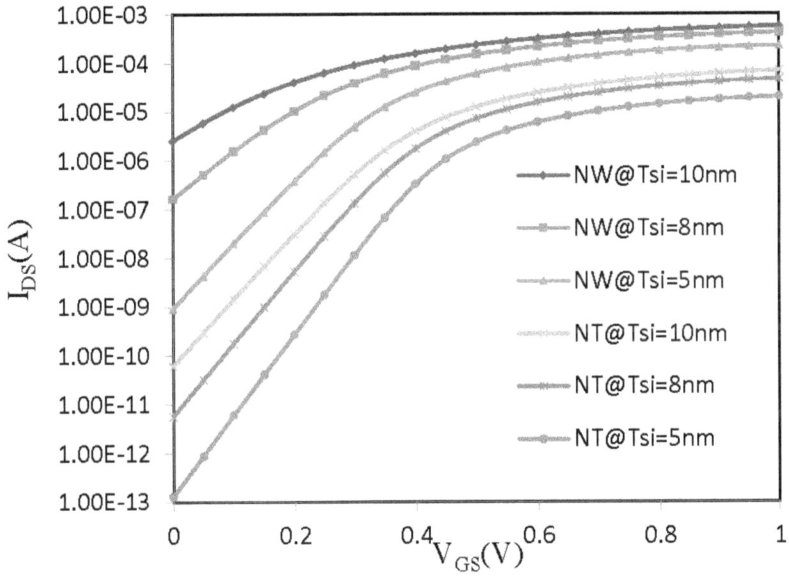

The thickness of nanowires and nanotubes is the critical design parameter to manage the carrier's transportation along the channel. So, it becomes necessary to analyze the performance of devices at different thickness. From Simulation results, it has been observed that subthreshold leakage decreases significantly

Figure 8. Comparison of Log I_{ds} vs V_{gs} characteristics of Nanowire and Nanotube MOSFET at L_g=20nm, t_{si}=5nm, V_{ds}=0.75V for different oxide thickness

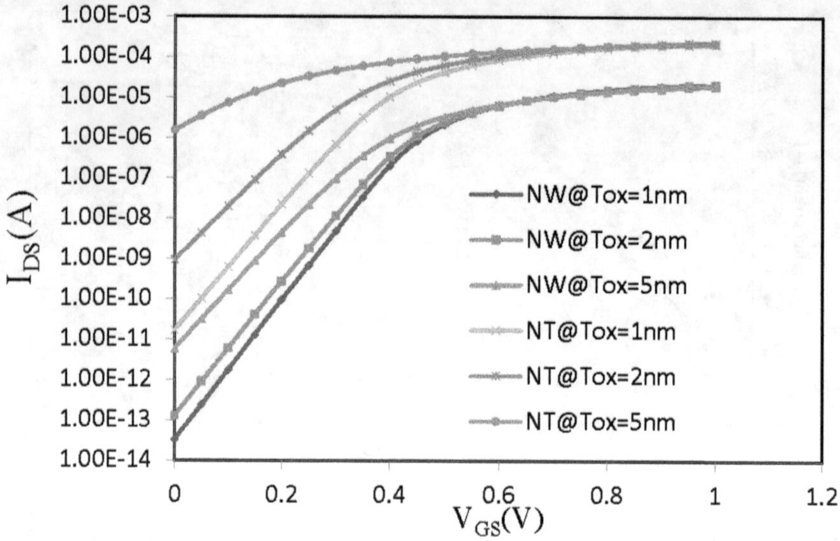

on the reduction of channel thickness as depicted in Fig.7. This is attributed to a strong 2-D quantization of the carriers in the channel. However, high drop-down in the ON-current can be seen as reduces the channel thickness, because, the strong electric field impacts conduction leading to generate hot carrier's effects. Thereby, subthreshold current is reduced by decreasing channel thickness.

Figure 8 depicts the drain current, I_{ds} vs gate voltage, V_{gs} of nanowire and nanotube MOSFETs at different oxide thickness (T_{ox}) and L_g=20nm, t_{si}=5nm, V_{ds}=0.75V. From simulations results, it has been investigated that subthreshold leakage decreases considerably as oxide thickness decreases for both NT and NW FETs, due to weak electrostatic hold of gates over the channel.

The significant increase in gate tunnelling has been observed as the gate oxide thickness decrease below 2 nm, leading to gate leakage current (GIDL). Thereby, the reliability and performance of the device degrade. Therefore, High-k dielectric gate oxide materials have been introduced to have high performance and can be considered for future CMOS technology.

CONCLUSION

In this work, by very rigorous mechanical simulation, NT-MOSFETs are found to be more reliable than GAA NW-FETs. The presence of the core inner gate causes full

Figure 9. (a) Comparison of Log I-V characteristics of Nanowire and Nanotube MOSFETs (b) Energy Band diagram for NW and NT at L_g=20nm, t_{ox}=1nm, Vds=0.75V, T_{si}=5nm for SiO_2 and HfO_2 as dielectric (c) Energy of NW FET with Air and Sio2 dielectric as a spacer

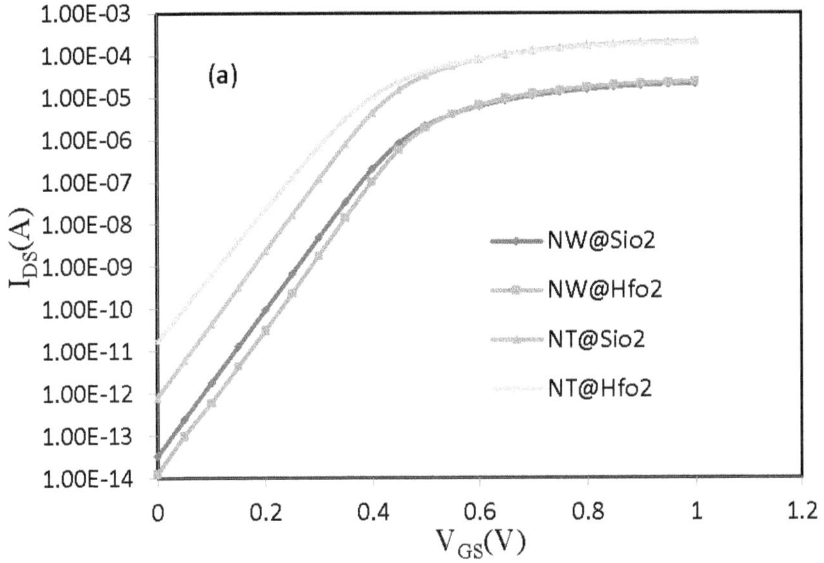

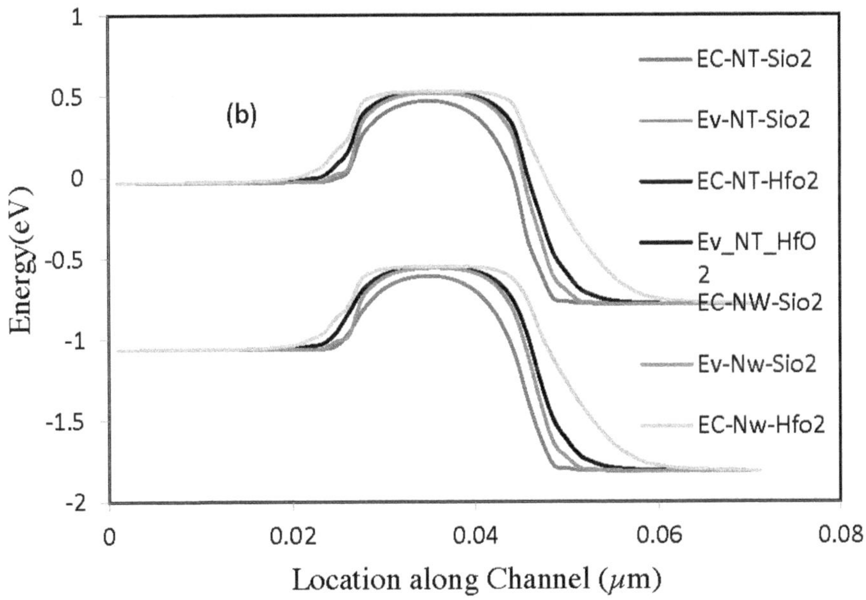

volume inversion of carriers in the ultrathin channel, leading to produce high drive current. Channel thickness found to be a very imperative design parameter to select because drive current decreases as channel thickness reduce for both NW and NT FETs. The high-k oxides can be considered to have less gate induced leakage, as the gate oxide thickness reduced below 2nm. The threshold voltage and drive current can also be increased by keeping high-k gate oxide. SiNT FETs with hafnium oxide as gate oxide produced excellent drive current as compared to GAA Si-NW FETs; make it the most suitable candidates for system-on-chip (SoC) design.

REFERENCES

Aaen, P., Plá, J. A., & Wood, J. (2007). *Modeling and characterization of RF and microwave power FETs*. Cambridge University Press. doi:10.1017/CBO9780511541124

Ahmad, H. M. N., Arifuzzman, A. K. M., Nowshin, N., & Akram, C. (2011). Effects of Series resistance, Effective Mobility and Output Conductance on Si-NWFET Based on Y-function Technique. *International Journal of Engineering Science and Technology*, *3*(1).

Autran, J. L., Nehari, K., & Munteanu, D. (2005). Compact modeling of the threshold voltage in silicon nanowire MOSFET including 2D-quantum confinement effects. *Molecular Simulation*, *31*(12), 839–843. doi:10.1080/08927020500314027

Baidya, A., Lenka, T. R., & Baishya, S. (2016). Mixed-mode simulation and analysis of 3D double gate junctionless nanowire transistor for CMOS circuit applications. *Superlattices and Microstructures*, *100*, 14–23. doi:10.1016/j.spmi.2016.08.028

Baral, K., Singh, P. K., Kumar, S., Chander, S., & Jit, S. (2020). Ultrathin body nanowire hetero-dielectric stacked asymmetric halo doped junctionless accumulation mode MOSFET for enhanced electrical characteristics and negative bias stability. *Superlattices and Microstructures*, *138*, 106364. doi:10.1016/j.spmi.2019.106364

Cao, G. (2004). *Nanostructures and nanomaterials: synthesis, properties and applications*. World scientific. doi:10.1142/p305

Colinge, J. P. (2008). The new generation of SOI MOSFETs. *Romanian Journal of Information Science and Technology*, *11*(1), 3–15.

Colinge, J. P. (2008). The new generation of SOI MOSFETs. *Romanian Journal of Information Science and Technology*, *11*(1), 3–15.

Colinge, J. P., Lee, C. W., Afzalian, A., Akhavan, N. D., Yan, R., Ferain, I., ... Kelleher, A. M. (2010). Nanowire transistors without junctions. *Nature Nanotechnology, 5*(3), 225-229.

Davies, P. R., & Roberts, M. W. (2007). *Atom resolved surface reactions: nanocatalysis*. Royal Society of Chemistry.

Dura, J., Martinie, S., Munteanu, D., Triozon, F., Barraud, S., Niquet, Y. M., & Autran, J. L. (2011, September). Analytical model of drain current in nanowire MOSFETs including quantum confinement, band structure effects and quasi-ballistic transport: device to circuit performances analysis. In *2011 International Conference on Simulation of Semiconductor Processes and Devices* (pp. 43-46). IEEE. 10.1109/SISPAD.2011.6035045

Fahad, H. M., & Hussain, M. M. (2012). Are nanotube architectures more advantageous than nanowire architectures for field effect transistors? *Scientific Reports*, *2*(1), 1–7. doi:10.1038rep00475 PMID:22741059

Fahad, H. M., Smith, C. E., Rojas, J. P., & Hussain, M. M. (2011). Silicon nanotube field effect transistor with core–shell gate stacks for enhanced high-performance operation and area scaling benefits. *Nano Letters*, *11*(10), 4393–4399. doi:10.1021/nl202563s PMID:21923117

Goel, E., Kumar, S., Singh, B., Singh, K., & Jit, S. (2017). Two-dimensional model for subthreshold current and subthreshold swing of graded-channel dual-material double-gate (GCDMDG) MOSFETs. *Superlattices and Microstructures, 106*, 147–155. doi:10.1016/j.spmi.2017.03.047

Gu, J. J., Wang, X. W., Wu, H., Shao, J., Neal, A. T., Manfra, M. J., ... Ye, P. D. (2012, December). 20–80nm Channel length InGaAs gate-all-around nanowire MOSFETs with EOT= 1.2 nm and lowest SS= 63mV/dec. In *2012 International Electron Devices Meeting* (pp. 27-6). IEEE.

Iñiguez, B., Fjeldly, T. A., Lázaro, A., Danneville, F., & Deen, M. J. (2006). Compact-modeling solutions for nanoscale double-gate and gate-all-around MOSFETs. *IEEE Transactions on Electron Devices*, *53*(9), 2128–2142. doi:10.1109/TED.2006.881007

Jain, A., Sharma, S. K., & Raj, B. (2016). Design and analysis of high sensitivity photosensor using Cylindrical Surrounding Gate MOSFET for low power applications. *Engineering Science and Technology, an International Journal, 19*(4), 1864-1870.

Kim, Y. B. (2010). Challenges for nanoscale MOSFETs and emerging nanoelectronics. *Transactions on Electrical and Electronic Materials*, *11*(3), 93–105. doi:10.4313/TEEM.2010.11.3.093

Kranti, A., Yan, R., Lee, C. W., Ferain, I., Yu, R., Akhavan, N. D., . . . Colinge, J. P. (2010, September). Junctionless nanowire transistor (JNT): Properties and design guidelines. In *2010 Proceedings of the European Solid State Device Research Conference* (pp. 357-360). IEEE.

Kumar, A., Bhushan, S., & Tiwari, P. K. (2017). A threshold voltage model of silicon-nanotube-based ultrathin double gate-all-around (DGAA) MOSFETs incorporating quantum confinement effects. *IEEE Transactions on Nanotechnology*, *16*(5), 868–875. doi:10.1109/TNANO.2017.2717841

Kumar, A., Bhushan, S., & Tiwari, P. K. (2017). A threshold voltage model of silicon-nanotube-based ultrathin double gate-all-around (DGAA) MOSFETs incorporating quantum confinement effects. *IEEE Transactions on Nanotechnology*, *16*(5), 868–875. doi:10.1109/TNANO.2017.2717841

Kumar, A., Bhushan, S., & Tiwari, P. K. (2017). Analytical modeling of subthreshold characteristics of ultra-thin double gate-all-around (DGAA) MOSFETs incorporating quantum confinement effects. *Superlattices and Microstructures*, *109*, 567–578. doi:10.1016/j.spmi.2017.05.038

Larson, J. M., & Snyder, J. P. (2006). Overview and status of metal S/D Schottky-barrier MOSFET technology. *IEEE Transactions on Electron Devices*, *53*(5), 1048–1058. doi:10.1109/TED.2006.871842

Peng, M. T. L. (2007). *Modeling the Effect of Velocity Saturation in Nanoscale Mosfet* (Doctoral dissertation). Universiti Teknologi Malaysia.

Raj, B., Saxena, A. K., & Dasgupta, S. (2009). Analytical modeling for the estimation of leakage current and subthreshold swing factor of nanoscale double gate FinFET device. *Microelectronics International*, *26*(1), 53–63. doi:10.1108/13565360910923188

Richter, C. A., Xiong, H. D., Zhu, X., Wang, W., Stanford, V. M., Hong, W. K., Lee, T., Ioannou, D. E., & Li, Q. (2008). Metrology for the electrical characterization of semiconductor nanowires. *IEEE Transactions on Electron Devices*, *55*(11), 3086–3095. doi:10.1109/TED.2008.2005394

Scarlet, S. P., Ambika, R., & Srinivasan, R. (2017). Effect of eccentricity on junction and junctionless based silicon nanowire and silicon nanotube FETs. *Superlattices and Microstructures*, *107*, 178–188. doi:10.1016/j.spmi.2017.04.015

Schaefer, H. E. (2010). *Nanoscience: the science of the small in physics, engineering, chemistry, biology and medicine*. Springer Science & Business Media. doi:10.1007/978-3-642-10559-3

Sharma, S., Raj, B., & Khosla, M. (2016). Comparative Analysis of MOSFET CNTFET and NWFET for High Performance VLSI Circuit Design. A Review. *Journal of VLSI Design Tools & Technology*, *6*(2), 19–32.

Sharma, S. K., Jain, A., & Raj, B. (2018). Analysis of triple metal surrounding gate (TM-SG) III–V nanowire MOSFET for photosensing application. *Opto-Electronics Review*, *26*(2), 141–148. doi:10.1016/j.opelre.2018.03.001

Sharma, S. K., Raj, B., & Khosla, M. (2015, October). Performance enhancement of junctionless nanowire FET with laterally graded channel doping and high-K spacers. In *2015 IEEE 4th Global Conference on Consumer Electronics (GCCE)* (pp. 556-559). IEEE. 10.1109/GCCE.2015.7398608

Sharma, S. K., Raj, B., & Khosla, M. (2016). A Gaussian approach for analytical subthreshold current model of cylindrical nanowire FET with quantum mechanical effects. *Microelectronics Journal*, *53*, 65–72. doi:10.1016/j.mejo.2016.04.002

Sharma, S. K., Raj, B., & Khosla, M. (2017). Subthreshold Performance of In 1–x Ga x As Based Dual Metal with Gate Stack Cylindrical/Surrounding Gate Nanowire MOSFET for Low Power Analog Applications. *Journal of Nanoelectronics and Optoelectronics*, *12*(2), 171–176. doi:10.1166/jno.2017.1961

Sharma, V. K., Pattanaik, M., & Raj, B. (2014). PVT variations aware low leakage INDEP approach for nanoscale CMOS circuits. *Microelectronics and Reliability*, *54*(1), 90–99. doi:10.1016/j.microrel.2013.09.018

Sharma, V. K., Pattanaik, M., & Raj, B. (2014). ONOFIC approach: Low power high speed nanoscale VLSI circuits design. *International Journal of Electronics*, *101*(1), 61–73. doi:10.1080/00207217.2013.769186

Skotnicki, T., Hutchby, J. A., King, T. J., Wong, H. S., & Boeuf, F. (2005). The end of CMOS scaling: Toward the introduction of new materials and structural changes to improve MOSFET performance. *IEEE Circuits and Devices Magazine*, *21*(1), 16–26. doi:10.1109/MCD.2005.1388765

Skotnicki, T., Hutchby, J. A., King, T. J., Wong, H. S., & Boeuf, F. (2005). The end of CMOS scaling: Toward the introduction of new materials and structural changes to improve MOSFET performance. *IEEE Circuits and Devices Magazine*, *21*(1), 16–26. doi:10.1109/MCD.2005.1388765

Song, J. Y., Choi, W. Y., Park, J. H., Lee, J. D., & Park, B. G. (2006). Design optimization of gate-all-around (GAA) MOSFETs. *IEEE Transactions on Nanotechnology*, *5*(3), 186–191. doi:10.1109/TNANO.2006.869952

Spanier, J. E. (2006). One-dimensional semiconductor and oxide nanostructures. *Nanotubes and Nanofibers*, 199-232.

Sze, S. M. (2008). *Semiconductor devices: physics and technology*. John Wiley & Sons.

Tekleab, D. (2014). Device performance of silicon nanotube field effect transistor. *IEEE Electron Device Letters*, *35*(5), 506–508. doi:10.1109/LED.2014.2310175

Tekleab, D., Tran, H. H., Sleight, J. W., & Chidambarrao, D. (2014). U.S. *Patent No. 8,871,576*. Washington, DC: U.S. Patent and Trademark Office.

Tiwari, P. K., Samoju, V. R., Sunkara, T., Dubey, S., & Jit, S. (2016). Analytical modeling of threshold voltage for symmetrical silicon nano-tube field-effect-transistors (Si-NT FETs). *Journal of Computational Electronics*, *15*(2), 516–524. doi:10.100710825-016-0819-0

Tiwari, P. K., Samoju, V. R., Sunkara, T., Dubey, S., & Jit, S. (2016). Analytical modeling of threshold voltage for symmetrical silicon nano-tube field-effect-transistors (Si-NT FETs). *Journal of Computational Electronics*, *15*(2), 516–524. doi:10.100710825-016-0819-0

Vilouras, A., Heidari, H., Gupta, S., & Dahiya, R. (2017). Modeling of CMOS devices and circuits on flexible ultrathin chips. *IEEE Transactions on Electron Devices*, *64*(5), 2038–2046. doi:10.1109/TED.2017.2668899

Vishvakarma, S. K., Agrawal, V., Raj, B., Dasgupta, S., & Saxena, A. K. (2007). Two dimensional analytical potential modeling of nanoscale symmetric double gate (SDG) MOSFET with ultra thin body (UTB). *Journal of Computational and Theoretical Nanoscience*, *4*(6), 1144–1148. doi:10.1166/jctn.2007.2390

Wan, Y., Sha, J., Chen, B., Fang, Y., Wang, Z., & Wang, Y. (2009). Nanodevices based on silicon nanowires. *Recent Patents on Nanotechnology*, *3*(1), 1–9. doi:10.2174/187221009787003348 PMID:19149750

Wiederrecht, G. (2010). *Handbook of nanoscale optics and electronics*. Academic Press.

Wolf, E. L. (2004). *Nanophysics and nanotechnology*. Weinaheim: WIEYaVCH.

Young, K. K. (1989). Short-channel effect in fully depleted SOI MOSFETs. *IEEE Transactions on Electron Devices*, *36*(2), 399–402. doi:10.1109/16.19942

Zheng, G., Lu, W., Jin, S., & Lieber, C. M. (2004). Synthesis and fabrication of high-performance n-type silicon nanowire transistors. *Advanced Materials*, *16*(21), 1890–1893. doi:10.1002/adma.200400472

Chapter 4
Diameter Dependent Ultrasonic Investigation of SiC Nanowires

Sudhanshu Tripathi
https://orcid.org/0000-0001-8865-7987
Amity School of Engineering and Technology, Amity University, Noida, India & University School of Information, Communication and Technology, Guru Gobind Singh Indraprastha University, Delhi, India

Rekha Agarwal
Amity School of Engineering and Technology, Amity University, Noida, India

Rashmi Vashisth
Amity School of Engineering and Technology, Amity University, Noida, India

Devraj Singh
Department of Physics, Institute of Physical Sciences for Study and Research, V.B.S.P. University, India

ABSTRACT

Despite various methods available theoretically and experimentally for material characterization, the ultrasonic characterization method has proven as one of important method because of its non-destructive evaluation (NDE) technique. This chapter explores the ultrasonic characterization of wide band gap semiconducting material SiC at nanoscale. In the present chapter, diameter dependent nonlinear elastic properties (second and third order elastic constants), thermophysical properties, and ultrasonic properties of single crystalline SiC-nanowires have been computed using Lennard-Jones potentials in the high temperature regime. The size dependent mechanical and ultrasonic properties of the SiC-nanowires and their mechanical stability has also been estimated using the higher order elastic constants

DOI: 10.4018/978-1-7998-6467-7.ch004

of the material. The correlation between size dependent ultrasonic attenuation and thermophysical properties have been established for the SiC-nanowires. Obtained results are discussed in correlation with available theoretical/experimental results for application in nanodevices.

INTRODUCTION

The rapid advancement in the technological domain leads to the faster development in the field of semiconductor devices to the smaller dimension for improvement of their performance-size, integration density, speed, multifunctionality requires high-quality insulators, semiconductors, and metals, which are well characterized (Liu, et al., 2020; Shen, et al., 2020). The importance of the material quality and device reliability assessment lies in the development of the fast, non-destructive, and accurate material technique for the determination of critical material parameters, e.g., elastic parameters, mechanical parameters, optical and electrical parameters etc., under the number of physical conditions (Pushpakaran, et al., 2013; Ponraj, Dhanabalan, Attolini, & Salviati, 2016).The characterization of material provides the knowledge of material properties at different physical conditions. The material structure, synthesis, composition, and processing impress the material properties. These material properties strongly depend on the structure of the material even though the material composition resides unchanged (Ponraj, Dhanabalan, Attolini, & Salviati, 2016). Thus, the microstructural relationship in the materials is of great importance. The primary material characterization techniques are destructive and non-destructive techniques (DT and NDT) in nature (Zheng, Huang, & Xu, 2020). Silicon carbide (SiC) is one of the relevant materials of the IV-IV group, widely used for high temperature (Spaziani & Lu, 2018), high frequency (Pushpakaran, et al., 2013), high radiation structural (Mandal, Muzykov, Krishna, Hayes, & Sudarshan, 2011), thermal and power electronics applications (Bland, et al., 2019). Shen, et al., 2020, have recently studied the properties of SiC-nanowires as a promising candidate for absorbing material with carbon-based absorbers, polymer composites, and ceramic compositing for different applications. SiC governs electrical and optical properties in semiconductor heterostructures due to lattice mismatch (Tse, et al., 2013). However, the discovery of more than two hundred polytypes of SiC crystals, only minimal types of SiC nanowires, has synthesized in a controllable manner. The SiC nanostructures for their excellent electrical, optical, mechanical, and thermal properties have broad application prospects in the field of microelectromechanical systems, nanoelectromechanical systems, biomedical engineering, microelectronics and nanoelectronics devices, etc. (Ponraj, Dhanabalan, Attolini, & Salviati, 2016; Cellini, Gao, & Riedo, 2019; Qi, Musetti, Fu, Zhu, & Huang, 2019). The knowledge about second-order elastic constants (SOECs) and their pressure derivatives, as well as third-order elastic constants (TOECs), enables to analyze the impact of stress due to crystal growth process and device processing mechanism of electronic devices (Łepkowski, Majewski, & Jurczak, 2005). The impurities also affect the thermal conductivity of SiC. As the differences in the atomic mass between the impurities and substituted atom increases, the thermal conductivity decreases (Kawamura, et al., 2008). According to (Zhou, Zuo, Wang, Zhang, & Pan, 2011), the electronic structure and stability of single-crystalline 2H-SiC at the nanoscale and their dependency on size, surface effects have been investigated. Rigorous studies have been done in recent years (Ma, Hao, & Hossain, 2019; He & Lilley, 2008; Liu, et al., 2020) to obtain mechanical properties such as surface effects, toughness, strength, etc. of SiC polytypes at the nanoscale (nanotubes and nanowires). At the nanoscale, SiC nanostructures combine the physical properties of bulk SiC. A variety of methods has developed for SiC nanocrystal fabrication. Recently SiC nanowires have been investigated for various applications. The variation of diameter (size) affecting the material stiffness denoting its mechanical characterization in the linear regime of mechanical deformation (Zeng & Zheng, 2007).

The present chapter deals with the discussion of material characterization using the theoretical model established on the ultrasonic nondestructive evaluation (NDE) technique. Initially, it discusses the brief introduction of ultrasonic NDE based material characterization. Then in the next section, it covers the formulation for the calculation of elastic constants of the material using the potential model approach. Later the concise study of ultrasonic velocity and its type of propagation through anisotropic medium have been discussed. Besides this, the theoretical concepts for the estimation of mechanical quantities related to ultrasonic parameters addressed. Finally, the mechanisms responsible for attenuation of the ultrasonic wave propagating through the crystalline/structured material has discussed. The present work explores the evaluation of the SOECs and TOECs and estimation of the size reliant ultrasonic velocity and the ultrasonic attenuation along with the related non-linear parameters (Grüneisen parameters, specific heat, thermal energy density, acoustic coupling constants, etc.) in SiC nanowires at room temperature (300K) and the establishment of the theories for the computation.

ULTRASONICS: NDE BASED TECHNIQUE FOR MATERIAL CHARACTERIZATION

The material characterization is a field involved with discovering new materials and enhancing previously known materials by improving a greater appreciation of properties under distinct physical conditions. Among the various experimental and theoretical material characterization techniques available, ultrasonic characterization is prevalent due to its non-destructive nature. The frequency of ultrasonic waves lies in the range of 25 kHz to 200 MHz (Pillai, 2010; Kittel, 2007). Ultrasonic waves utilized for both destructive and non-destructive characterization of fluids/ materials at bulk as well as nanoscale. The high frequency and low-intensity ultrasonic waves are used for destructive characterization, whereas low frequency and high-intensity ultrasonics used for the characterization of materials in a non-destructive manner (Pandey & Pandey, 2010). The ultrasonic properties of the materials are related to the material properties under varying physical status like temperature, pressure, size variation at the nanoscale, magnetization, etc. Thus, the computation of the ultrasonic properties is significant for material characterization. To ascertain the inherent properties of the substance under varying physical conditions using ultrasonic NDT, it is required to evaluate the ultrasonic parameters for the characterization of the materials at bulk/nanoscale. The ultrasonic NDT is being used at a wide scale at industries level either during the materials manufacturing process or after manufacturing for inspection and quality control (Pandey, Singh, & Yadav, 2007; Pandey & Pandey, 2010).

The estimation of ultrasonic velocity and ultrasonic attenuation are the two necessary physical quantities used for material characterization via ultrasonic NDT. The ultrasonic velocity deals with the parameters affecting the speed of the phonon propagating through the medium. In comparison, the ultrasonic attenuation provides information about the cause for energy loss during ultrasonic wave propagation (Pandey & Pandey, 2010). The ultrasonic velocity is an important parameter as it directly depends on the materials density, SOECs, and the angle of the propagation of ultrasonic waves with the unique axis of SiC-nanowires. Out of these the materials density and SOECs highly depends on the material composition/ concentration and the physical conditions (e.g., size, temperature, pressure, etc.) (Tripathi, Agarwal, & Singh, Size dependent elastic and thermophysical properties of zinc oxide nanowires, 2019a; Yadav, Yadav, Pandey, & Singh, 2008). Thus, the non-destructive evaluation of ultrasonic velocity not only provides information about the structural properties of the material but also gives a direct understanding of elastic and mechanical behavior of the material under different physical conditions. During the propagation of the ultrasonic waves over the material, the loss in the energy occurs termed as ultrasonic attenuation. Ultrasonic attenuation in the material is correlated with several thermophysical properties (ultrasonic velocities, elastic constants, Grüneisen number, thermal conductivity, thermal energy density, thermal relaxation time, etc.) and occurs as a result of different mechanisms occurs during wave propagation, e.g., electron-phonon interaction (EPI) mechanism, thermal loss, scattering, absorption, phonon-phonon interaction (PPI) mechanism, magnon-phonon interaction (MPI) mechanism, etc. Thus, evaluation of the ultrasonic attenuation is one crucial property for the material characterization (Pandey & Pandey, 2010; Yadav, Yadav, Pandey, & Singh, 2008).

Theory for Computation of SOECs and TOECs

The determination of elastic properties of the chosen material is essential for the application in different fields where the mechanical stability, capability, and durability are of significant interest. The material property, due to which after the removal of external forces acting on it, resumes its original shape, is termed as elasticity. For anisotropic crystal, the stress tensor obtained is proportional to strain tensor, and their coefficients termed as second-order elastic constants (SOECs) or material stiffness constants. Similarly, the energy density of the material may be expanded in the powers of strain. According to Cauchy's theorem, the nth order strain derivative of the free energy density provides nth order elastic constants. Therefore, SOECs and TOECs for the material can be obtained via the second and third-order strain derivatives of energy density, respectively. These elastic constants provide a better understanding of their non-linear elastic behavior. The SOECs and TOECs of hexagonal SiC have been determined using the Lennard Jones potential between the atoms. In contrast, the energy density obtained by consideration of atomic interaction as defined by Lennard Jones potential. Here, the atomic interactions are considered up to second-order neighbourhood (Pillai, 2010; Kittel, 2007).

The nth-order elastic constants of hexagonal closed packed (hcp) crystalline material can be described as nth order partial derivative of the free energy density ($\mathcal{F}$) of the material (Mori & Hiki, 1978) given as

$$C_{ijklmn} = \left(\frac{\partial^n \mathcal{F}}{\partial \eta_{ij} \partial \eta_{kl} \partial \eta_{mn}} \right) \tag{1}$$

The overall free energy density may be expressed in terms of the Lagrangian strain tensor components, using the Taylors series expansion as

$$\mathcal{F} = \sum_{n=0}^{\infty} \mathcal{F}_n = \sum_{n=0}^{\infty} \frac{1}{n!} \left(\frac{\partial^n \mathcal{F}}{\partial \eta_{ij} \partial \eta_{kl} \partial \eta_{mn}} \right) \eta_{ij} \eta_{kl} \eta_{mn} \tag{2}$$

Now restricting the overall free energy density till the 3rd order term, Eq. (2) maybe written as

$$\mathcal{F}_2+\mathcal{F}_3=\frac{1}{2!}\left(\frac{\partial^2\mathcal{F}}{\partial\eta_{ij}\partial\eta_{kl}}\right)\eta_{ij}\eta_{kl}+\frac{1}{3!}\left(\frac{\partial^3\mathcal{F}}{\partial\eta_{ij}\partial\eta_{kl}\partial\eta_{mn}}\right)\eta_{ij}\eta_{kl}\eta_{mn}$$

$$=\frac{1}{2!}C_{ijkl}\eta_{ij}\eta_{kl}+\frac{1}{3!}C_{ijklmn}\eta_{ij}\eta_{kl}\eta_{mn} \tag{3}$$

At finite temperature T, the free energy density $\mathcal{F}$ is the sum of elastical energy density ($\mathcal{U}$) and the vibrational portion of the free energy density. The elastical energy density per unit cell ($\mathcal{U}$) of the material up to the second neighborhood in terms atoms in the basal plane (k) and atoms below and above the basal plane (l) can be written as

$$\mathcal{U}_2+\mathcal{U}_3=\sum_{k=1}^{6}\mathcal{U}_s(k)+\sum_{l=1}^{6}\mathcal{U}_s(l) \tag{4}$$

Where s represents the undeformed state of the crystal. Under homogeneous deformation, the interatomic vector between the deformed state (s') and undeformed state (s) is related to cartesian components of the vectors and Lagrangian strain components η_{ij} given as

$$(s')^2-(s)^2=2\varepsilon_i\varepsilon_j\eta_{ij}=2\rho$$

The elastical energy density has been extended in terms of unit cell volume V_H, material density ρ and interaction potential ϕ_s as (Ghate, 1965)

$$\mathcal{U}_n(\varsigma)=\frac{1}{2V_H}\Sigma\frac{1}{n!}\rho^n\Delta^n\phi_s(\varsigma) \tag{5}$$

Where $\Delta^n=\frac{1}{s}\frac{d}{ds}$, and $\phi_s\left(=\frac{p_0}{s^m}+\frac{q_0}{s^n}\right)$. The interaction potential depends on integers (m, n) and the values of constants (p_0, q_0), respectively. The value of state vector s is the function of the lattice parameters a and c of hcp structured wurtzite type materials. The energy density equation up to the third-order term (Mori & Hiki, 1978) can be simplified using Eq. (4) and Eq. (5) as

$$\mathcal{U}=\mathcal{U}_2+\mathcal{U}_3=\frac{1}{2V_H}\left\{\left[\sum_{k=1}^{6}\frac{1}{2!}\rho^2(k)\Delta^2\phi_s(k)+\sum_{l=1}^{6}\frac{1}{2!}\rho^2(l)\Delta^2\phi_s(l)\right]\right.$$
$$\left.+\left[\sum_{k=1}^{6}\frac{1}{3!}\rho^3(k)\Delta^3\phi_s(k)+\sum_{l=1}^{6}\frac{1}{3!}\rho^3(l)\Delta^3\phi_s(l)\right]\right\} \tag{6}$$

On comparing Eq. (3) with Eq. (6), crystal symmetry provides an expression for six SOECs and ten TOECs, respectively (Pandey, Singh, & Yadav, 2007; Yadav, Yadav, Pandey, & Singh, 2008). The SOECs and TOECs mathematical relations for hcp structured materials are given in Eq. (7) and Eq. (8) respectively.

$$\left.\begin{array}{ll} C_{11}=24.1p^4C & C_{12}=5.918p^4C \\ C_{13}=1.925p^6C & C_{33}=3.464p^8C \\ C_{44}=2.309p^4C & C_{66}=9.851p^4C \end{array}\right\} \tag{7}$$

$$\left.\begin{array}{l} C_{111}=126.9p^2C''+8.853p^4C';\;\; C_{112}=19.168p^2C''-1.61p^4C' \\ C_{113}=1.924p^4C''+1.155p^6C'';\;\; C_{123}=1.617p^4C''-1.155p^6C' \\ C_{133}=3.695p^6C'';\;\; C_{155}=1.539p^4C'';\;\; C_{144}=2.309p^4C'';\;\; C_{344}=3.464p^6C'' \\ C_{222}=101.039p^2C''+9.007p^4C';\;\; C_{333}=5.196p^8C'' \end{array}\right\} \tag{8}$$

Where the axial ratio is given as $p=c/a$. The harmonic and anharmonic properties (χ,ψ) of the materials are the function of lattice parameters, the Lennard-Jones potential parameter (b_0), (m, n) respectively, given as

$$\chi=(8a)^{-(n+4)}nb_0(n-m);\qquad \psi=-\chi\left[6a^2(m+n+6)\right]^{-1}.$$

Also the constants $C'=\chi ap^{-5;}$ $C''=\psi(ap^{-1})^3$ are computed using lattice parameters, harmonic, and anharmonic parameters, respectively.

Theory for Computation of Ultrasonic Velocity

An ultrasonic velocity is a vital tool for material characterization, where the different types of ultrasonic velocities, i.e., longitudinal and shear/transverse wave velocity, have been utilized to study the anisotropic and mechanical parameters of the crystalline material. The ultrasonic velocity of the structured material depends on SOECs, materials density (ρ), and the wave propagation angle with the unique axis of the material (θ). The analysis of the ultrasonic velocity of hcp materials provides information about their anisotropic behavior. For anisotropic materials, different types of ultrasonic velocities associated with each direction: one longitudinal velocity V_L and two shear velocities V_{S1} and $V_{S2,}$ which correspond to transverse acoustics (Yadav, Pandey, & Singh, 2019). The expressions for ultrasonic velocities along the three directions are as follows (Tripathi, Agarwal, & Singh, Size dependent elastic and thermophysical properties of zinc oxide nanowires, 2019a).

$$V_L = \left\{ \left\{ C_{33} cos^2\theta + C_{11} sin^2\theta + C_{44} + \left[C_{11} sin^2\theta - C_{33} cos^2\theta + C_{44} \left(cos^2\theta - sin^2\theta \right) \right]^2 \right.\right.$$
$$\left.\left. +4cos^2\theta sin^2\theta \left(C_{13} + C_{44} \right)^2 \right\} / 2\rho \right\}^{\frac{1}{2}} \tag{9}$$

$$V_{S1} = \left\{ \left\{ C_{33} cos^2\theta + C_{11} sin^2\theta + C_{44} - \left[C_{11} sin^2\theta - C_{33} cos^2\theta + C_{44} \left(cos^2\theta - sin^2\theta \right) \right]^2 \right.\right.$$
$$\left.\left. +4cos^2\theta sin^2\theta \left(C_{13} + C_{44} \right)^2 \right\} / 2\rho \right\}^{\frac{1}{2}} \tag{10}$$

$$V_{S2} = \left\{ \left\{ C_{44} cos^2\theta + C_{66} sin^2\theta \left(C_{13} + C_{44} \right)^2 \right\} / 2\rho \right\}^{\frac{1}{2}} \tag{11}$$

Depending on the value of the angle (θ), the direction of the propagation of ultrasonic wave changes. For $\theta=0°$, the ultrasonic waves propagate along <001> direction and correspondingly Eqs. (9)-(11) reduced to

$$V_L = \{C_{33} / \rho\}^{\frac{1}{2}} ; V_{S1} = V_{S1} = \{C_{44} / \rho\}^{\frac{1}{2}} \tag{12}$$

Similarly, for $\theta=90°$, the ultrasonic waves propagate along 100 direction and correspondingly Eqs. (9)-(11) reduced to

$$V_L = \{C_{11}/2\rho\}^{\frac{1}{2}}; V_{S1} = \{C_{44}/\rho\}^{\frac{1}{2}}; V_{S2} = \{(C_{11}-C_{12})/2\rho\}^{\frac{1}{2}} \tag{13}$$

Thus, the ultrasonic velocity can be evaluated theoretically via the information of SOECs and the material density. The Debye average velocity (V_D) (Oligschleger, Jones, Reimann, & Schober, 1996) obtained via ultrasonic velocities as a weighted average of shear and longitudinal velocities. V_D is a crucial parameter as it averages the sound velocity, which is indirectly related to SOECs of the chosen materials through the ultrasonic velocity. The crystallographic texture information and other lattice thermal phenomena can obtained by utilization of Debye average velocity (V_D), given as

$$\mathrm{V_D} = 3^{-1}\left[V_L^{-3} + V_{S1}^{-3} + V_{S2}^{-3}\right]^{-\frac{1}{3}} \tag{14}$$

The Debye average velocity (V_D) also affects the Debye temperature (θD$_)$ of the material. The Debye temperature is also an influential criterion indicating the temperature at which all the vibrational modes are excited and supports in the further estimation of thermophysical parameters like thermal energy density and specific heat. The Debye temperature varies in direct proportion to Debye average velocity given as

$$\theta_D = \hbar V_D \left(6\pi^2 N_A\right)^{1/3} / k_B \tag{15}$$

Here k_B is Boltzmann constant, $\hbar$ is the reduced-order Plank's constant, and N_A is the Avogadro number.

Theory of Mechanical Properties

The strength, hardness and stability analysis of the material can be analyzed by the mechanical properties like the anisotropy factor (EA_F), Poisson's ratio (υ), shear modulus (G), Young's modulus (Y), Pugh's ratio (B/G), bulk modulus (B), and Vicker's hardness (H_V). The shear modulus and bulk modulus have obtained using the SOECs via Voigt-Reuss and Hill (Voigt, 1928; Reuss, 1929; Hill, 1963) approximation techniques. The expressions for the shear modulus (G_V and G_R) and the bulk modulus (B_V and B_R) using the Voigt and Reuss approximation technique given as

$$\left.\begin{aligned} G_V &= \left(C_{11}+C_{12}+2C_{33}-4C_{13}+12C_{44}+12C_{66}\right)/30 \\ G_R &= 2.5\text{C}/\left[3B_V C_{44}C_{66}+\left(\left(C_{11}+C_{12}\right)C_{33}-2C_{13}^2\right)\left(C_{44}+C_{66}\right)\right] \\ \text{C} &= \left[\left(C_{11}+C_{12}\right)C_{33}-2C_{13}^2\right]C_{44}C_{66} \end{aligned}\right\} \quad (16)$$

Similarly, the lower and upper limits for bulk modulus of the hexagonal structured crystal is given as

$$\left.\begin{aligned} B_R &= \left[\left(C_{11}+C_{12}\right)C_{33}-2C_{13}^2\right]/\left[C_{11}+C_{12}+2C_{33}-4C_{13}\right] \\ B_V &= \left[2\left(C_{11}+C_{12}\right)+C_{33}+4C_{13}\right]/9 \end{aligned}\right\} \quad (17)$$

The average of Voigt and Reuss limits termed as Voigt-Reuss-Hill (VRH) theory (Hill, 1963). The average value for the bulk modulus and the shear modulus using VRH theory given as

$$B=(B_V+B_R)/2;\ G=(G_V+G_R)/2 \quad (18)$$

The Poisson's ratio (PR) is the fundamental thermodynamic parameters of an isotropic elastic medium (Greaves, Greer, Lakes, & Rouxel, 2011). The negative ratio of the lateral strain to axial strain for an axially strained sample determines the PR (υ), which also specified in terms of B and G, as

$$\upsilon = [3(B/G) - 2] / [6(B/G) + 2] \quad (19)$$

Similarly, the B/G ratio can also be applied to determine the Young modulus (Y), given as

$$Y = [9(B/G)] / [3(B/G) + 1] \quad (20)$$

The anisotropy factor (EA_F) explores the degree of isotropy of the material (Lau & McCurdy, 1998), has a significant impact on the substantial properties such as the elastic instability, crack propagation, and anisotropic deformations. The EA_F value close to 1 indicates the isotropic character of the materials plane. The proportionate anisotropy factor for the bulk modulus (EA_B) and the shear modulus (EA_G) (Chung

& Buessem, 1967) and universal anisotropy factor (EA_U) (Ranganathan & Ostoja-Starzewski, 2008) also measures the anisotropic degree of the material. For isotropic material, the EA_B, EA_G, and EA_U must be zero, while the departure from zero measures the anisotropic degree of the material. The expression for evaluation of EA_P, EA_B, EA_G and EA_U given as

$$EA_F = \frac{C_{44}\left(C_{11} + 2C_{13} + C_{33}\right)}{C_{11}C_{33} - C_{13}{}^2} \tag{21}$$

$$EA_B\left(\%\right) = \frac{B_V - B_R}{B_V + B_R} \times 100;\ EA_G\left(\%\right) = \frac{G_V - G_R}{G_V + G_R} \times 100;\ EA_U = 5\frac{G_V}{G_R} + \frac{B_V}{B_R} - 6 \tag{22}$$

The ratio of B to G, i.e., $Pu=B/G$ known as Pugh's indicator, identifies the ductile and brittle nature of the material (Pugh, 1954). The material behaves in a ductile manner for $Pu>1.75$ and in fragile nature for $Pu<1.75$. The PR (υ) also gives an indication about the ductility and brittleness behavior of the material (Zagorac, Zagorac, Djukic, Jordanov, & Matović, 2019), and when $\upsilon>0.26$, the material has ductile nature otherwise the material behaves as brittle. The Pettifor's rule (Pettifor, 1992) defines Cauchy Pressure (CP) as $C_{P100} = C_{13} - C_{44}$ for <100> plane and $C_{P001} = C_{12} - C_{66}$ for <001> plane of hcp crystal. The positive/negative values of CP also indicate the ductility/brittleness of the material. The hardness is an influential criterion of a material which tells about the resistance of materials to penetration or deformation under applied abrasion. Because of the absence of a standard scale, there are some empirical formulas relating to tensile strength and hardness. (Chen, Niu, Li, & Li, 2011) have proposed the Vicker's hardness by considering Pugh's bitterness standard given as

$$H_V = 2\left(G/\left(Pu\right)^2\right)^{0.585} - 3\left(GPa\right) \tag{23}$$

The mechanical stability of the hcp structured materials can be analysed using the Born stability criteria (Born & Huang, 1954) defined based on SOECs as

$$C_{11} > \left|C_{12}\right|; \left(C_{11} + C_{12}\right)C_{13} > 2C_{13}^2; C_{44} > 0; C_{66} > 0 \tag{24}$$

METHOD OF ULTRASONIC ATTENUATION AND RELATED THERMAL PARAMETERS

The ultrasonic attenuation in crystalline material has evaluated using the Mason and Bateman approach (Mason W. P., 1965a; Mason & Bateman,1964). The rate of decay energy of the ultrasonic wave is called ultrasonic attenuation. The loss in strength of ultrasonic waves during its propagation through the material is observed, which is due to the various phenomenon, e.g., scattering, absorption, diffraction, etc. The major component of energy is lost dissipated in the form of heat. The ultrasonic attenuation in different types of materials may arise due to several well-known causes, which depend on the physical properties of the concerned medium. Various possible causes of ultrasonic attenuation (Mason W. P., Effect of impurities and phonon processes on the ultrasonic attenuation of germanium, crystal quartz, and silicon, 1965a; Mason & Bateman, Ultrasonic-wave propagation in pure silicon and germanium, 1964; Mason & Baerwald, Piezoelectric crystals and their applications to ultrasonics, 1951) may be classified as follows:

- Losses due to grain boundaries
- Losses due to lattice imperfection
- Thermo-elastic relaxation mechanism losses
- Phonon-viscosity mechanism losses
- Electron-phonon mechanism losses

Losses Due to Grain Boundaries

The loss due to the grain boundaries (ultrasonic attenuation) mostly occurs in polycrystalline materials and generally not found in perfect/single-crystalline materials. During the propagation of ultrasonic waves of small wavelength comparable to the grain size of the material, the regular reflections occur due to discontinuity of the elastic modulus at the boundaries, which causes loss in energy (Granato & Lücke, 1956). This type of energy loss is mainly dependent on the grain diameter, the anisotropy of the crystal, and the wavelength of ultrasonic waves. Thus, for the ultrasonic wave of small wavelength, the losses are frequency dependent while for the larger wavelength ultrasonic waves, Rayleigh scattering takes place.

Losses Due to Lattice Imperfection

The departure from the regular, symmetrical, and periodic arrangement of atoms within the crystal termed as lattice imperfection. The imperfection gives rise to the absorption of the ultrasonic waves. The attenuation of ultrasonic waves can arise due to edge or screw dislocation (Bhalla, Singh, Jain, & Kumar, 2016), which is due to the interaction of phonons and the vibrational energy of impurity atoms or dislocation. During the ultrasonic wave propagation through the crystal lattice, the energy levels of normal atoms periodically raise/lowers, causes continuous relaxation jumps between the energy states, and hence contribute to the ultrasonic attenuation. The exchange of energy between the ultrasonic wave and dislocation brought under forced vibration causes a relaxation absorption.

Thermo-Elastic Relaxation Mechanism Losses

The thermo-elastic loss within the material is the result of heat energy flow between the regions of rarefied and compressed zone created due to the propagation of ultrasonic longitudinal waves throughout the crystal lattice. In the case of the shear wave, there is no compression or rarefaction, so no loss occurs due to this mechanism. The expression for the thermo-elastic relaxation mechanism loss (Pandey & Pandey, 2010; Mason & Baerwald, 1951; Mason W. P1965a; Mason W., 1955b) given as

$$\left(\frac{\alpha}{f^2}\right)_{TE} = \frac{4\pi^2 \kappa T \left\langle \gamma_i^j \right\rangle^2}{2\rho V_L^5} \tag{25}$$

Here κ signifies the thermal conductivity, $\left\langle \gamma_i^j \right\rangle$ the Grüneisen parameters (GPs). The GPs for hexagonal crystals obtained via SOECs and TOECs (Bala, Singh, Pandey, & Yadav, 2020; Nandanpawar & Rajagopalan, 1982). The thermal conductivity (Clarke, 2003) of the material is given by

$$\kappa = k_B V_D \left(\frac{M}{n\rho N_A}\right)^{-\frac{2}{3}} \tag{26}$$

Also, the thermal relaxation time is the reestablishment time taken by the thermal phonons to achieve thermal equilibrium due to ultrasonic wave propagation, termed as thermal relaxation time (Tripathi, Agarwal, & Singh, 2020b).

$$\tau_{th} = 3\kappa / C_V V_D^2 \tag{27}$$

Where the specific heat per unit volume of the material $C_{V,}$ has been obtained via θ_D/T Tables of AIP Handbook (Gray, 1972). The thermal relaxation time (τ_{th}) for the longitudinal wave (τ_l) and shear wave (τ_s) have related as

$$\tau_{th} = \frac{1}{2}\tau_l = \tau_S \tag{28}$$

Phonon-Viscosity Mechanism Losses

The phonon-viscosity mechanism was first proposed by Akhieser (Yadav, Yadav, Pandey, & Singh, 2008; Dhawan, Wan, Verma, Pandey, & Yadav, 2015), explaining that the interaction of thermal phonons and acoustical phonons in solids are the main reason behind the ultrasonic attenuation at higher temperature regime. Akhieser stated that the ultrasonic wave propagation through material tends to modulate the distribution of thermal phonons of the medium, which attains thermal equilibrium by phonon-phonon interaction (PPI). During the thermal relaxation process, the loss in the ultrasonic energy obtained termed as ultrasonic attenuation. Later the modified form of Akhieser theory was proposed (Oligschleger, Jones, Reimann, & Schober, 1996; Wang, Wang, & Wu, 2010) by consideration of finite thermal relaxation time (τ). The expression for evaluation of ultrasonic attenuation coefficient (Yadav & Singh, 2001; Tosi, 1964) is given by

$$\alpha_{Akh} = \alpha_{p-p} = \frac{\omega^2\tau_{th}\left(DE_0/3\right)}{2\rho V^3\left(1+\omega^2\tau_{th}^2\right)} \Rightarrow \left(\alpha/f^2\right)_{Akh} = \frac{4\pi^2\tau_{th}\left(DE_0/3\right)}{2\rho V^3\left(1+\omega^2\tau_{th}^2\right)} \tag{29}$$

Here E_0 is the thermal energy density of the chosen material, and D is the acoustical coupling constant, which measures transformation between the acoustical energy and thermal energy. The acoustic coupling constant

$$D = 3\frac{\Delta C}{E_0} = 3\left(3E_0\left\langle\left(\gamma_i^j\right)^2\right\rangle - \left\langle\gamma_i^j\right\rangle^2 C_V T\right)/E_0,$$

depends on the thermal parameters like the thermal energy density of the chosen material $E_{0.}$ The parameter ΔC, represents the change in the elastic modulus.

Electron-Phonon Mechanism Losses

This type of ultrasonic attenuation due to electron-phonon interaction (EPI) generally observed in the material at low temperatures (< 100K). The ultrasonic attenuation is a result of the dissipation of the acoustical energy due to the coupling between the ultrasonic wave propagating through the material and the conduction electrons of material. According to the Debye theory of specific heat at low temperatures, the exchange of energy occurs between crystal lattice and free electrons. The mean free path of the conduction band electrons is comparable to the wavelength of the acoustical phonons (Morse, Myers, & Walker, 1960; Mason W., 1955b; Beyer & Letcher, 1969; Bömmel, 1954), hence the probability of the interaction occurs between acoustical phonons and free electrons. The experimental investigation (Bala & Singh, Elastic and ultrasonic properties of fermium monopnictides, 2020; Jassby & Vreeland Jr, 1970) also reveals that the primary cause of acoustical wave attenuation at low temperatures is due to the interaction of phonons with electrons.

Based on the above described theories of ultrasonic attenuation, it is clear that for a perfect non-ferromagnetic, non-ferroelectric, and perfect single crystal, only three reasons are predominantly responsible for ultrasonic attenuation. The three dominant factors are EPI, PPI, and ultrasonic attenuation due to the thermoelastic relaxation mechanism. Thus, The total ultrasonic attenuation $(\alpha/f^2)_{total,}$ is the overall impact of ultrasonic attenuation due to Akhieser losses observed in longitudinal, shear mode of propagation, and thermoelastic loss, respectively. $(\alpha/f^2)_{total}$ is given as

$$(\alpha/f^2)_{total} = (\alpha/f^2)_{TE} + (\alpha/f^2)_{Akh.Long} + (\alpha/f^2)_{Akh.Shear} \tag{30}$$

Where $(\alpha/f^2)_{TE}$ represents thermoelastic loss, $(\alpha/f^2)_{Akh.Shear}$ and $(\alpha/f^2)_{Akh.Long}$ are the shear and longitudinal components of Akhieser losses.

NON-DESTRUCTIVE EVALUATION OF 2H SIC-NANOWIRES

Only limited pieces of literature are available, which explores the non-linear properties of 2H-SiC nanowires (2H SiC-NWs). As per the authors' knowledge, despite the number of theoretical and experimental work done for SiC at the nanoscale, various nonlinear elastic, mechanical, thermophysical, and ultrasonic properties remain untouched. Therefore, the present work describes the size-reliant ultrasonic characterization of 2H SiC-NWs in the hexagonal phase, which includes the determination of its elastic, ultrasonic, mechanical, and thermal, properties.

The hexagonal phase of 2H-SiC at the nanoscale, particularly the wurtzite (wz) structure, with a bandgap energy of 3.33 eV, is a promising candidate for numerous

Figure 1. The cross-sectional view of three diameters of 2H-SiC nanowires (H1, H2, H3)

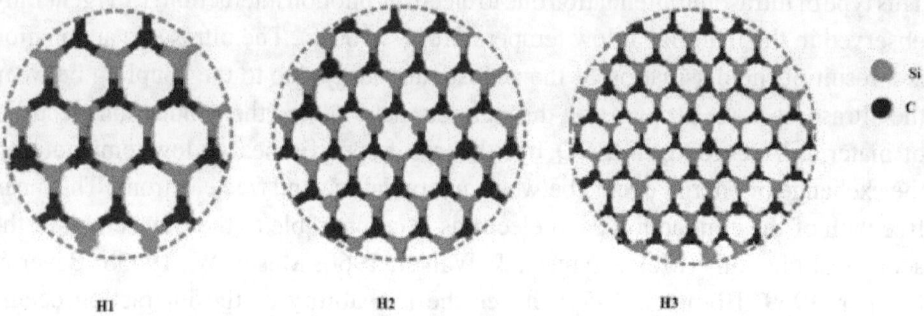

electronic applications. Figure 1 spectacles the cross-section of 2H-SiC nanowires (Miranda & Pérez, 2016; Li, Li, Li, Gong, & Fang, 2017) in the growth direction <001> with different diameters defined concerning carbon atoms position at the surface, marked as H1 (24Si and 24 C atoms per cell), H2 (54Si and 54 C atoms per cell) and H3 (96 Si and 96 C atoms per cell). The diameters of 2H SiC nanowires understudy marked as H1, H2, H3 are d_1=9.38 Å, d_2=15.48 Å, and d_3= 21.62 Å respectively. The SOECs and TOECs of 2H-SiC nanowires have evaluated using the lattice parameters obtained from the literature (Jones & Menon, 2014) and the assessed value of the Lennard Jones parameter under the equilibrium condition using Eqs. (7) - (8), and their computed values are depicted in Table 1 at 300K and compared with the reported values of SiC (Liang, Upmanyu, & Huang, 2005; Rich, et al., 2009). The obtained negative values of TOECs are justified by the previously reported similar wz-structured materials (Zheng, Huang, & Xu, 2020; Spaziani & Lu, 2018; Mandal, Muzykov, Krishna, Hayes, & Sudarshan, 2011; Tse, et al., 2013; Chung & Buessem, 1967). The negative values of TOECs due to the strain components of nanowire showing negative nature. The higher values of TOECs for longitudinal moduli (C_{111}, C_{222}, and C_{333}) than their corresponding shear moduli TOECs (C_{144}, C_{155}) indicates that external stimuli will be less effective in shear mode. $C_{333} < C_{111}$ suggests that anisotropy along the c-axis will be less prominent along the ab-plane of SiC nanowires. The obtained values of higher-order elastic constants for 2H-SiC nanowires are within range of 10-20% variation of previously reported values of similar structured SiC at bulk scale, validating the theoretical approach for the computation of SOECs and TOECs for SiC nanowires.

As depicted in Table 1, the SOECs values for different diameters of 2H-SiC nanowires also satisfy the Born's stability criterion given in Eq. (24), indicating their mechanical stability. The mechanical properties of 2H-SiC nanowires can be

Table 1. SOECs and TOECs (GPa) of 2H SiC-NWs with diameter d1,d2,d3 at 300K

SOECs & TOECs	H1	H2	H3	*a	*b
C_{11}	518	508	506	493	541
C_{12}	127	124	124	187	117
C_{13}	111	109	108	91	61
C_{33}	540	531	528	547	586
C_{44}	133	131	130	137	162
C_{66}	211	208	206	153	212
C_{111}	-2812	-2768	-2765	-2676	----
C_{112}	-1339	-1312	-1312	-1040	----
C_{113}	-1089	-1056	-1055	-1009	----
C_{123}	-1256	-1178	-1177	-1199	----
C_{133}	-1253	-1218	-1218	-1219	----
C_{344}	-423	-408	-408	-392	----
C_{144}	-104	-98	-97	-97	----
C_{155}	-28	-27	-27	-16	----
C_{222}	-2225	-2165	-2161	-2086	----
C_{333}	-2303	-2284	-2280	-2107	----

*a- (Jones & Menon, 2014), *b- (Sarasamak, Limpijumnong, & Lambrecht, 2010)

Table 2. B_V, B_R, B and G_V, G_R, G, Y, H_V, Cp along <100> and <100> planes (in GPa), EA_P, EA_B, EA_G, EA_U, Pu and υ for 2H SiC-NWs at 300 K.

SiC NWs	B_V	B_R	B	G_V	G_R	G	Y	H_V
H1	253.02	253	253.01	177.61	171.91	174.76	426.16	23.59
H2	248.02	248.01	248.02	174.11	168.52	171.03	417.75	23.28
H3	247.26	247.25	247.26	173.58	168	170.79	416.48	23.18
	$C_{p<001>}$	$C_{p<100>}$	EA_F	EA_B	EA_G	EA_U	Pu	υ
H1	-84.53	-22.24	0.64	0.0017	1.63	0.16	1.62	0.21
H2	-82.85	-21.80	0.64	0.0018	1.63	0.16	1.62	0.21
H3	-82.63	-21.73	0.64	0.0017	1.63	0.16	1.62	0.21

found out based on evaluated values of SOECs using the relation described in Eqs. (16)-(23) and depicted in Table 2. (Liang, Upmanyu, & Huang, 2005) reported that the non-linear core elasticity and surface modulus are the two crucial factor which governs the size-reliant elasticity of the nanowires. Based on the ratio of the surface atoms to total atoms of 2H-SiC nanowires for the range of diameters understudy, the difference is quite significant in comparison to the nanowires of diameter greater than 21.62 $\dot{A}$. Thus for 2H-SiC nanowires, the surface modulus becomes the dominant factor that governs the elasticity of nanowires. The obtained values of bulk modulus (*B*) for 2H-SiC nanowires are 250±10 GPa, lying within the range of experimentally reported values 241±17 GPa of SiC nanowires using high-pressure X-ray diffraction (Rich, et al., 2009). The shear modulus (*G*) is essential to analyze the material hardness. Once the values of *B* and *G* for the material obtained, Young's modulus (*Y*) for the material can easily be obtained. The *Y* of the 2H-SiC nanowire, as depicted in Table 2, decreases with an increase in the size of the nanowire, a similar result has been reported for SiC nanowires using the molecular dynamic method (Ma, Hao, & Hossain, 2019). The positive value of Poisson's ratio indicates the expansion of the material in the other two directions perpendicular to the direction of force applied. The value of Poisson's ratio obtained is 0.21, stating the nature of bonding force within the atom is non-central. The obtained value of Pugh's indicator shows the brittleness of the 2H-SiC nanowires at room temperature. This brittleness behavior of the nanowires also justified by the evaluated negative values of Cauchy pressure (C_p). The anisotropy factor (EA_F) utilized to evaluate the anisotropy degree of wz-structured materials. The value of EA_F close to 1 indicates the mechanical behavior prone to isotropy. For isotropic crystals, the values of EA_P, EA_B, and EA_G must be zero. It has observed that 2H-SiC nanowires with different diameters have positive values of EA_P, EA_B, and EA_G. However, it is clear from Table 2 that EA_G>EA_B, indicates that shear anisotropy is more significant than the compressibility anisotropy for 2H-SiC nanowires. It is depicted from Table 2 that, the 2H-SiC nanowires are elastic anisotropic in nature and shows a small anisotropy of the bulk modulus but high anisotropic degree of shear modulus. The Vicker's hardness (*Hv*) predicts the hardness of the material, evaluated using Eq. (23). The material's hardness shows its ability to resist the elastic and plastic deformations under the influence of external forces applied. It depends on the elastic constants, ductility, plasticity, strain, etc. the low value of Hv indicates weak covalent bonds, which may also be predicted by the large valued Pugh's ratio. The crystallographic texture of 2H-SiC nanowires has analyzed using the ultrasonic velocity. Figure 2 illustrates the angle-dependent ultrasonic longitudinal velocity for 2H-SiC-NWs for the three diameters under study. A slight change in the value of the ultrasonic velocities with diameter observed as a result of its relationship with materials elastic constants and lattice parameter, which changes slightly with a change in diameter. As depicted in Fig. 2, the minima of V_L, is observed at an angle of 50°. The angle-dependent ultrasonic quasi-shear wave velocity for 2H-SiC-NWs for different diameters understudy shown in Fig. 3 and it has been observed that maxima for V_{S1}, is at an angle 45°. Figure 4 shows the angle-dependent ultrasonic shear velocity for 2H-SiC-NWs for the different sizes,

Figure 2. Ultrasonic longitudinal velocity VL for 2H SiC-NWs (H1, H2, H3) at 300K.

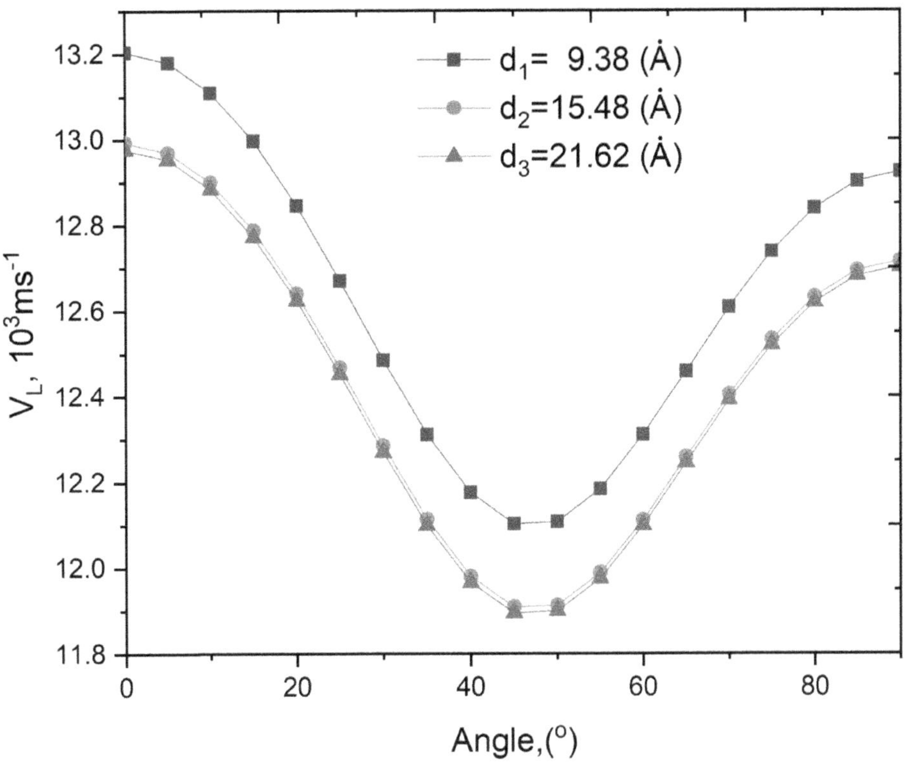

and it is clear that V_{S2} increases with an increase in angle for each 2H-SiC-NWs under study (i.e., H1, H2, H3). It is also apparent from Fig. 5, that the Debye average velocity V_D has its maxima for an ultrasonic wave propagating at an angle 55° with a unique axis. The maxima V_D occurs due to the joint effect of constituent velocities V_L, V_{S1}, and V_{S2}, as defined in Eq. (14).

The ultrasonic velocities vales have evaluated using Eqs. (9)-(11). The ultrasonic velocities (V_L, V_{S1}, and V_{S2}: here L stands for longitudinal wave, S1 for shear wave, and S2 for quasi shear wave) at an angle parallel to the unique axis has been computed with the help of the SOECs depicted in Table 1 for different diameters. The obtained values of V_L, V_S are in good accord with the existing values reported in the literature for wz-structured SiC (Sarasamak, Limpijumnong, & Lambrecht, 2010; Kamitani, et al., 1997), which validates the ultrasonic theoretical model used for the characterization of 2H-SiC nanowires. The values of E_0, C_V have been computed from the AIP Handbook (Gray, 1972). The values of E_0, C_V, D_L, D_S, and

Figure 3. Ultrasonic quasi-shear velocity VS1 for 2H SiC-NWs (H1, H2, H3) at 300K.

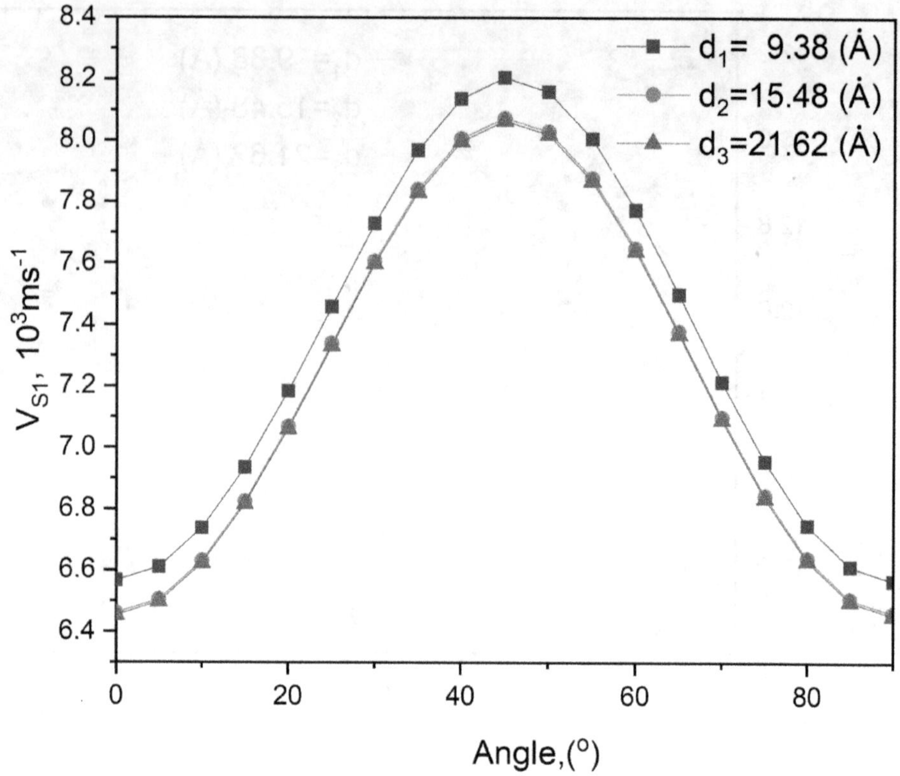

ultrasonic velocities given in Table 3. The slight change in the values of E_0, C_V with diameter observed due to its relationship with V_D and Debye temperature θ_D. The obtained values D_L and D_S for 2H-SiC nanowires are higher than that of bulk. Also, it is evident from Table 3; that $D_L > D_S$ indicates the dominancy of surface modulus in 2H-SiC nanowires. $D_L > D_S$ implies more acoustical energy will converted to thermal energy for the ultrasonic wave propagating along the length of the nanowire.

The thermal conductivity values for 2H-SiC nanowires have been determined using Eq. (26) and are depicted in Table 4. It is evident from the relation described that the material parameters and Debye average velocity (V_D) are the governing factors for the thermal conductivity of the given material. V_D, C_V and κ jointly affects the thermal relaxation time (τ_{th}) due to which the reestablishment time of thermally generated phonons will be minimum when the ultrasonic wave is propagating at an angle of 55° with a unique axis. The order of τth_is in pico-seconds shows that the equilibrium position will be retained by the thermal phonon distribution in ~10-1^{2s}.

Figure 4. Ultrasonic shear velocity VS2 for 2H SiC-NWs (H1, H2, H3) at 300K.

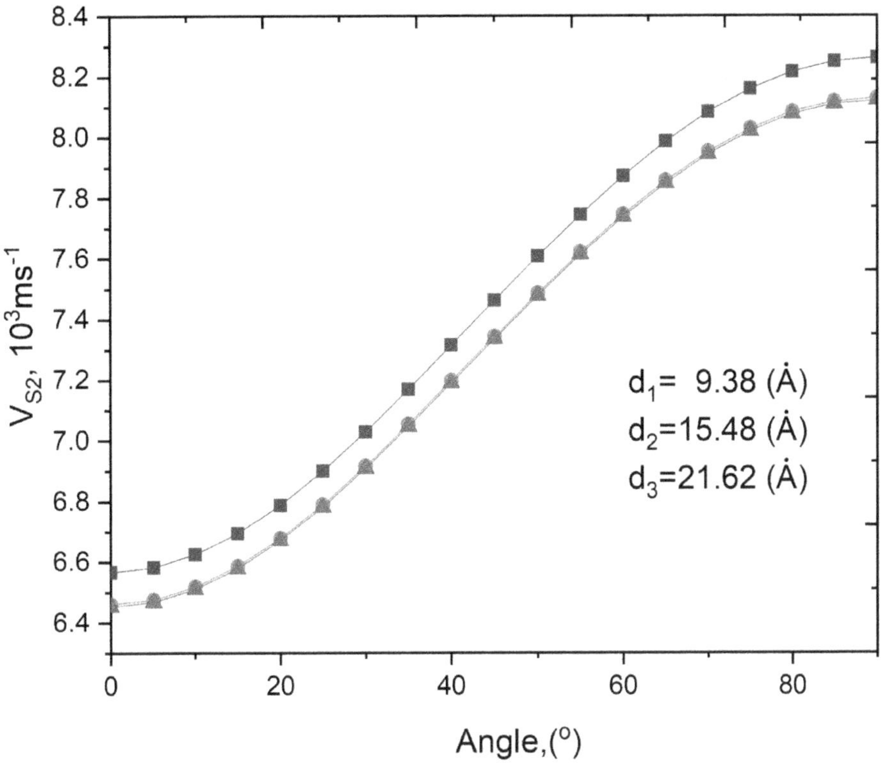

It is crucial to determine the ultrasonic absorption over the squared frequency (α/f^2) to evaluate the ultrasonic properties of nanowires. Also, an assortment of mechanisms responsible for total ultrasonic attenuation such as Akhieser loss, thermo-elastic loss, etc., and their values have computed using the relation described in Eq. (25) and Eq. (29). All components, along with total ultrasonic absorption, have been depicted in Table 4. Table 4 demonstrates the values of the ultrasonic attenuation for the shear wave $(\alpha/f^2)_{Akh.Shear}$ and longitudinal wave $(\alpha/f^2)_{Akh.Long}$ with the assumption $\omega\tau_{th} \ll 1$. As explained that ultrasonic attenuation at any temperature is strongly dependent on τ_{th}/κ, non-*l*inearity parameter, and ultrasonic velocity. Thus, the dominancy of the ultrasonic attenuation due to longitudinal wave propagation is observed over the ultrasonic attenuation due to shear wave propagation. Since the longitudinal wave for each SiC nanowires has its maxima at an angle of 50° with unique axis Fig. 2 and Eq. (29) shows that $(\alpha/f^2)_{Akh.Long}$ is proportional to V_L^{-3}, hence the total ultrasonic attenuation in these materials will be maximum during ultrasonic

Figure 5. Debye average velocity VS for 2H SiC-NWs (H1, H2, H3) at 300K.

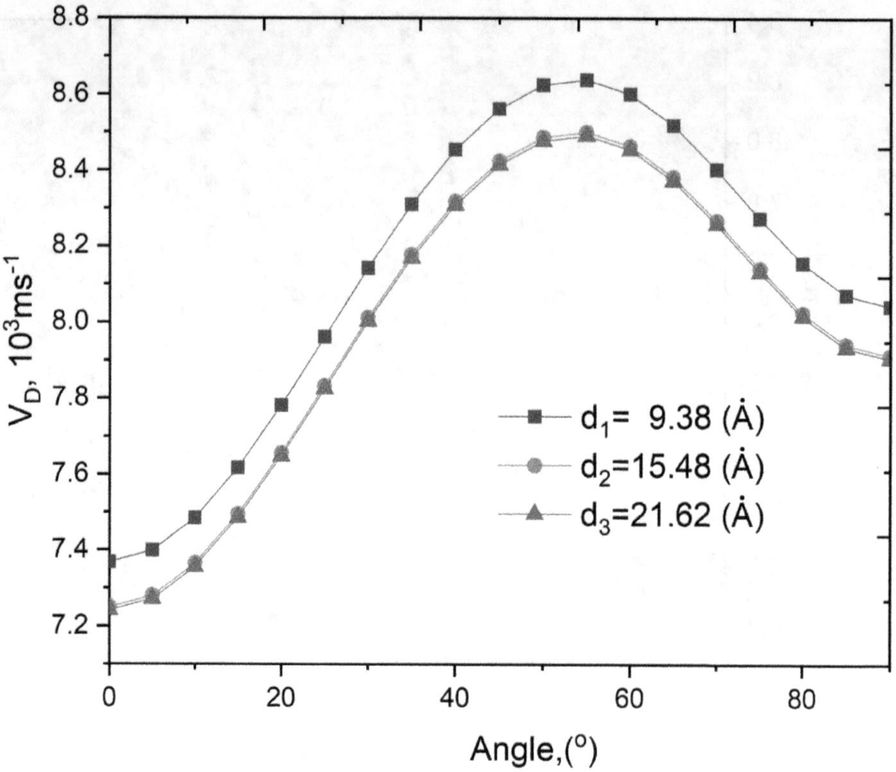

Table 3. C_V (10^5 Jm^{-3}), E_0 (10^6 Jm^{-3}K^{-1}), D_L, D_S, and ultrasonic velocities (10^3ms^{-1}) of 2H SiC-NWs at 300K.

SiC-NWs	C_V	E_0	D_L	D_S	V_L	$V_{S1}=V_{S2}$	V_D
H1	0.76	68.74	54.38	1.59	13.20	6.56	7.36
H2	0.79	78.96	54.99	1.60	12.99	6.46	7.25
H3	0.79	78.96	55.01	1.60	12.97	6.45	7.24

wave propagation at 50° with a unique axis. Table 4 depicts that the PPI process is dominant over the thermoelastic relaxation process to produce total ultrasonic attenuation in wz-structured 2H-SiC nanowires. The $(\alpha/f^2)_{total}$ as described in Eq. (30) depends on ultrasonic attenuation due to Akhieser-type losses (PPI) and attenuation due to the thermal relaxation process. It is also apparent from Table 4

Table 4. κ ($Wm^{-1}K^{-1}$), τ_{th}(ps) and ultrasonic attenuation ($10^{-18}Nps^2m^{-1}$) of 2H SiC-NWs at 300K.

SiC-NWs	τt_h	κ	$(\alpha/f2)A_{kh.Long}$	$(\alpha/f2)A_{kh.Shear}$	$(\alpha/f2)T_E$	$(\alpha/f2)T_{otal}$
H1	0.60	8.399	4.20	0.5024	0.0279	5.24
H2	0.67	9.46	5.63	0.6806	0.0373	7.02
H3	0.68	9.45	5.66	0.6686	0.0339	7.03

that the ultrasonic attenuation value increases with the increasing size of the 2H-SiC nanowire while we found that the change in the thermal conductivity of the nanowires is almost negligible for with diameter variation. So, we can say that the ultrasonic attenuation is not only directly depends on thermal conductivity but also dependent on all associated parameters such as elastic constants, ultrasonic velocity, Grüneisen parameters, non-linearity parameters of 2H-SiC nanowires.

As per the authors' knowledge, no theoretical or experimental data has been found in the available literature for lattice thermal conductivity and ultrasonic attenuation under the suggested theory for the wurtzite 2H-SiC nanowires for direct comparison of the same. The understanding of these mechanical and thermo-elastic properties such as sound velocities, Pugh's indicator, thermal conductivity, and SOECs and TOECs along with ultrasonic attenuations may become fundamental to many innovative applications and for theoretical advancement.

CONCLUSION

We have presented an ultrasonic characterization of 2H SiC-NWs with varying diameters, and determined the elastic, thermophysical, and the ultrasonic properties of 2H SiC-NWs with varying size, using a simple interaction potential model. The obtained data and discussion elaborated for 2H SiC-NWs approves our theoretical approach using Lennard -Jones potential model for assessment of the higher-order elastic constants and the ultrasonic parameters. The description of SOECs and TOECs is similar to wz-structured materials. The Born's criteria for mechanical stability have been satisfied for 2H-SiC nanowire. The Pugh's ratio confirms the brittle nature of 2H SiC-NWs. The thermal relaxation time monitored is in the picoseconds. The ultrasonic velocity characteristics reveal that for the ultrasonic wave propagating at an angle 55° the distribution of thermal phonons restoration time has been found least. $(\alpha/f2)Akh.^{L}o_{ng>(\alpha/}f2)Akh.S^{h}e_{ar\,valida}$tes the extended contribution of the ultrasonic attenuation for the wave propagation along with the longitudinal mode. The obtained results indicate that the ultrasonic attenuation not only depends on thermal conductivity but also relies on associated parameters. The acquired elastic, thermophysical, and ultrasonic properties in correlation with other physical properties may be utilized for the industrial application of 2H SiC-NWs. This study improves the understanding of material behavior under varying physical conditions. The characteristic traits related to elastic properties and size-dependent ultrasonic properties of 2H SiC-NWs along with other established features might be used for the characterization and application in nanoscale devices.

REFERENCES

Bala, J., & Singh, D. (2020, June 4). Elastic and ultrasonic properties of fermium monopnictides. Eng*ineering and Applied Science Research, 47(2)*, 182–189.

Bala, J., Singh, D., Pandey, D., & Yadav, C. (2020, February 26). Mechanical and Thermophysical Properties of ScM (M: Ru, Rh, Pd, Ag) Intermetallics. *International Journal of Thermophysics*, *41*(4), 1–13. doi:10.100710765-020-02624-9

Beyer, R. T., & Letcher, S. V. (1969). *Physical Ultrasonics*. Academic Press.

Bhalla, V., Singh, D., Jain, S., & Kumar, R. (2016, March 17). Ultrasonic attenuation in rare-earth monoarsenides. *Pramana -. Journal of Physics*, *86*(6), 1355–1367. doi:10.100712043-015-1183-5

Bland, H. A., Thomas, E. L., Klemencic, G. M., Mandal, S., Morgan, D. J., Papageorgiou, A., ... Williams, O. A. (2019). superconducting Diamond on silicon Nitride for Device Applications. *Scientific Reports*, *9*(1), 1–9. doi:10.103841598-019-39707-z PMID:30626917

Bömmel, H. E. (1954, October 1). Ultrasonic attenuation in superconducting lead. *Physical Review*, *96*(1), 220–221. doi:10.1103/PhysRev.96.220

Born, M., & Huang, K. (1954). *Dynamical theory of crystal lattices. U.K.* Clarendon press.

Cellini, F., Gao, Y., & Riedo, E. (2019, March 11). Å-Indentation for non-destructive elastic moduli measurements of supported ultra-hard ultra-thin films and nanostructures. *Scientific Reports*, *9*(1), 1–16. doi:10.103841598-019-40636-0 PMID:30858472

Chen, X.-Q., Niu, H., Li, D., & Li, Y. (2011, September). Modeling hardness of polycrystalline materials and bulk metallic glasses. *Intermetallics*, *19*(9), 1275–1281. doi:10.1016/j.intermet.2011.03.026

Chung, D., & Buessem, W. (1967). The elastic anisotropy of crystals. *Journal of Applied Physics*, *38*(5), 2010–2012. doi:10.1063/1.1709819

Clarke, D. R. (2003, January 30). Materials selection guidelines for low thermal conductivity thermal barrier coatings. *Surface and Coatings Technology*, *163-164*, 67–74. doi:10.1016/S0257-8972(02)00593-5

Dhawan, P. K., Wan, M., Verma, S., Pandey, D., & Yadav, R. (2015, February 20). Effect of diameter and surface roughness on ultrasonic properties of GaAs nanowires. *Journal of Applied Physics*, *117*(7), 074307. doi:10.1063/1.4913289

Ghate, P. (1965, August 30). Third-order elastic constants of alkali halide crystals. *Physical Review*, *139*(5A), A1666–A1674. doi:10.1103/PhysRev.139.A1666

Granato, A., & Lücke, K. (1956). Theory of mechanical damping due to dislocations. *Journal of Applied Physics*, *27*(6), 583–593. doi:10.1063/1.1722436

Gray, D. E. (1972). *American Institute of Physics handbook*. McGraw-Hill.

Greaves, G. N., Greer, A., Lakes, R. S., & Rouxel, T. (2011, October 24). Poisson's ratio and modern materials. *Nature Materials*, *10*(11), 823–837. doi:10.1038/nmat3134 PMID:22020006

He, J., & Lilley, C. M. (2008, May 30). Surface effect on the elastic behavior of static bending nanowires. *Nano Letters*, *8*(7), 1798–1802. doi:10.1021/nl0733233 PMID:18510370

Hill, R. (1963, September). Elastic properties of reinforced solids: Some theoretical principles. *Journal of the Mechanics and Physics of Solids*, *11*(5), 357–372. doi:10.1016/0022-5096(63)90036-X

Jassby, K. M., & Vreeland, T. Jr. (1970). An experimental study of the mobility of edge dislocations in pure copper single crystals. *Philosophical Magazine*, *21*(174), 1147–1168. doi:10.1080/14786437008238500

Jones, S., & Menon, C. S. (2014, April 04). Non-linear elastic behavior of hexagonal silicon carbide. *Physica Status Solidi. B, Basic Research*, *251*(6), 1186–1191. doi:10.1002/pssb.201451024

Kamitani, K., Grimsditch, M., Nipko, J., Loong, C.-K., Okada, M., & Kimura, I. (1997, June). The elastic constants of silicon carbide: A Brillouin-scattering study of 4H and 6H SiC single crystals. *Journal of Applied Physics*, *82*(6), 3152–3154. doi:10.1063/1.366100

Kawamura, T., Hori, D., Kangawa, Y., Kakimoto, K., Yoshimura, M., & Mori, Y. (2008, December 19). Thermal conductivity of SiC calculated by molecular dynamics. *Japanese Journal of Applied Physics*, *47*(12, 12R), 8898–8901. doi:10.1143/JJAP.47.8898

Kittel. (2007). *Introduction to solid state physics* (7th ed.). Wiley India Pvt. Limited.

Lau, K., & McCurdy, A. (1998, October 1). Elastic anisotropy factors for orthorhombic, tetragonal, and hexagonal crystals. *Physical Review B: Condensed Matter and Materials Physics*, *58*(14), 8980–8984. doi:10.1103/PhysRevB.58.8980

Łepkowski, S., Majewski, J., & Jurczak, G. (2005, December 1). Nonlinear elasticity in III-N compounds: Ab initio calculations. *Physical Review B: Condensed Matter and Materials Physics*, *72*(24), 245201. doi:10.1103/PhysRevB.72.245201

Li, Y.-J., Li, Y.-L., Li, S.-L., Gong, P., & Fang, X.-Y. (2017). Structural, electronic, and optical properties of hexagonal and triangular SiC NWs with different diameters. *Chinese Physics B*, *26*(4), 047309. doi:10.1088/1674-1056/26/4/047309

Liang, H., Upmanyu, M., & Huang, H. (2005, June 8). Size-dependent elasticity of nanowires: Nonlinear effects. *Physical Review B: Condensed Matter and Materials Physics*, *71*(24), 241403. doi:10.1103/PhysRevB.71.241403

Liu, Y., Dong, L., Lu, J., Huo, W., Du, Y., Zhang, W., & Zhang, Y. (2020, April 5). Microstructure and mechanical properties of SiC nanowires reinforced titanium matrix composites. *Journal of Alloys and Compounds*, *819*, 152953. doi:10.1016/j.jallcom.2019.152953

Ma, L., Hao, T., & Hossain, Z. M. (2019, October 22). Size-dependent toughness and strength in defective 3C-SiC nanowires. *Journal of Applied Physics*, *126*(16), 164303. doi:10.1063/1.5120034

Mandal, K. C., Muzykov, P. G., Krishna, R., Hayes, T., & Sudarshan, T. S. (2011, April). Thermally stimulated current and high temperature resistivity measurements of 4H semi-insulating silicon carbide. *Solid State Communications*, *151*(7), 532–535. doi:10.1016/j.ssc.2011.01.019

Mason, W. (1955b, January 15). Ultrasonic attenuation due to lattice-electron interaction in normal conducting metals. *Physical Review*, *97*(2), 557–558. doi:10.1103/PhysRev.97.557

Mason, W. P. (1965a). Effect of impurities and phonon processes on the ultrasonic attenuation of germanium, crystal quartz, and silicon. In *Physical Acoustics* (Vol. 3, pp. 235–286). Elsevier. doi:10.1016/B978-0-12-395669-9.50013-8

Mason, W. P., & Baerwald, H. (1951). Piezoelectric crystals and their applications to ultrasonics. *Physics Today*, *4*(5), 23–24. doi:10.1063/1.3067231

Mason, W. P., & Bateman, T. (1964). Ultrasonic-wave propagation in pure silicon and germanium. *The Journal of the Acoustical Society of America*, *36*(4), 644–652. doi:10.1121/1.1919031

Miranda, A., & Pérez, L. A. (2016, January). Electronic properties of fluorinated silicon carbide nanowires. *Computational Materials Science*, *111*, 294–300. doi:10.1016/j.commatsci.2015.09.028

Mori, S., & Hiki, Y. (1978, May). Calculation of the third-and fourth-order elastic constants of alkali halide crystalsCalculation of the third-and fourth-order elastic constants of alkali halide crystals. *Journal of the Physical Society of Japan*, *45*(5), 1449–1456. doi:10.1143/JPSJ.45.1449

Morse, R., Myers, A., & Walker, C. (1960, June 15). Fermi surfaces of gold and silver from ultrasonic attenuation. *Physical Review Letters*, *4*(12), 605–607. doi:10.1103/PhysRevLett.4.605

Nandanpawar, M., & Rajagopalan, S. (1982). Grüneisen numbers in hexagonal crystals. *The Journal of the Acoustical Society of America*, *71*(6), 1469–1472. doi:10.1121/1.387844

Oligschleger, C., Jones, R., Reimann, S., & Schober, H. (1996, March 1). Model interatomic potential for simulations in selenium. *Physical Review B: Condensed Matter and Materials Physics*, *53*(10), 6165–6173. doi:10.1103/PhysRevB.53.6165 PMID:9982015

Pandey, D., Singh, D., & Yadav, R. (2007, July). Ultrasonic wave propagation in IIIrd group nitrides. *Applied Acoustics*, *68*(7), 766–777. doi:10.1016/j.apacoust.2006.04.004

Pandey, D. K., & Pandey, S. (2010). Ultrasonics: A technique of material characterization. In D. Dissanayake (Ed.), Acoustic Waves (pp. 397–430). Academic Press.

Pettifor, D. (1992). Theoretical predictions of structure and related properties of intermetallics. *Materials Science and Technology*, *8*(4), 345–349. doi:10.1179/mst.1992.8.4.345

Pillai, S. O. (2010). *Solid State Physics* (7th ed.). New Age Science.

Ponraj, J. S., Dhanabalan, S. C., Attolini, G., & Salviati, G. (2016, May 17). SiC nanostructures toward biomedical applications and its future challenges. *Critical Reviews in Solid State and Material Sciences*, *41*(5), 430–446. doi:10.1080/10408436.2016.1150806

Pugh, S. (1954). XCII. Relations between the elastic moduli and the plastic properties of polycrystalline pure metals. *The London, Edinburgh and Dublin Philosophical Magazine and Journal of Science*, *45*(367), 823–843. doi:10.1080/14786440808520496

Pushpakaran, B. N., Hinojosa, M., Bayne, S. B., Veliadis, V., Urciuoli, D., El-Hinnawy, N., Borodulin, P., Gupta, S., & Scozzie, C. (2013). High temperature unclamped inductive switching mode evaluation of SiC JFET. *IEEE Electron Device Letters*, *34*(4), 526–528. doi:10.1109/LED.2013.2247020

Qi, C., Musetti, S., Fu, L.-H., Zhu, Y.-J., & Huang, L. (2019, May 13). Biomolecule-assisted green synthesis of nanostructured calcium phosphates and their biomedical applications. *Chemical Society Reviews*, *48*(10), 2698–2737. doi:10.1039/C8CS00489G PMID:31080987

Ranganathan, S. I., & Ostoja-Starzewski, M. (2008, August 1). Universal elastic anisotropy index. *Physical Review Letters*, *101*(5), 055504. doi:10.1103/PhysRevLett.101.055504 PMID:18764407

Reuss, A. (1929). Calculation of the flow limits of mixed crystals on the basis of the plasticity of monocrystals. *Zeitschrift für Angewandte Mathematik und Mechanik*, *9*, 49–58. doi:10.1002/zamm.19290090104

Rich, R. M., Stelmakh, S., Patyk, J., Wieligor, M., Zerda, T., & Guo, Q. (2009, June 1). Bulk modulus of silicon carbide nanowires and nanosize grains. *Journal of Materials Science*, *44*(11), 3010–3013. doi:10.100710853-009-3431-x

Sarasamak, K., Limpijumnong, S., & Lambrecht, W. R. (2010, July 1). Pressure-dependent elastic constants and sound velocities of wurtzite SiC, GaN, InN, ZnO, and CdSe, and their relation to the high-pressure phase transition: A first-principles study. *Physical Review B: Condensed Matter and Materials Physics*, *82*(3), 035201. doi:10.1103/PhysRevB.82.035201

Shen, Z., Chen, J., Li, B., Li, G., Zhang, Z., & Hou, X. (2020, January 30). Recent progress in SiC nanowires as electromagnetic microwaves absorbing materials. *Journal of Alloys and Compounds*, *815*, 152388. doi:10.1016/j.jallcom.2019.152388

Spaziani, L., & Lu, L. (2018). Silicon, GaN and SiC: There's room for all: An application space overview of device considerations. In *2018 IEEE 30th International Symposium on Power Semiconductor Devices and ICs (ISPSD)* (pp. 8-11). IEEE. doi:10.1109/ISPSD.2018.8393590

Tosi, M. P. (1964). Cohesion of ionic solids in the Born model. In F. S. Turnbull (Ed.), *Solid state physics* (Vol. 16, pp. 1–120). Academic Press., doi:10.1016/S0081-1947(08)60515-9

Tripathi, S., Agarwal, R., & Singh, D. (2019a, July). Size dependent elastic and thermophysical properties of zinc oxide nanowires. *Johnson Matthey Technology Review*, *63*(3), 166–176. doi:10.1595/205651319X15514400132039

Tripathi, S., Agarwal, R., & Singh, D. (2020b, March 7). Size-Dependent Ultrasonic and Thermophysical Properties of Indium Phosphide Nanowires. *Zeitschrift fur Naturforschung. Section A. Physical Sciences*, *75*(4), 373–380. doi:10.1515/zna-2019-0351

Tse, G., Pal, J., Monteverde, U., Garg, R., Haxha, V., Migliorato, M., & Tomić, S. (2013, August). Non-linear piezoelectricity in zinc blende GaAs and InAs semiconductors. *Journal of Applied Physics*, *114*(7), 073515. doi:10.1063/1.4818798

Voigt, W. (1928). *Lehrbuch der kristallphysik* (Vol. 962). Teubner Leipzig.

Wang, R., Wang, S., & Wu, X. (2010, August). On third-order elastic constants for ductile rare-earth intermetallic compounds: A first-principles study. *Intermetallics*, *18*(8), 1653–1658. doi:10.1016/j.intermet.2010.04.021

Yadav, A., Yadav, R., Pandey, D., & Singh, D. (2008, June 30). Ultrasonic study of fission products precipitated in the nuclear fuel. *Materials Letters*, *62*(17-18), 3258–3261. doi:10.1016/j.matlet.2008.02.036

Yadav, C. P., Pandey, D. K., & Singh, D. (2019, February 8). Ultrasonic study of Laves phase compounds ScOs 2 and YOs 2. *Indian Journal of Physics*, *93*(9), 1147–1153. doi:10.100712648-019-01389-8

Yadav, R. R., & Singh, D. (2001). Ultrasonic attenuation in lanthanum monochalcogenides. *Journal of the Physical Society of Japan*, *70*(6), 1825–1832. doi:10.1143/JPSJ.70.1825

Zagorac, D., Zagorac, J., Djukic, M., Jordanov, D., & Matović, B. (2019, October). Theoretical study of AlN mechanical behaviour under high pressure regime. *Theoretical and Applied Fracture Mechanics*, *103*, 102289. doi:10.1016/j.tafmec.2019.102289

Zeng, D.-J., & Zheng, Q.-S. (2007, August 15). Resonant frequency-based method for measuring the Young's moduli of nanowires. *Physical Review B: Condensed Matter and Materials Physics*, *76*(7), 075417. doi:10.1103/PhysRevB.76.075417

Zheng, J., Huang, J., & Xu, S. (2020, April). Multiscale micro-/nanostructures on single crystalline SiC fabricated by hybridly polarized femtosecond laser. *Optics and Lasers in Engineering*, *127*, 105940. doi:10.1016/j.optlaseng.2019.105940

Zhou, R., Zuo, R., Wang, L., Zhang, B., & Pan, B. (2011, April 20). Size-and surface-dependent electronic structures of crystalline SiC nanotubes. *Journal of Applied Physics*, *109*(8), 084318. doi:10.1063/1.3567114

Chapter 5
III-V-Based Gate-All-Around Cylindrical Nanowire Junctionless Field Effect Transistor

Pooja Shilla
UIET, Panjab University, Chandigarh, India

Raj Kumar
UIET, Panjab University, Chandigarh, India

Arvind Kumar
UIET, Panjab University, Chandigarh, India

ABSTRACT

This chapter represents some essential aspects of nanowires and their transport properties. Scaling of MOSFETs becomes a huge problem for industries due to short channel effects (SCEs) and sub-threshold leakage current. So, nanowires become a good solution to SCEs due to their structure. This chapter is divided into three parts. The first part gives a brief introduction of nanowire and different materials that can replace Si (channel material) and SiO2 (oxide material) due to their superior performance over Si. In the second part, the device structure and device structural measurement is discussed. In the third part, transport properties are discussed. This chapter shows the behavior of nanowire on changing different device materials and device dimensions. Electrical characteristics of Si and III-V based nanowires FETs will be analyzed and compared. High-k dielectric as oxide material also helps in improving device performance. HfO2 shows improvement in device characteristics over SiO2 taken as an oxide material. Junctionless nanowire MOSFET has also been designed and analyzed.

DOI: 10.4018/978-1-7998-6467-7.ch005

INTRODUCTION

Electronic devices are an essential part of today's life. It is hard to think, construct an electronic device without complementary metal oxide semiconductor (CMOS) technology. CMOS technology has played an important role in the integrated circuit industry. According to CMOS technology, device size can be reduced by increasing on-chip transistors with every generation. Tapered device ensures improved performance, high drivability and increased speed. And also reduce cost and packaging density (Neamen, 2012)(Sze & Lee, 2008)(Kim, 2010)(Larson & Snyder, 2006). This minimization of device is known as scaling and this downscaling is done by following Moore's law (Kim, 2010). Now the device scaling reached in nanometer regime, a number of problems obstruct normal behaviour of MOSFETs. Short channel effects are introduced due to reduced size which increases threshold voltage because of two dimensional (2D) electrostatic charge confinements in the active region (channel) of the MOSFET(Gupta, 2018)(Young, 1989). Due to increment in threshold voltage, power consumption also increased. Due to reduced gate length heat generation is high due to increased power consumption at high- frequency operations. Power consumption can be reduced by lowering supply voltage which results in less operational speed of MOSFETs. So power management becomes a big task during scaling of MOSFETs. Short channel requires ultra-shallow junction(Skotnicki et al., 2005) (Nader Shehata et al., 2015)(Division et al., 2014)(P. Razavi et al., 2011) formation which is a great challenge for fabrication labs.

Downscaling of MOSFETs is the most effective way to enhance performance as well as current density of the device [16]. To extend Moore's law further chip density should be increased to enhance operating speed of device which can be achieved by reducing the size of device (Sze & Lee, 2008)(Kim, 2010). Reduction in channel length introduce some short channel effects(SCEs) like DIBL, hot carriers, velocity saturation etc.(Bradley et al., 2003). Further scaling of conventional MOSFET becomes a huge challenge due to the formation of the junctions in the device (Division et al., 2014)(P. Razavi et al., 2011). The introduction of multigate devices becomes a necessity to overcome issues like SCEs, leakage current. MOSFET with different device geometry like Double gate, Tri gate, SOI (Silicon on Insulator) MOSFET, Fin FET and with different mechanism like junctionless MOSFET were designed to improve performance of device. In double gate MOSFETs, presence of two gates (top gate and bottom gate) makes strong control over the channel region through the both gates and leads to reduction in SCEs of device. Doubling of gate to channel coupling helps in SCEs reduction (Wong et al., 1997)(Rahman & Lundstrom, 2002). Further scaling of device is also possible by using DG-MOSFET and also number of transistor per chip can be increased. DG-MOSFET provide decrease in threshold voltage, good carrier mobility of electrons, better switching property in the device, small leakage and near ideal sub-threshold slope with less power consumption (Tsormpatzoglou et al., 2008)(Wu et al., 2013) (Jiang et al., 2015).

Trigate MOSFETs consists of three gates placed around the channel region. The gates are placed on both sides as well as top of MOSFET. Scaled gate length results in SCEs enhancement with high controllability of gate over channel region. In Trigate MOSFET, three gates are fabricated around the channel to increases the gate controllability. The presence of the three gates allows the electrons to travel on three-time enhanced surface area. Trigate MOSFET reduces leakage current and consumes less power than the conventional MOSFET (Nirmal et al., 2013) (Holtij et al., 2015)(Ávila-Herrera et al., 2016).

Parasitic capacitance is also an important factor which effects device performance. Silicon on insulator (SOI) MOSFET technology resolves parasitic capacitance issue by utilizing layered silicon–insulator–silicon substrate instead of conventional silicon substrate during semiconductor fabrication. Layered silicon helps in parasitic capacitance reduction which improves speed of the device [6]. The implementation of SOI technology allows further downscaling in the device. SOI MOSFET provides many advantages over the conventional MOSFET such as device with lower parasitic capacitance due to substrate isolation, less power consumption due to lower parasitic capacitance, higher performance. Isolation from the substrate also gives advantage of high power efficiency with lower leakage current (J. P. Colinge, 2008)(Chaudhry & Kumar, 2004)(Reddy et al., 2005).

FinFETs are 3D 'fin' like structures that arise above the substrate. These 'fins' effectively establish the source and drain which authorized more volume than a traditional planar transistor for the same area. The gates wrapped around the fin grant more control over channel region due to sufficient length. However, extended channel region originates minute leakage current through the body during off state. This also permits to utilize low threshold voltages which give better performance as well as lower power dissipation (Leung & Chui, 2011).

Some obstacles faced during scaling of MOSFET devices are overcomes by developing a novel technology called Junctionless technology. Junctionless transistor is easy to fabricate due to the absence of junction (Tai et al., 2010)(Chiang, 2012) (C. et al., 2013)(Xiao et al., 2016). Unlike the conventional MOSFET, JLTs are uniformly heavily doped offers full depletion in the OFF state. Gate oxide with a large work-function is required for operation of device(Pratap et al., 2016)(Sahay & Kumar, 2017)(Jazaeri et al., 2014)(Oproglidis et al., 2017). This device exhibit bulk conduction rather than surface conduction. This device works in depletion mode. As high doping concentration already insures a large current drive so no need to goes in accumulation mode (Holtij et al., 2015)(Jiang et al., 2015).

In comparison to conventional MOSFETs, Junctionless MOSFETs have considerably simple fabrication process due to absence of junction as no doping concentration gradient is present. In JLTs, the doping of source, drain and channel are uniform and of same type (Jean Pierre Colinge, Lee, Afzalian, et al., 2010). In JLTs, for current to be flow in the channel region, source, drain and channel should be heavily doped (J. P. Colinge et al., 2011)(Jean Pierre Colinge, Lee, Afzalian, et al., 2010a). Further reduction can be possible with JLTs.

INTRODUCTION TO JUNCTIONLESS TECHNOLOGY

Downscaling helps to improve device performance by increasing power with current density (Pratap et al., 2016)(Sahay & Kumar, 2017). But shortened channel length allows SCEs like saturation velocity, hot carriers effect, surface scattering, leakage currents etc. (Skotnicki et al., 2005)(Kim, 2010). Due to increase in SCEs it is difficult to stay with Moore's law (to increase performance of integrated circuits, chip density should be doubled in every two years) with MOSFET. So, junctionless technology was introduced to satisfy the Moore's Law. Junctionless transistor provides reduction in leakages, decrease in power supply, decrease in power consumption, faster device performance due to reduction in parasitic capacitance, improvement in performance of the device (J. P. Colinge et al., 2011)(Jean Pierre Colinge, Lee, Afzalian, et al., 2010a)(Dehdashti Akhavan et al., 2011)(Jean Pierre Colinge, Lee, Ferain, et al., 2010).

Junctionless Field Effect Transistor

Junctionless field effect transistor can be considered as a resistor with uniform doping throughout the device. Uniform doping solve problem of diffusion and ultrasharp junction due to absence of conc. gradient. Operation of junctionless device was possible by using work function difference and heavy doping profile (Moon et al., 2013)(Jin et al., 2014)(Baidya et al., 2016)(Song et al., 2014). The main factor to operate a junctionless device is (i) work function difference (ii) heavy doping. Work Function difference helps to achieve full depletion. So created potential barrier ensures no conduction between source to drain regions (Tai et al., 2010)(Chiang, 2012). Applied gate voltage takes device in partial depletion, potential barrier starts decreasing and conduction will start between source and drain. Device reached in full depletion region from the partial depletion region after applying gate voltage and attains flat band (Chiang, 2012)(C. et al., 2013)(Xiao et al., 2016)(Sahay & Kumar, 2017). We can increase gate voltage above flat band voltage, but there no need of it. We can achieve full conduction before the flat band voltage. This device shows better performance with high ON current, better switching (ON-OFF current ratio) and better sub-threshold slope. We can say this device gives better device performance (S. K. Sharma et al., 2016)(S. Sharma, 2017).

Junctionless Nanowire FET

To get better control on channel, here we choose cylindrical geometry. This cylindrical geometry is called cylindrical Gate-All-Around (GAA) MOSFET. In this electric field created by gate is more enhanced and uniform as compared to bottom and top gated devices (Pratap et al., 2016)(Baidya et al., 2017)(Jazaeri et al., 2014).

GAA MOSFET is like a wire with ends source and drain and can be swap able. A wired shaped channel with nano dimension is formed between source and drain with the increment in gate voltage. As scaling of device reaches to nm. So device is called nanowire FET. As it is junctionless, that's why junctionless nanowire FET (Wu et al., 2013; Baidya et al., 2016).

Figure 1. Structure of Junctionless nanowire MOSFET in cylindrical domain

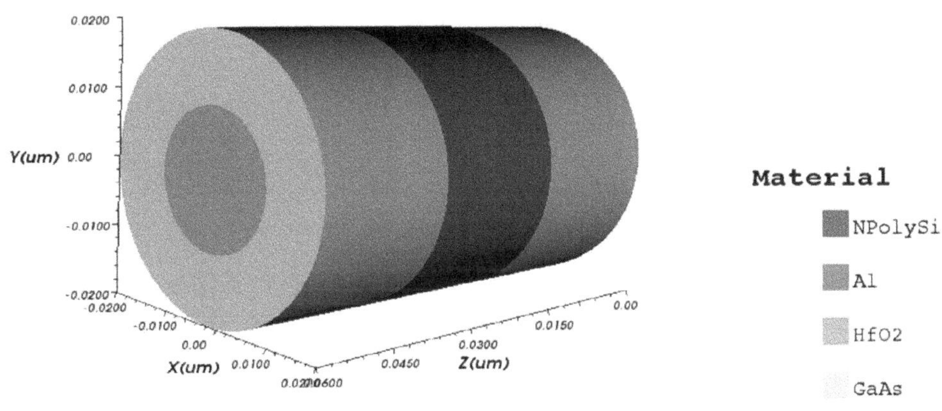

CONDUCTION MECHANISM INSIDE JUNCTIONLESS NANOWIRE FET

As there are two keys points to fabricate a junctionless device (i) use of nano scale dimensions and (ii) high doping concentration. The junctionless MOSFET is a fully depleted below threshold voltage due to uniformly heavily doping throughout the device. Increased gate voltage enhances electron concentration in channel region and when peak value of electron concentration is reached to doping concentration N_D, device reached to threshold voltage. Further increment in gate voltage (above threshold voltage) takes the device to partially depletion from fully depletion and arrived to flat-band voltage where the device cross-section becomes neutral. As the device already acquires full conduction and provides large current drive, so no need to take it to accumulation mode. So from this we can conclude that junctionless device works in depletion mode and provides full conduction at flat band voltage (Jean Pierre Colinge, Lee, Afzalian, et al., 2010)(J. P. Colinge et al., 2011).

Figure 2. Conduction mode in junctionless device.

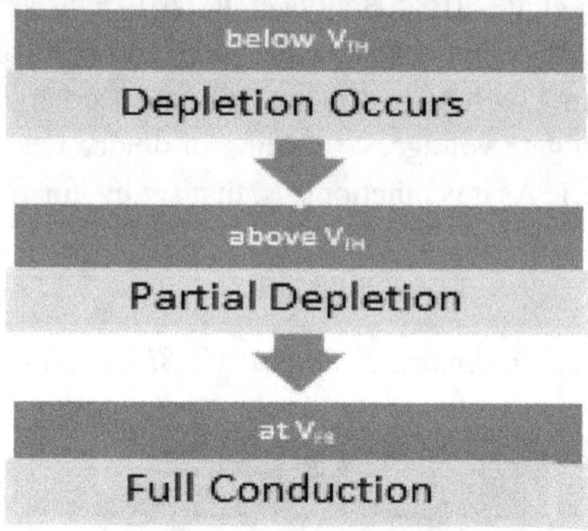

All conduction modes are shown in fig.2 through which conduction occurs in junctionless device.

Figure 3. Conduction in junctionless GAA MOSFET with energy band diagram

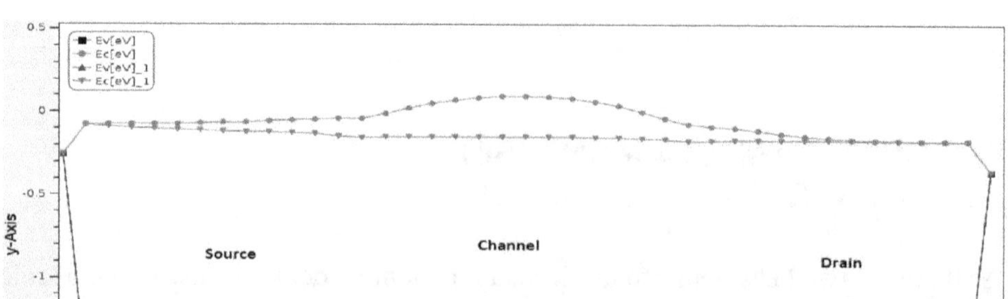

Device conduction is shown in Figure 3 with simulated energy band diagram. This figure shows the device off state (at 0V) in which device is fully depleted and on state (at 1V) in which all the bands are flat. One thing is to be noticed that the device radius must be thin enough to be fully depleted. When the device is fully depleted the band structure is looks like in fig. 3. On applying gate voltage the device energy level again aligned at the same level and the device is turned ON.

Figure 4. Electron concentration in n-type junctionless nanowire transistor

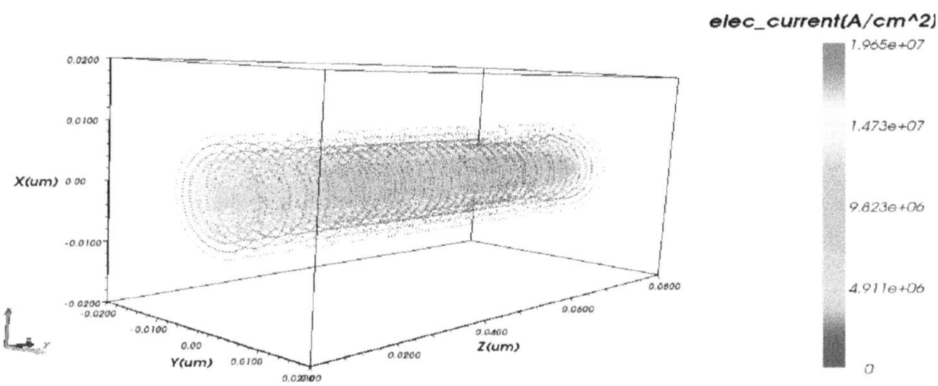

Figure 4 depicts electronic concentration of n-type junctionless MOSFET. Junctionless device exhibits bulk conduction rather than surface conduction. Whole conduction process is controlled by gate voltage as below threshold (V_{th}) the channel region is depleted of electrons. Neutral channel of junctionless device starts conducting and connects source to channel after reaching at threshold voltage. Further increment in threshold voltage expands width and thickness of channel. Channel region behaves as simple resistor in flat energy band situation (Gautam et al., 2013).

Advantage of JLTs over Conventional MOSFET

- Junctionless devices have the potential to operate at faster and use less energy than the conventional transistors(Sahay & Kumar, 2017)(Baidya et al., 2016).
- JLTs shows low leakage currents with near-ideal sub threshold slope and less dependency of mobility on gate voltage as well as temperature in comparison of junctional transistors(S. K. Sharma et al., 2016).
- Absence of concentration gradients in junctionless transistor takes advantage and makes fabrication process easier and greatly simplified compared to junction transistor(Baidya et al., 2016)(Baidya et al., 2017).

GATE OXIDE MATERIALS

As continues scaling of devices demanding high speed processors, SiO_2 do not with stand scaling due to its insufficient dielectric constant (k). Reduced oxide thickness introduce high energetic electrons called 'hot electrons' which enhance excess charge in oxide after tunneling oxide layer, may affect the device performance and can be a cause of device failure with time(Wu et al., 2013)(Dowdy et al., 2012)(Song et al., 2014). The oxide capacitance (C_{ox}) comprises a capacitor with parallel plates divided by gate oxide thickness (T_{ox}). Now the gate metal electrode behaves as parallel plates and a conducting channel is formed in the device (semiconductor region). Current flow in device depends inversely on thickness of oxide. Hence, gate voltage control decreases due to tunneling of electron on decreasing oxide thickness (Baidya et al., 2016)(Baidya et al., 2017). Oxide capacitance calculation makes oxide thickness an important factor as reduction in oxide thickness also results in high leakage current. Thin oxide layer problem can be resolved by using high-k dielectrics as gate oxide instead of SiO_2. Here the concept of EOT occurs. EOT is the gate oxide thickness of the SiO_2 layer of a transistor that would be required to achieve similar capacitance density as the high-k material used (Song et al., 2014)(Kumar et al., 2017).

$$EOT = T_{high\text{-}k}\left(\frac{\text{k of SiO2}}{\text{k of high}-\text{k}}\right) \tag{1}$$

As stated earlier, scaling down MOS transistors introduces many challenges in device design, manufacturing and analysis; alternate processes of devices' performance improvement are being indispensable day by day. Though there is no hard line boundary for device scaling limit, complexities associated with device scaling are increasing noticeably. This warns that the device design methodology is about to reach the scaling limit and about to cross the fundamental limits of semiconductor physics. Therefore exploring for new techniques without scaling down has become a top topic of research through the last few years (Skotnicki et al., 2005)(Kim, 2010).

Exploring for techniques of device performance improvement has divided the research

Mainstream into two major groups:

- Material Engineering
- Structural Modification

The first way to improve device performance without scaling is to change the materials of substrate, gate dielectric, contact etc. of the device. Many research works are being continuously carried on to explore for new materials as a replacement of the conventional silicon and silicon-dioxide(Young, 1989)(Kim, 2010)(Skotnicki et al., 2005). Changeable band gap, higher mobility, direct semiconductor nature, wide range of variability in composition etc. make III-V compounds an excellent choice for next generation MOSFETs (Dowdy et al., 2012)(Pedram Razavi & Fagas, 2013). By using compound semiconductors and carbon nanotubes, MOS operational characteristics can be improved considerably without crossing scaling limit.

Due to its lower threshold voltages, Si MOSFETs have an exponent rise in diffusion or sub-threshold current which becomes a major factor of leakage current. As device scaling reached into nanometer regime, electric field increases in channel region and also increment in tunneling components like BTBT, source/drain tunneling take place exponentially. An effective trend of low effective mass has been observed with respect to bandgap and dielectric constant. Bandgap goes rapidly down and dielectric constant increases when a low effective mass or high mobility material taken as channel material. III-V materials have low effective mass and high mobility compared to Si andGe. Thus, III-V materials may be a prominent candidate for highly scaled n-MOSFET (Sahay & Kumar, 2016)(Bala & Khosla, 2018)(Oktyabrsky & Ye, 2010)(Das, 2019). A comparison is shown between III-V materials in table (1).

Table 1. Different III-V materials with their properties

Bulk materials	Bulk bandgap (eV)	Effective mass (bulk) (me*/m0)	Bulk mobility (cm2/V s)	Direct/indirect	Bulk crystal structure
Si	1.12	0.19	1400	Indirect	Diamond structure
Ge	0.661	0.082	3900	Direct	Diamond structure
GaN	3.47	0.2	2000	Direct	Wurtzite
GaAs	1.424	0.067	8500	Direct	Cubic zinc blende
InGaAs	0.75	0.041	12,000		Cubic zinc blende
InAs	0.354	0.023	40,000	Direct	Cubic zinc blende
InSb	0.17	0.014	80,000	Direct	Cubic zinc blende
Graphene (suspended)	0	0	200,000	Gapless	2D atomic thin layer

Materials which show poor conduction of electricity are known as dielectric materials. Polarization in a dielectric can be achieved by applying electric field and also the effective follower of electrostatic fields. Electrostatic field can be stored energy by reducing the current between oppositely charged poles without disturbing the electrostatic lines of flux. Same property is applicable for capacitors. Essential properties of a dielectric are its capability and dielectric constant. Dielectric effectiveness is inversely proportional to energy lost as heat (Jazaeri et al., 2014) (Dowdy et al., 2012).

Most of the dielectrics are solid in nature. For example: ceramic, silicon dioxide, hafnium oxide, glass, plastics etc. Various liquids and gases can also be considered in the category of good dielectric materials. Air acts as a tremendous dielectric and used in variable capacitors. Generally, Metal oxides have high dielectric constants (Song et al., 2014). High capacitance with low physical volume makes high-dielectric-constant substances more efficient and popular. However, these high-k dielectric materials usually not able to tolerate intense electrostatic fields that much a low-k dielectric substances like air can. Very high voltage across a dielectric material allows too intense electrostatic field results in sudden current conduction (Kumar et al., 2017) and this phenomenon is known as dielectric breakdown.

HIGH-K DIELECTRIC MATERIALS

Dielectric materials with k value high offers high value of oxide capacitance (C_{ox}) which may affect the threshold voltage (V_T) as well as device performance. As the k value is depends upon the depositing behavior, the dielectric constants can be determined by depositing nature of material over the Si substrate. Leakage can be reduced by using higher electrical permittivity over thicker films and reliability of dielectric layer with their electrical thickness to ultrathin SiO_2 layer tells about improvement of device (Baidya et al., 2016)(Baidya et al., 2017)(Song et al., 2014) (Kim, 2010).

Need of High-K Dielectric Materials

From a decade, for the fabrication of MOSFETs Silicon dioxide (SiO_2) are taken as a gate oxide material. To enhance the number of on-chip transistor, downscaling of device should be necessary. During scaling, enhancement of gate capacitance depends inversely on oxide thickness of transistors. Thus enhancement in device performance can be achieved by drive current. As oxide thickness scales down below 2 nm, an abrupt increment in leakage currents has been analyzed due to tunneling, this leads to high power consumption and makes device less reliable. Thus leakage current can be reduced by replacing gate oxide material SiO_2 with high-k dielectric material due to high gate capacitance (Dowdy et al., 2012)(Song et al., 2014)(Baidya et al., 2016)(Baidya et al., 2017).

In this work, GaAs based nanowire junctionless transistor has been proposed, as electron mobility through crystalline structure of GaAs is very high. Hafnium Oxide (HfO_2) as gate oxide material used instead of SiO_2 to enhance the device performance. Simulation of GaAs-nanowire junctionless n-MOSFETs has been done by using Visual TCAD. Performance of GaAs-NW-JLT has been analyzed by varying the design parameter such as Oxide thickness, Channel thickness and drain bias to check its credibility for future CMOS applications.

Device Structure

3D structural view of junctionless GaAs-nanowire MOSFET is shown in Fig. 5. It has uniform doping throughout the device and source/drain as different ends which can be interchangeable. Fig. 6 shows different regions using circular cross-sectional view of cylindrical junctionless GAA device. The gate is wrapped over channel region providing excellent electrostatic control. Fig. 7 shows 2D cross-sectional view of device. Design parameters taken in simulation setup are described in Table 2.

Figure 5. 3D structural view of junctionless GaAs-nanowire MOSFET

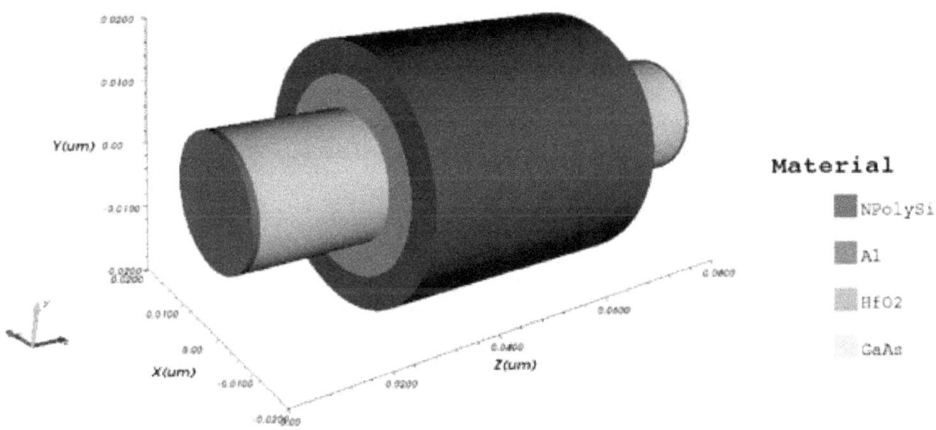

Figure 6. (a) cylindrical view of GaAs nanowire MOSFET (b) 2D front circular cross-sectional view

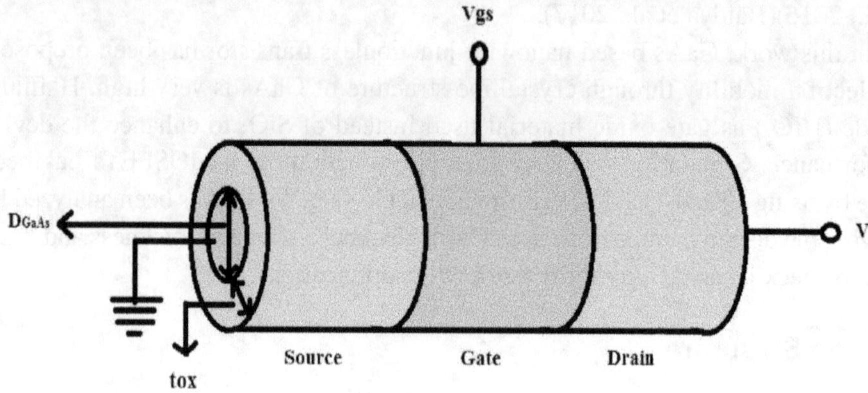

Figure 7. 2D Side cross–sectional view of junctionless GaAs nanowire MOSFET

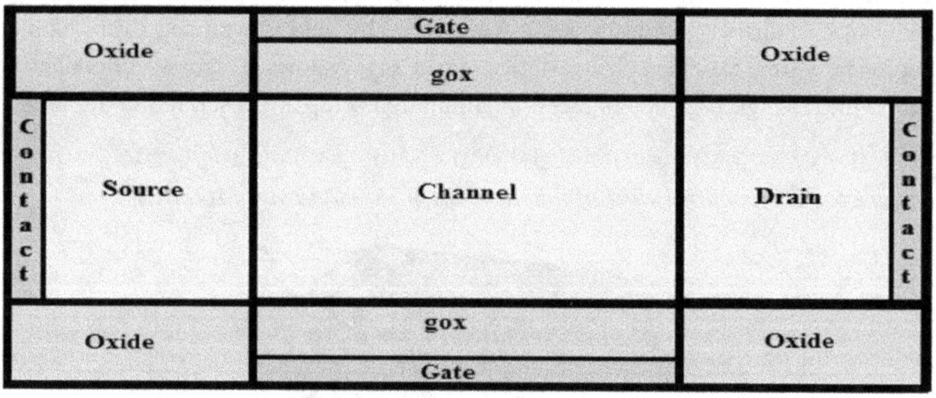

Table 2. Dimensions of different design parameters used for simulation

S. No.	Design Parameter	Values
1	Channel thickness(t_d)	10 nm
2	Oxide thickness(t_{ox})	5nm
3	Gate length(L_g)	20 nm
4	Source/drain length	20 nm
5	Gate oxide material	HfO_2
6	Channel material	GaAs
7	Doping conc.	$1\times10^{19}cm^{-3}$
8	Work function	4.8eV

RESULTS AND DISCUSSIONS

This section represents simulated results in the form of I-V plots to represents transfer characteristics of junctionless GaAs-Nanowire MOSFET by altering different dimensional design parameters. Different device design models have been used in cylindrical co-ordinates to solve Poisson's equation using 3D Visual TCAD simulator.

Figure 8. Comparison of Log (I_{ds}) vs V_{gs} characteristics of JL GaAs-nanowired FET between SiO_2 and HfO_2

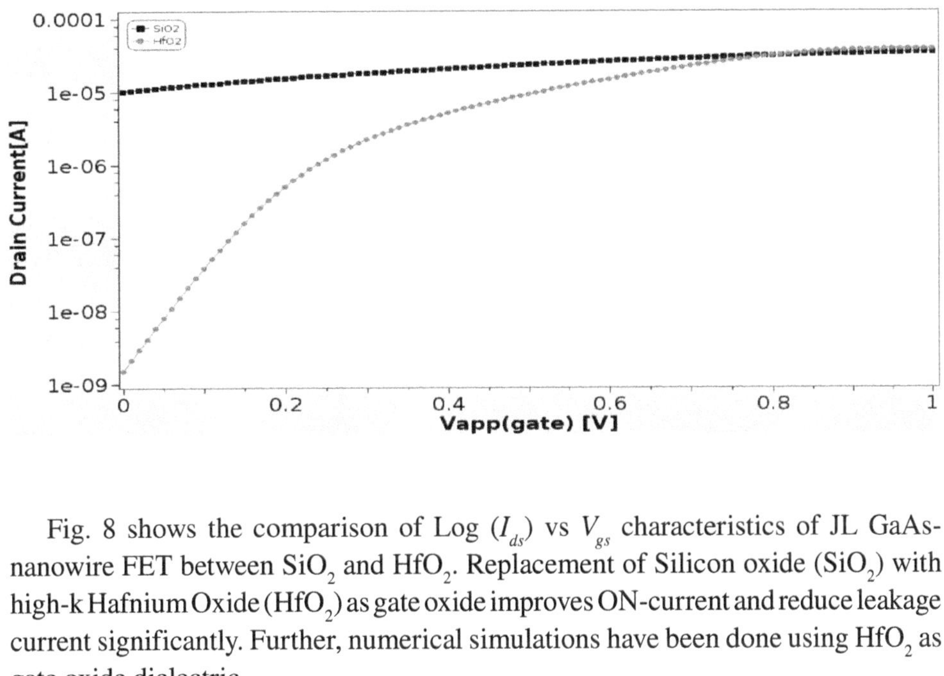

Fig. 8 shows the comparison of Log (I_{ds}) vs V_{gs} characteristics of JL GaAs-nanowire FET between SiO_2 and HfO_2. Replacement of Silicon oxide (SiO_2) with high-k Hafnium Oxide (HfO_2) as gate oxide improves ON-current and reduce leakage current significantly. Further, numerical simulations have been done using HfO_2 as gate oxide dielectric.

In Fig. 9, Log I_{ds} vs V_{gs} transfer characteristics curves of JL GaAs-nanowire FET for various drain voltages (V_{ds}) was depicted. It is analyzed that increase in drain voltage increases the Subthreshold current (OFF-current) attributed to drain induced barrier lowering (DIBL) effect. Direct Band to Band tunneling (BTBT) produce leakage current which leads to static power consumption.

Now, the variations in Log I_{ds} vs V_{gs} can also be analyzed with different gate oxide thickness of device through Fig. 10. The leakage current reduced significantly and the drive current increase with decrease in oxide thickness. This is because reduction in gate oxide thickness provides strong control of gate over the channel region of nanowire.

Figure 9. Log I_{ds} vs V_{gs} characteristic curves of JL GaAs-nanowired FET by varying drain voltages t_{ox} =5nm, t_{GaAs} =10nm and L_{ch} = 20nm.

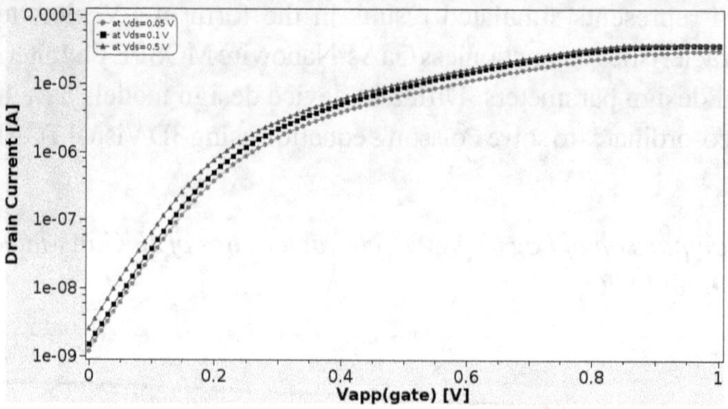

Figure 10. Comparison of Log I_{ds} vs V_{gs} characteristics of junctionless GAA MOSFET by oxide thickness with t_{ch}=10nm, V_{ds} =0.1v and L_{ch}= 20nm

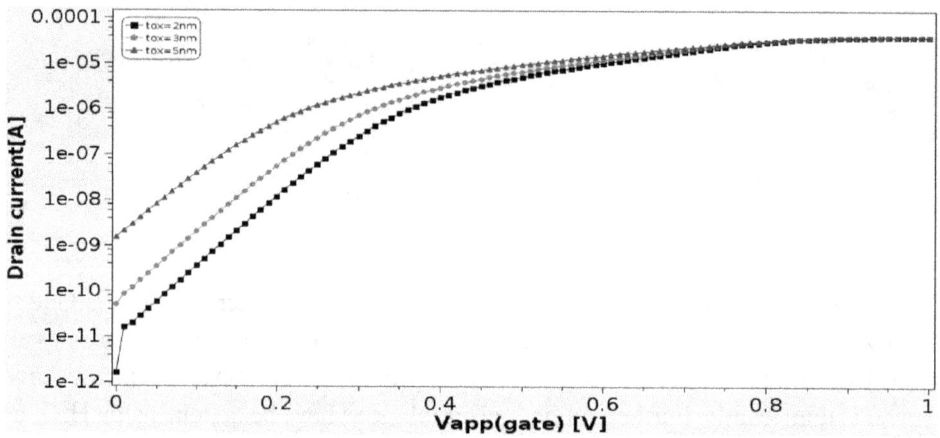

The channel thickness is leading design parameter of device to study the carrier transport properties. Now, it required to analyze the effect of channel thickness on device performance. Fig. 11 illustrates Log I_{ds} vs V_{gs} characteristics plot on varying different channel thickness. It was observed that Leakage current decreases manifold with decrease in channel thickness attributed to strong gate control on bulk conduction mechanism. Therefore, the electrical performances of junctionaless cylindrical GAA nanowire MOSFET can be improved by selecting suitably small channel thickness because the defect density in the channel also decreases with small size of device.

However, conduction mechanism is directly related with channel thickness. Due to strong electric field effect on conduction mechanism, ON-current decreases with channel thickness which also responsible for hot carrier's effects. Hence, small nanowire thickness shows less subthreshold leakage.

The subthreshold slope (SS) is a parameter to estimate the subthreshold characteristics of nanoscale short-channel MOSFET devices. It gives estimation of gate voltage required to raise output current per decade current. It tells about the ON-OFF switching behavior of device. The SS of the *I–V* characteristics can be extracted as

$$SS= \frac{\Delta Vgs}{\Delta \log(Id)} \tag{2}$$

Figure 11. Comparison of Log I_{ds} vs V_{gs} of junctionless GAA MOSFET by GaAs-nanowire thickness with t_{ox} =5nm, V_{ds} =0.1v and L_{ch}= 20nm

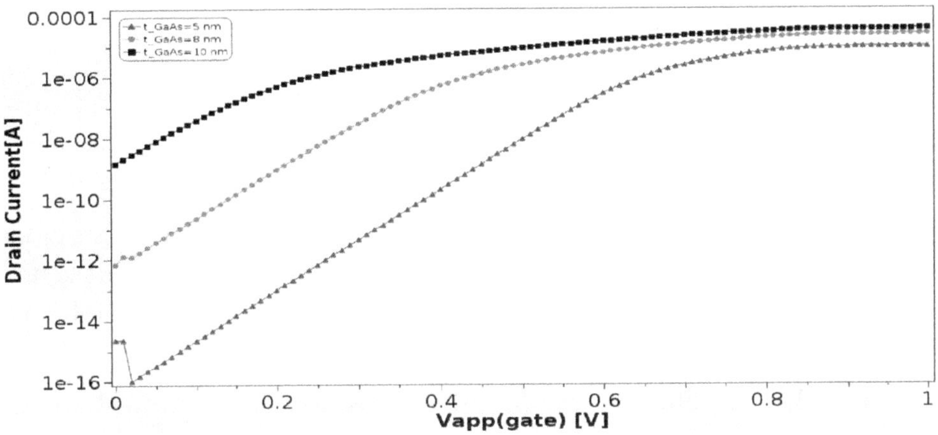

One more important factor is I_{ON} /I_{OFF} ratio and calculated for showing device performance. I_{ON} should be high as it is device ON current and I_{OFF} should be low as it is subthreshold leakage current. So for good performance of device I_{ON} /I_{OFF} should be high. To get higher ON/OFF ratio SS should be small.

Table 3. Calculated I_{ON} /I_{OFF} and SS for various channel thickness

Sr. No.	Channel thickness (nm)	I_{ON} /I_{OFF}	SS(mV/dec)
1.	5	1E+11	40
2.	8	1E+11	50
3.	10	1E+10	60

CONCLUSION

In this paper, junctionless Gate All around (GAA) nanowired MOSFET is presented to determine scalaing limits of these devices when uniform doping is present. GaAs has been used as substrate material, gives better performance as compared to Silicon. It is observed that Hafinum oxide(HfO_2) is optimal replacement of Silicon oxide (SiO_2) as gate oxide material to enhanced device performance. Electrical performances of junctionaless cylindrical nanowire transistors can be enhanced by having suitable gate oxide and channel thickness. III-V materials with suitable value of design parameters enhance device performance in terms of ON-Current, OFF-current, SS and I_{ON} /I_{OFF}. However, DIBL is responsible for subthreshold leakage current with the increment in drain bias. Junctionless GaAs-Nanowire MOSFET is also a promising candidate for future CMOS applications.

REFERENCES

Ávila-Herrera, F., Paz, B. C., Cerdeira, A., Estrada, M., & Pavanello, M. A. (2016). Charge-based compact analytical model for triple-gate junctionless nanowire transistors. *Solid-State Electronics*, *122*, 23–31. doi:10.1016/j.sse.2016.04.013

Baidya, A., Baishya, S., & Lenka, T. R. (2017). Impact of thin high-k dielectrics and gate metals on RF characteristics of 3D double gate junctionless transistor. *Materials Science in Semiconductor Processing*, *71*(April), 413–420. doi:10.1016/j.mssp.2017.08.031

Baidya, A., Lenka, T. R., & Baishya, S. (2016). Mixed-mode simulation and analysis of 3D double gate junctionless nanowire transistor for CMOS circuit applications. *Superlattices and Microstructures*, *100*, 14–23. doi:10.1016/j.spmi.2016.08.028

Bala, S., & Khosla, M. (2018). Design and simulation of nanoscale double-gate TFET/tunnel CNTFET. *Journal of Semiconductors*, *39*(4), 044001. Advance online publication. doi:10.1088/1674-4926/39/4/044001

Bradley, K., Gabriel, J. C. P., Star, A., & Grüner, G. (2003). Short-channel effects in contact-passivated nanotube chemical sensors. *Applied Physics Letters*, *83*(18), 3821–3823. doi:10.1063/1.1619222

C., L., Y, Z., S, D., & Han, R. (2013). Subthreshold behavior models for nanoscale short-channel junctionless cylindrical surrounding-gate MOSFETs. *IEEE Transactions on Electron Devices, 60*(11), 3655–3662.

Chaudhry, A., & Kumar, M. J. (2004). Controlling Short-Channel Effects in Deep-Submicron SOI MOSFETs for Improved Reliability: A Review. *IEEE Transactions on Device and Materials Reliability*, *4*(1), 99–109. doi:10.1109/TDMR.2004.824359

Chiang, T. K. (2012). A new quasi-2-D threshold voltage model for short-channel junctionless cylindrical surrounding gate (JLCSG) MOSFETs. *IEEE Transactions on Electron Devices*, *59*(11), 3127–3129. doi:10.1109/TED.2012.2212904

Colinge, J. P. (2008). The new generation of soi mosfets. *Romanian Journal of Information Science and Technology*, *11*(1), 3–15.

Colinge, J. P., Kranti, A., Yan, R., Lee, C. W., Ferain, I., Yu, R., Akhavan, N. D., & Razavi, P. (2011). Junctionless Nanowire Transistor (JNT): Properties and design guidelines. *Solid-State Electronics*, *65–66*(1), 33–37. doi:10.1016/j.sse.2011.06.004

Colinge, J. P., Lee, C.-W., Afzalian, A., Akhavan, N. D., Yan, R., Ferain, I., Razavi, P., O'Neill, B., Blake, A., White, M., Kelleher, A.-M., McCarthy, B., & Murphy, R. (2010a). Nanowire transistors without junctions. *Nature Nanotechnology*, *5*(3), 225–229. doi:10.1038/nnano.2010.15 PMID:20173755

Colinge, J. P., Lee, C.-W., Afzalian, A., Akhavan, N. D., Yan, R., Ferain, I., Razavi, P., O'Neill, B., Blake, A., White, M., Kelleher, A.-M., McCarthy, B., & Murphy, R. (2010b). Nanowire transistors without junctions. *Nature Nanotechnology*, *5*(3), 225–229. doi:10.1038/nnano.2010.15 PMID:20173755

Colinge, J. P., Lee, C.-W., Ferain, I., Akhavan, N. D., Yan, R., Razavi, P., Yu, R., Nazarov, A. N., & Doria, R. T. (2010). Reduced electric field in junctionless transistors. *Applied Physics Letters*, *96*(7), 23–26. doi:10.1063/1.3299014

Das, S. R. (2019). *Advances in InSb and InAs Nanowire Based Nanoelectronic Field Effect Transistors*. Academic Press.

Dehdashti Akhavan, N., Ferain, I., Razavi, P., Yu, R., & Colinge, J. P. (2011). Improvement of carrier ballisticity in junctionless nanowire transistors. *Applied Physics Letters*, *98*(10), 2011–2014. doi:10.1063/1.3559625

Division, C. T., Company, M., & Mc, T. S. M. (2014). *Mu ultigate transistors*. Academic Press.

Dowdy, R., Walko, D. A., Fortuna, S. A., & Li, X. (2012). Realization of unidirectional planar GaAs. *IEEE Electron Device Letters*, *33*(4), 522–524. doi:10.1109/LED.2012.2186115

Gautam, R., Saxena, M., Gupta, R. S., & Gupta, M. (2013). Gate all around MOSFET with vacuum gate dielectric for improved hot carrier reliability and RF performance. *IEEE Transactions on Electron Devices*, *60*(6), 1820–1827. doi:10.1109/TED.2013.2256912

Gupta, J. B. (2018). *Electronic Devices and Circuits*. Pearson Education.

Holtij, T., Kloes, A., & Iñíguez, B. (2015). 3-D compact model for nanoscale junctionless triple-gate nanowire MOSFETs, including simple treatment of quantization effects. *Solid-State Electronics*, *112*, 85–98. doi:10.1016/j.sse.2015.02.002

Jazaeri, F., Barbut, L., & Sallese, J. M. (2014). Trans-capacitance modeling in junctionless gate-all-around nanowire FETs. *Solid-State Electronics*, *96*, 34–37. doi:10.1016/j.sse.2014.04.022

Jiang, C., Liang, R., Wang, J., & Xu, J. (2015). A two-dimensional analytical model for short channel junctionless double-gate MOSFETs. *AIP Advances*, *5*(5), 057122. Advance online publication. doi:10.1063/1.4921086

Jin, X., Liu, X., Lee, J. H., & Lee, J. H. (2014). Modeling of subthreshold characteristics of short channel junctionless cylindrical surrounding-gate nanowire metal-oxide-silicon field effect transistors. *Physica Scripta*, *89*(1), 015804. Advance online publication. doi:10.1088/0031-8949/89/01/015804

Kim, Y.-B. (2010). Challenges for Nanoscale MOSFETs and Emerging Nanoelectronics. *Transactions on Electrical and Electronic Materials*, *11*(3), 93–105. doi:10.4313/TEEM.2010.11.3.093

Kumar, S., Kumari, A., & Das, M. K. (2017). Modeling gate-all-around Si/SiGe MOSFETs and circuits for digital applications. *Journal of Computational Electronics*, *16*(1), 47–60. doi:10.100710825-016-0941-z

Larson, J. M., & Snyder, J. P. (2006). Overview and status of metal S/D Schottky-barrier MOSFET technology. *IEEE Transactions on Electron Devices*, *53*(5), 1048–1058. doi:10.1109/TED.2006.871842

Leung, G., & Chui, C. O. (2011). Variability of inversion-mode and junctionless finfets due to line edge roughness. *IEEE Electron Device Letters*, *32*(11), 1489–1491. doi:10.1109/LED.2011.2164233

Moon, D., Choi, S.-J., Duarte, J. P., & Choi, Y.-K. (2013). Investigation of silicon nanowire gate-all-around junctionless transistors built on a bulk substrate. *IEEE Transactions on Electron Devices*, *60*(4), 1355–1360. doi:10.1109/TED.2013.2247763

Neamen, D. A. (2019). *Semiconductor physics and devices: basic principles*. doi:10.1017/CBO9781107415324.004

Nirmal, D., & Kumar, P. V. J. (2013). *Nanoscale tri gate MOSFET for ultra low power applications using high-k dielectrics*. Academic Press.

Oktyabrsky, S., & Ye, P. D. (2010). Fundamentals of III-V semiconductor MOSFETs. *Fundamentals of III-V Semiconductor MOSFETs*. doi:10.1007/978-1-4419-1547-4

Oproglidis, T. A., Karatsori, T. A., Barraud, S., Ghibaudo, G., & Dimitriadis, C. A. (2017). Leakage current conduction in metal gate junctionless nanowire transistors. *Solid-State Electronics*, *131*, 20–23. doi:10.1016/j.sse.2017.02.003

Pratap, Y., Haldar, S., Gupta, R. S., & Gupta, M. (2016). Gate-Material-Engineered Junctionless Nanowire Transistor (JNT) with Vacuum Gate Dielectric for Enhanced Hot-Carrier Reliability. *IEEE Transactions on Device and Materials Reliability*, *16*(3), 360–369. doi:10.1109/TDMR.2016.2583262

Rahman, A., & Lundstrom, M. S. (2002). A compact scattering model for the nanoscale double-gate MOSFET. *IEEE Transactions on Electron Devices*, *49*(3), 481–489. doi:10.1109/16.987120

Razavi, P., & Fagas, G. (2013). Electrical performance of III-V gate-all-around nanowire transistors. *Applied Physics Letters*, *103*(6), 1–4. doi:10.1063/1.4817997

Razavi, P., Fagas, G., Ferain, I., Akhavan, N. D., Yu, R., & Colinge, J. P. (2011). Performance investigation of short-channel junctionless multigate transistors. *2011 12th International Conference on Ultimate Integration on Silicon, ULIS 2011*, 122–125. 10.1109/ULIS.2011.5758005

Reddy, G. V., Kumar, M. J., & Member, S. (2005). *A New Dual-Material Double-Gate (DMDG)*. Academic Press.

Sahay, S., & Kumar, M. J. (2016). Insight into Lateral Band-to-Band-Tunneling in Nanowire Junctionless FETs. *IEEE Transactions on Electron Devices*, *63*(10), 4138–4142. doi:10.1109/TED.2016.2601239

Sahay, S., & Kumar, M. J. (2017). Spacer Design Guidelines for Nanowire FETs from Gate-Induced Drain Leakage Perspective. *IEEE Transactions on Electron Devices*, *64*(7), 3007–3015. doi:10.1109/TED.2017.2702067

Sharma, S. (2017). An Extensive Evaluation of Futuristic Gate All Around Junctionless Nanowire MOSFET Using Numerical Simulation. *International Journal for Research in Applied Science and Engineering Technology*, *5*(8), 1974–1979. doi:10.22214/ijraset.2017.8279

Sharma, S. K., Raj, B., & Khosla, M. (2016). A Gaussian approach for analytical subthreshold current model of cylindrical nanowire FET with quantum mechanical effects. *Microelectronics Journal*, *53*, 65–72. doi:10.1016/j.mejo.2016.04.002

Shehata, N., Gaber, A.-R., Naguib, A., Selmy, A. E., Hassan, H., Shoeer, I., Ahmadien, O., & Nabeel, R. (2015). 3D Multi-gate Transistors: Concept, Operation, and Fabrication. *J. of Electrical Engineering*, *3*(1), 1–14. doi:10.17265/2328-2223/2015.01.001

Skotnicki, T., Hutchby, J. A., & King, T. (2005). *Toward the Introduction of New Materials and Structural Changes to Improve MOSFET Performance*. Academic Press.

Song, Y., Zhang, C., Dowdy, R., Chabak, K., Mohseni, P. K., Choi, W., & Li, X. (2014). III-V junctionless gate-all-around nanowire MOSFETs for high linearity low power applications. *IEEE Electron Device Letters*, *35*(3), 324–326. doi:10.1109/LED.2013.2296556

Sze, S. M., & Lee, M. K. (2008). *Semiconductor devices: physics and technology*. Academic Press.

Tai, C. H., Lin, J. T., Eng, Y. C., & Lin, P. H. (2010). A novel high-performance junctionless vertical MOSFET produced on bulk-si wafer. *ICSICT-2010 - 2010 10th IEEE International Conference on Solid-State and Integrated Circuit Technology Proceedings*, *3*, 108–110. doi:10.1109/ICSICT.2010.5667835

Tsormpatzoglou, A., Dimitriadis, C. A., Clerc, R., Pananakakis, G., & Ghibaudo, G. (2008). Threshold voltage model for short-channel undoped symmetrical double-gate MOSFETs. *IEEE Transactions on Electron Devices*, *55*(9), 2512–2516. doi:10.1109/TED.2008.927394

Wong, H. S. P., Chan, K. K., & Taur, Y. (1997). Self-aligned (top and bottom) double-gate MOSFET with a 25 nm thick silicon channel. *Technical Digest - International Electron Devices Meeting, IEDM*, 427–430. 10.1109/iedm.1997.650416

Wu, M., Jin, X., Kwon, H., Chuai, R., Liu, X., & Lee, J. (2013). The optimal design of junctionless transistors with double-gate structure for reducing the effect of band-to-band tunneling. *JSTS: Journal of Semiconductor Technology and Science, 13*(3), 245–251. doi:10.5573/JSTS.2013.13.3.245

Xiao, Y., Zhang, B., Lou, H., Zhang, L., & Lin, X. (2016). A Compact Model of Subthreshold Current with Source/Drain Depletion Effect for the Short-Channel Junctionless Cylindrical Surrounding-Gate MOSFETs. *IEEE Transactions on Electron Devices, 63*(5), 2176–2181. doi:10.1109/TED.2016.2535247

Young, K. K. (1989). Short-Channel Effect in Fully Depleted SOI MOSFET's. *IEEE Transactions on Electron Devices, 36*(2), 399–402. doi:10.1109/16.19942

Section 3
Nanowire-Based Circuits Design

Chapter 6
Design and Simulation Analysis of NWFET for Digital Application

Shashi Bala
Chandigarh Engineering College, Landran, India

Raj Kumar
University Institute of Engineering and Technology, Panjab University, Chandigarh, India

Jeetendra Singh
National Institute of Technology, Sikkim, India

Sanjeev Kumar Sharma
National Institute of Engineering and Technology, Jalandhar, India

ABSTRACT

This chapter presents the design and simulation analysis nanowire-based FET (NWFET) for best possible Ig-Vgs characteristics. A NWFET is a device in which channel is wire-like structure with diameter or lateral dimension in nanometer (10-9 m) range. Performance analysis has been done for various design and process parameters variation to propose optimized parameter for best performance. Although a lot of focus has been put on homogenous Si based NWFETs, there has been a rising interest in III-V NWFETs. This is mainly due to the excellent carrier transport properties are provided by these materials. NWFETs have ability to suppress SCEs and are also good in suppressing OFF-current (IOFF), because of gate all around (GAA) configuration. Secondly, NWFETs have large ON-current (ION) due to quasi one-dimensional (1D) conduction of NWs, and as a result of low carrier scattering, conduction of NWs-based devices is very large.

DOI: 10.4018/978-1-7998-6467-7.ch006

INTRODUCTION

Over the last few years, the devices have scaled down from micrometre to nanometre regime to increase the packaging density and processing speed of the modern ICs. However, the switching characteristic of the transistor degrades attributed to keeping large number of devise in chip by scaling down its dimensions. The feature size of device has been scaled down to nanometre technology node by following the scaling rules. CMOS devices in nanometre regime have shown severe degradation in the subthreshold characteristics limits, Short Channel Effects (SCEs) comprising less drive current, increase leakage current, with the ongoing scaling of conventional MOSFETs (Tanaka et al.,1993), (Bricout et al.,1996), (Young et al.,1989), (Roy et al.,2005). Such SCEs have severe task to continually scale down the device without effecting device performance. To deal with these problems numerous semiconductor devices have been proposed by researches such as Double Gate Field Effect Transistors (DG-FET), triple gate Field Effect Transistors (TG-FET) Fin FET (Park et al., 2001), (Crupi et al., 2006), (Vega et al.,2009), (Park et al., 2003), Cylindrical Gate Metal-Oxide-Field-Effect-Transistor (CG-MOSFET) and Nanowire-Field-Effect-Transistor (NWFETs). NWFETs provide better electrostatic control and sharp sub threshold slope as compared to others (Kranti et al., 2001), (Ghosh et al., 2012), (Kaur et al., 2007), (Takato et al., 1991), (Lu et al., 2006), (Sharma et al., 2004).

The MOSFETs made off from NWs are usually entitled as Nanowire-Field-Effect-Transistor (NWFETs). So for improvement circuit application, different material is used for device channel. III-V compound materials based devices show better performance in terms of ON current and transconductances as compared to Si based transistors (Jain et al., 2016).. In recent times, devices, circuits and mechanism there have been investigated in literature (Sharma et al., 2015), (Sharma et al., 2017), (Gao et al., 2011), (Huang et al., 2001), (Sharma et al., 2016), (Yuan et al., 2009), (Guang-Xi et al., 2010), (Vishvakarma et al., 2010). Indium-Gallium-Arsenide ($In_{1-x}Ga_xAs$) has an auspicious channel material for upcoming nano devices (Lubow et al., 2010). Today, there is great demand for less cost, greatly sensitive sensors for different applications such as biomedical (Ouyang et al., 2006), optical interconnects (Gan et al., 2000), (Cho et al., 2004) optical storage [29, 30] etc.

NANOWIRES AND THEIR TYPES

A NW is very thin wire, the diameter of nanowire in nanoscale range (10^{-9} meters) and the length of the wire is in microns (10^{-6} m), as shown in figure 1

There are numerous categories of NWs like, oxides, metals, semiconductors. Table 1 précises the several kinds of NWs beside using their examples.

Figure 1. Nanowire

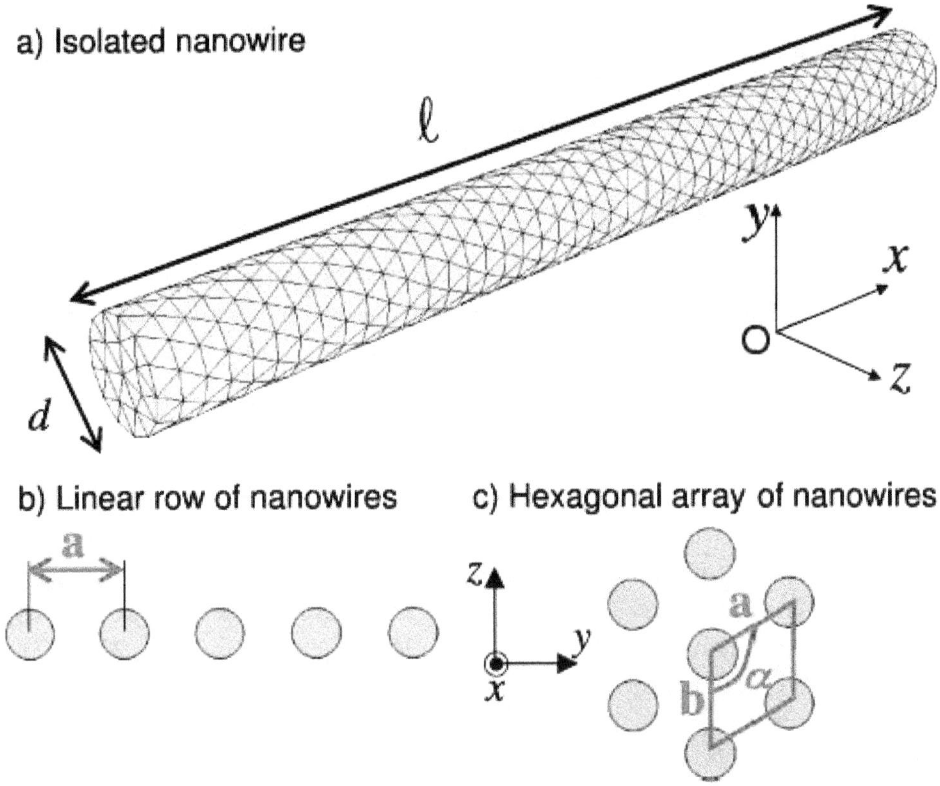

Table 1. NWs types and their Examples

S No.	Types of nanowires	Examples
1	Elemental	Si, Ge, B, Sn, Sb, Bi, Se, Te, Au, Ag, Cu, Ni
2	Metal Oxide	MgO, Al_2O_3, TiO_2,GeO_2, Mn_3O4, ZnO
3	Metal Nitride	BN, AIN, GaN, InN, Si_3N_4, Cu_xO
4	Metal Carbide	BC & SiC
5	Metal Chalcogenide	CdS, CdSe, CdTe, Pbs, Zns, Znse, Nbs_2
6	III-V compound	GaAs, InAs, InGaAs.

Nanowire FETs

In NWFET device channel is consist of nanowire having diameter in nanometer. Si based NWFET devices are difficult to fabricated, so a lot of attention has been placed on homogenous Si based NWFETs and III-V NWFETs have been introduced because these materials have superb carrier transport. NWFET have many benefits over conventional CMOS devices as (Iwai et al., 2009). Firstly, NW based devices have capability to reduced SCEs, low OFF-current (IOFF), due to Gate all around structure. Secondly, NW based devices have high ON-current (ION) because of quasi One Dimensional (1D) conduction of NWs. Due to 1D transport in nanowire conduction is very high and less carrier scattering (Sakaki, 1980). The band structure of nanowire is different as compared planer device; the sub-bands seem close to lowest band in nanowire. These distinct sub bands enhance near the conduction band (Ohno et al., 1992). The fabrication steps of nanowire based FET is less as compare to conventional CMOS devices (Lauhon et al., 2002).

The Structures of Double-Material-Gate NWFET and Simulation Parameters

The structure of NWFET has been designed by using Silvaco ATLAS. Figure 2 shows the 3D and 2d cut section of Double-Material-Gate NWFET with device dimension. The device channel, source, drain length is 20 nm and Diameter (D) is 5 nm. The dielectric used is Hfo_2 for high K and Sio_2 for low k (Chaturvedi et al., 2014). The work function (Φ_{M1}) of gate metal near source is 4.83eV and work function of metal at drain side is 4.41eV are used to get the preferred threshold voltage. The length of overall channel is equal to $L=20(L_1+L_2)$. Thickness of SiO_2 and HfO_2 is 0.4nm and 2.1nm.The device structure parameters are given in table 2.

Figure 2. (a) 3D and (b) 2D cut plane view of Double-Material-Gate NWFET

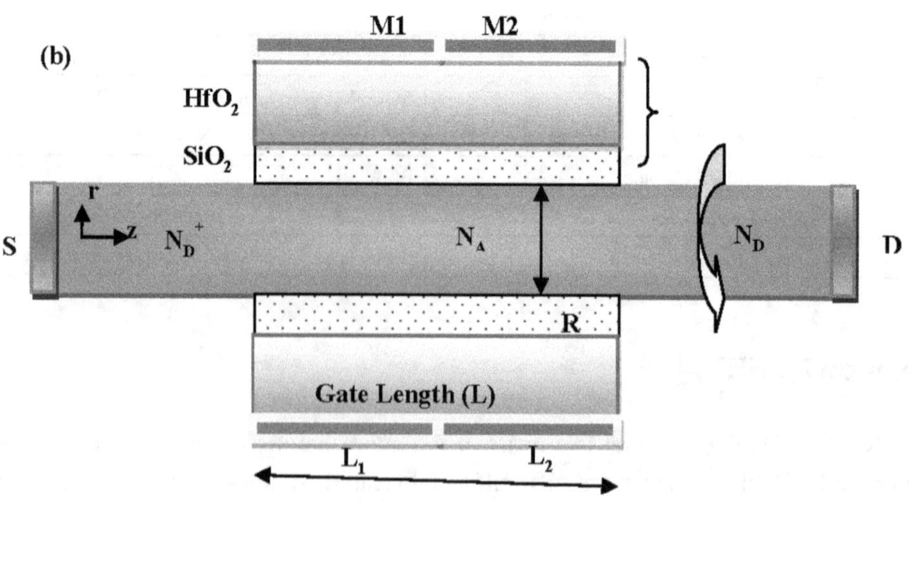

Table 2. Device parameter description for Double-Material-Gate NWFET

Device Case	Semiconducting material	Channel doping (N_A) (cm^{-3})	Source/Drain doping (N_D^+) (cm^{-3})	Gate metal work function in (eV)
Double-Material-Gate NWFET	$In_{0.53}Ga_{0.47}As$	2x10^{16}	1x10^{20}	$M_1$4.83, $M_{2=}$4.41

SIMULATION RESULTS AND DISCUSSION

To check the accuracy of the device, Firstly single -Material-Gate NWFET device design using 3D simulator (Silvaco ATLAS) with parameter taken from (Marin et al., 2015), the gate length of the device is 1 μm, oxide thickness t_{ox} is 1.5 nm, work function 5.05 eV of gate metal. The doing of channel is 1x10^{14} cm^{-3} and doping of source and drain doping (N_D^+) =5x10^{17} cm^{-3}. The simulation results have closed agreement with existing work (Marin et al., 2015), as given in Figure 3 and Figure 4.

Figure 3. I_d-V_{ds} characteristics of Single -Material-Gate NWFET using Silvaco ATLAS simulator

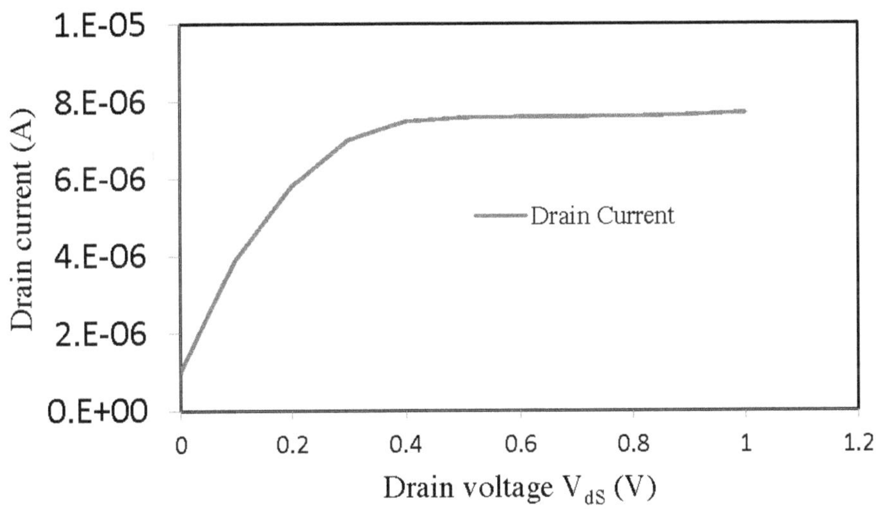

Effect of Channel Length Scaling

This section shows the behavior of Double-Material-Gate NWFET device by changing the length of the channel. Figure 4 shows the discrepancy in I_d-V_{ds} characteristics

Figure 4. I_d-V_{ds} characteristics of Single -Material-Gate NWFET [36] using Graph Digitizer tool

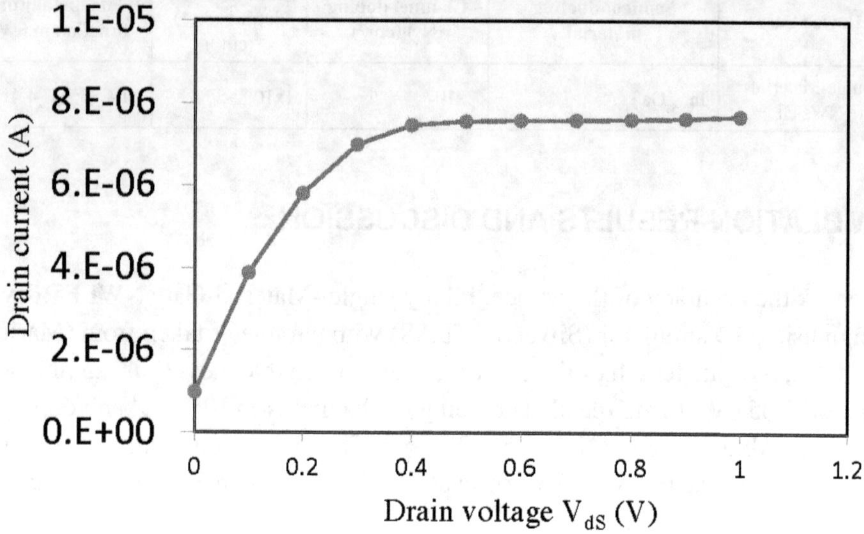

Figure 5. I_d-V_{ds} characteristics of Double-Material-Gate NWFET for different channel lengths

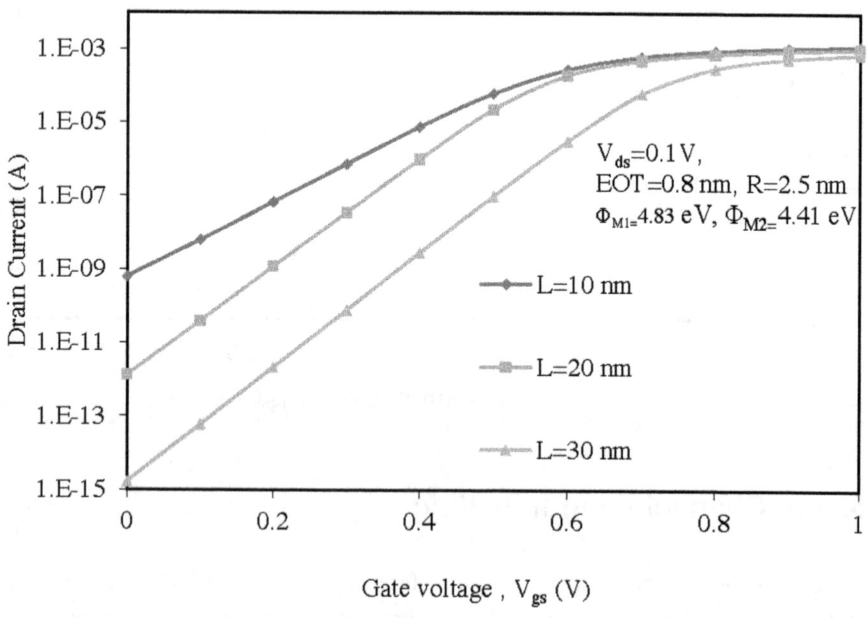

while reducing the length of channel from 30 to 10 nm. The channel radius is taken as constant in each case is 2.5 nm. The device performance degrade by reducing the channel length. This is due to the DIBL effect.

Figure 5 shows effect on SS by varying the channel length. It is observed that by increasing the channel length the SS reduces. The SS reduces due to enhancement in the electrostatic integrity of the device. The potential barrier at source region increases by increasing channel length. So electrons have not conduct at subthreshold region and SS decreases.

Figure 6. Effect on subthreshold slope by varying channel length channel length for Double-Material-Gate NWFET

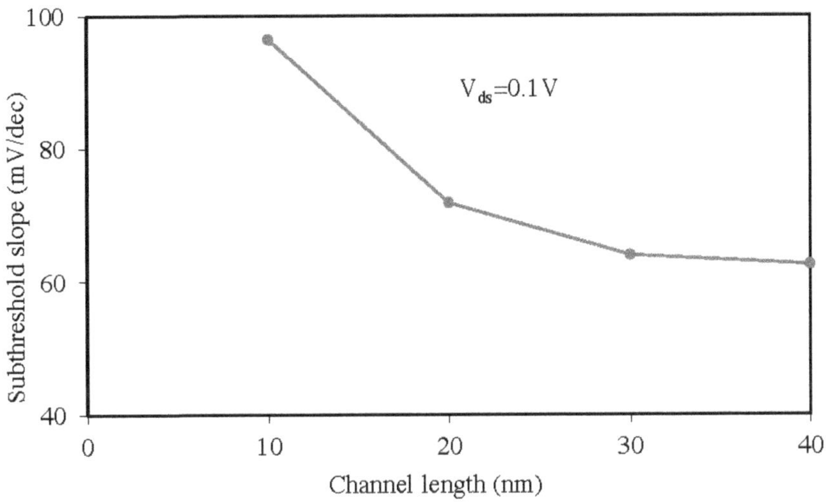

NANOWIRE FET BASED SENSORS

NWs as Semiconductor have extended abundant attention due to their unique electronic, photonic, mechanical, and electrical properties (Elfström et al., 2007), (Soci et al., 2013), (Brubaker et al., 2013) (Krogstrup et al., 2013). Semiconductor NWs using III-V compound (e.g. In1-xGaxAs) have more efficient in absorbing light energy as compare to Si nanowire, this unique property of helpful to design high sensitive photosensor. Sensors made by Semiconductor NWs using III-V compound (e.g. $In_{1-x}Ga_xAs$) have great sensitivity and low power consumption because of more surface area of NWs, The structure of NWFET as a sensor application is given in Figure 6.

Figure 7. NWFET for sensor application

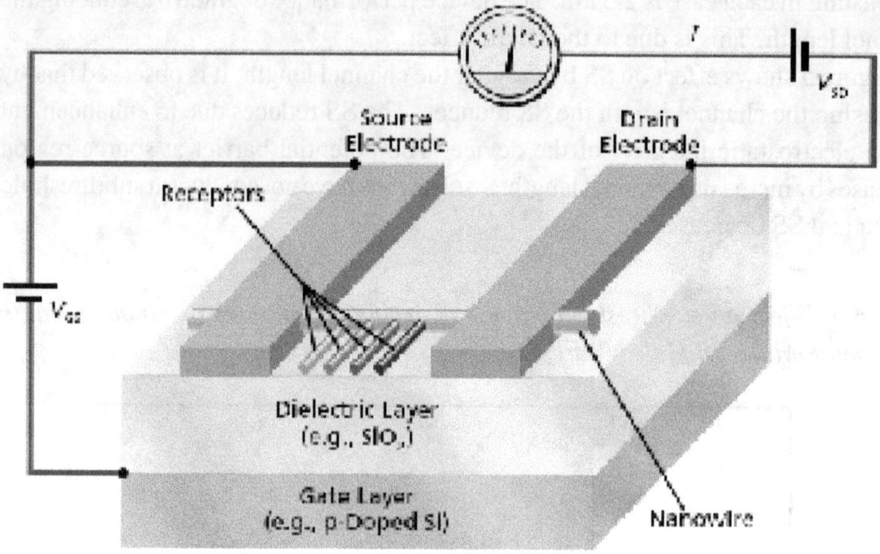

Optical Beam

Silvaco ATLAS-3D simulator used to extract device characteristics under dark and under incident radiation (Marin et al., 2015) includes advance LUMINOUS-3D optical simulator to extract device characteristics under incident radiation which uses Ray Trace method for calculating the photogeneration rate at defined mesh points. LUMINOUS-3D module is set the radiation intensity, optical parameters, location and wavelength with incorporated of BEAM keyword.

The light path from air to channel of NWFET through at four interfaces; first air to gate, second gate to oxide (HfO_2), third interface is gate to oxide (SiO_2), last interface is oxide to channel of NWFET. If reflection accrues, this interfaces supports to examining the performance of device underneath happening radiation. Reflection coefficients of these interfaces are given in (1).

$$R_c = \frac{(n_c - n_{c+1})^2 + k_{c+1}{}^2}{(n_c + n_{c+1})^2 + k_{c+1}{}^2} \tag{1}$$

where, n1 is refractive index in air, n2 is refractive index in gate, n3 is refractive index in oxide and n4 is refractive index in semiconductor. These all are the real values. Here c value lie 1 to 4, and R1, R2, R3 and R4 are at different four interfaces reflection coefficients. kc's are refractive indices of imaginary part. These imaginary and real values are given in (SOPRA) SOPRA database.

When light is incident at channel the Electrons and Hole Pairs are generated. Absorption coefficient (α), define the amount of light fall on the channel.

Electrons and Holes Pair generation rate determined by incident radiation absorbed channel. The generation rate depends on channel material (In0.53Ga0.47As). The generation rate of Electrons and Holes Pair is taken from (Simpson et al., 1999).

$$G_r = \alpha\phi_d \tag{2}$$

ϕ_d is the incident spectral photon flux density and is calculate as

$$\phi_d = I_o \frac{h}{c\lambda}(1-R_1)(1-R_2)(1-R_3) \tag{3}$$

where, h = Plank's constant, I= intensity of incident radiation, c = light speed

Figure7 shows that maximum value of photocurrent occurs at 0.450 μm, the reason behind this at this wavelength maximum light is absorbed by channel material.

The most essential parameters for photosensor are Quantum Efficiency (Qe) and Responsivity (R_e) which can be calculated by (4) and (5) (Xie et al., 2012).

$$\text{Responsivity}(R_e) = \frac{\text{Photocurrent}, (I_{ph})}{P_o(\text{Watts})} \tag{4}$$

$$\text{Quantum Efficiency}(Q_e) = R_e \times \frac{hc}{q\lambda} \tag{5}$$

Where, h plank's constant, q charge of electron, λ is Wavelength of incident radiations.

Table 3. Précises the comparative analysis of $In_{0.53}Ga_{0.47}$ as Double-Material-Gate NWFET photosensor with literature work. **Comparison has been done at** different performance parameters such as Quantum Efficiency (Q_e), Responsivity (R_e), device dimension and photosensitivity. It is founded that $In_{0.53}Ga_{0.47}As$ Double-Material-Gate NWFET photosensor has less size and high photosensitivity.

Figure 8. Photocurrent Vs Wavelength for Double-Material-Gate NWFET

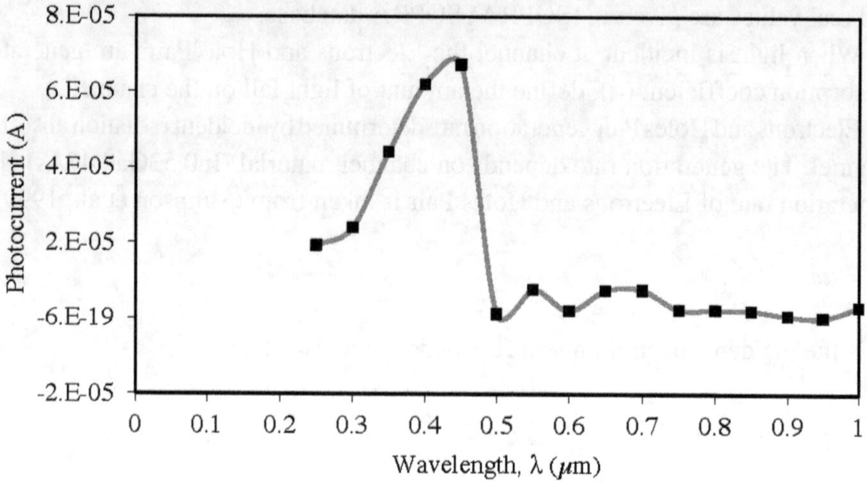

Table 3. Comparison of proposed $In_{0.53}Ga_{0.47}As$ DMGS CG/SG NWFET device with existing reported sensors in literature.

Photosensor	Responsivity (R_e) A W^{-1}	Quantum Efficiency(Q_e)	Photosensitivity	Dimensions
ZnSe nanowire (Salfi et al., 2006)	22	6810	-	80 nm (dia.) _ 4 mm (L)
CdS nanobelt (Gao et al., 2005)	-	-	1.5 $x10^3$	200 nm (width)
CdS nanowire (Li et al., 2005)	-	-	39	200 nm (dia.) array
CdSe nanowire (Kung et al., 2005)	-	5	10-100	200 nm (width) x 60 nm (height)
CdTe nanoribbon (Xie et al., 2012)	7.8 $x10^2$	2.4 $x10^5$	2	-
CdTe nanowire (Shaygan et al., 2014)	80.1	2.5 $x10^4$	19	49.1 nm (dia.) x 4.39 (L) mm
$In_{0.53}Ga_{0.47}As$ Double-Material-Gate NWFET	4.3 mA W^{-1}	11.2	1.36x10^9	5 nm (dia.) x 20 (L) nm

CONCLUSION

In this chapter, we have discussed about NWs and their types. Also design and analyed of III-V compound based Double-Material-Gate NWFET for best transfer characteristics. III-V compound material $In_{0.53}Ga_{0.47}As$ based Double-Material-Gate

NWFET has been design. Double-Material-Gate NWFET performance has been analyzed at different channel length. This chapter also has demonstrated $In_{0.53}Ga_{0.47}As$ based Double-Material-Gate NWFET as a photosensor and observed its performance. The Double-Material-Gate NWFET based photosensor have potential for abundant performance, $In_{0.53}Ga_{0.47}As$ sensors as tunable band gaps may be used as different application like indoor communication, and optical sensors.

REFERENCES

Bohr, M. T. (2002). Nanotechnology goals and challenges for electronic applications. *IEEE Transactions on Nanotechnology*, *1*(1), 56–62. doi:10.1109/TNANO.2002.1005426

Bricout, P. H., & Dubois, E. (1996). Short-channel effect immunity and current capability of sub-0.1-micron MOSFET's using a recessed channel. *IEEE Transactions on Electron Devices*, *43*(8), 1251–1255. doi:10.1109/16.506776

Brubaker, M. D., Blanchard, P. T., Schlager, J. B., Sanders, A. W., Roshko, A., Duff, S. M., Gray, J. M., Bright, V. M., Sanford, N. A., & Bertness, K. A. (2013). On-chip optical interconnects made with gallium nitride nanowires. *Nano Letters*, *13*(2), 374–377. doi:10.1021/nl303510h PMID:23324057

Chaturvedi, P., & Kumar, M. J. (2014). Impact of gate leakage considerations in tunnel field effect transistor design. *Japanese Journal of Applied Physics*, *53*(7), 074201. doi:10.7567/JJAP.53.074201

Cho, H., Kapur, P., & Saraswat, K. C. (2004). Power comparison between high-speed electrical and optical interconnects for interchip communication. *Journal of Lightwave Technology*, *22*(9), 2021–2033. doi:10.1109/JLT.2004.833531

Crupi, F., Kaczer, B., Degraeve, R., Subramanian, V., Srinivasan, P., Simoen, E., Dixit, A., Jurczak, M., & Groeseneken, G. (2006). Reliability comparison of triple-gate versus planar SOI FETs. *IEEE Transactions on Electron Devices*, *53*(9), 2351–2357. doi:10.1109/TED.2006.880824

Elfström, N., Juhasz, R., Sychugov, I., Engfeldt, T., Karlström, A. E., & Linnros, J. (2007). Surface charge sensitivity of silicon nanowires: Size dependence. *Nano Letters*, *7*(9), 2608–2612. doi:10.1021/nl0709017 PMID:17691849

Gan, F., Hou, L., Wang, G., Liu, H., & Li, J. (2000). Optical and recording properties of short wavelength optical storage materials. *Materials Science and Engineering B*, *76*(1), 63–68. doi:10.1016/S0921-5107(00)00400-1

Gao, A., Lu, N., Dai, P., Li, T., Pei, H., Gao, X., Gong, Y., Wang, Y., & Fan, C. (2011). Silicon-nanowire-based CMOS-compatible field-effect transistor nanosensors for ultrasensitive electrical detection of nucleic acids. *Nano Letters*, *11*(9), 3974–3978. doi:10.1021/nl202303y PMID:21848308

Gao, T., Li, Q. H., & Wang, T. H. (2005). CdS nanobelts as photoconductors. *Applied Physics Letters*, *86*(17), 173105. doi:10.1063/1.1915514

Ghosh, P., Haldar, S., Gupta, R. S., & Gupta, M. (2012). An analytical drain current model for dual material engineered cylindrical/surrounded gate MOSFET. *Microelectronics Journal*, *43*(1), 17–24. doi:10.1016/j.mejo.2011.10.001

Guang-Xi, H., Ling-Li, W., Ran, L., Ting-Ao, T., & Zhi-Jun, Q. (2010). Quantum-mechanical study on surrounding-gate metal-oxide-semiconductor field-effect transistors. *Communications in Theoretical Physics*, *54*(4), 763–767. doi:10.1088/0253-6102/54/4/33

Hu, L., & Chen, G. (2007). Analysis of optical absorption in silicon nanowire arrays for photovoltaic applications. *Nano Letters*, *7*(11), 3249–3252. doi:10.1021/nl071018b PMID:17927257

Huang, Y., Duan, X., Cui, Y., Lauhon, L. J., Kim, K. H., & Lieber, C. M. (2001). Logic gates and computation from assembled nanowire building blocks. *Science*, *294*(5545), 1313–1317. doi:10.1126cience.1066192 PMID:11701922

Iwai, H. (2009). Roadmap for 22 nm and beyond. *Microelectronic Engineering*, *86*(7-9), 1520–1528. doi:10.1016/j.mee.2009.03.129

Jain, A., Sharma, S. K., & Raj, B. (2016). Design and analysis of high sensitivity photosensor using Cylindrical Surrounding Gate MOSFET for low power applications. *Engineering Science and Technology, an International Journal, 19*(4), 1864-1870.

Kaur, H., Kabra, S., Bindra, S., Haldar, S., & Gupta, R. S. (2007). Impact of graded channel (GC) design in fully depleted cylindrical/surrounding gate MOSFET (FD CGT/SGT) for improved short channel immunity and hot carrier reliability. *Solid-State Electronics*, *51*(3), 398–404. doi:10.1016/j.sse.2007.01.025

Kranti, A., Haldar, S., & Gupta, R. S. (2001). An accurate 2D analytical model for short channel thin film fully depleted cylindrical/surrounding gate (CGT/SGT) MOSFET. *Microelectronics Journal*, *32*(4), 305–313. doi:10.1016/S0026-2692(01)00008-8

Krogstrup, P., Jørgensen, H. I., Heiss, M., Demichel, O., Holm, J. V., Aagesen, M., Nygard, J., & Fontcuberta i Morral, A. (2013). Single-nanowire solar cells beyond the Shockley–Queisser limit. *Nature Photonics, 7*(4), 306–310. doi:10.1038/nphoton.2013.32

Kung, S. C., van der Veer, W. E., Yang, F., Donavan, K. C., & Penner, R. M. (2010). 20 μs photocurrent response from lithographically patterned nanocrystalline cadmium selenide nanowires. *Nano Letters, 10*(4), 1481–1485. doi:10.1021/nl100483v PMID:20334354

Lauhon, L. J., Gudiksen, M. S., Wang, D., & Lieber, C. M. (2002). Epitaxial core–shell and core–multishell nanowire heterostructures. *Nature, 420*(6911), 57–61. doi:10.1038/nature01141 PMID:12422212

Li, Q., & Penner, R. M. (2005). Photoconductive cadmium sulfide hemicylindrical shell nanowire ensembles. *Nano Letters, 5*(9), 1720–1725. doi:10.1021/nl050994x PMID:16159212

Lin, C., & Povinelli, M. L. (2009). Optical absorption enhancement in silicon nanowire arrays with a large lattice constant for photovoltaic applications. *Optics Express, 17*(22), 19371–19381. doi:10.1364/OE.17.019371 PMID:19997158

Lu, W., & Lieber, C. M. (2006). Semiconductor nanowires. *Journal of Physics. D, Applied Physics, 39*(21), R387–R406. doi:10.1088/0022-3727/39/21/R01

Lubow, A., Ismail-Beigi, S., & Ma, T. P. (2010). Comparison of drive currents in metal-oxide-semiconductor field-effect transistors made of Si, Ge, GaAs, InGaAs, and InAs channels. *Applied Physics Letters, 96*(12), 122105. doi:10.1063/1.3367708

Marin, E. G., Ruiz, F. G., Schmidt, V., Godoy, A., Riel, H., & Gámiz, F. (2015). Analytic drain current model for III–V cylindrical nanowire transistors. *Journal of Applied Physics, 118*(4), 044502. doi:10.1063/1.4927041

Marin, E. G., Ruiz, F. G., Schmidt, V., Godoy, A., Riel, H., & Gámiz, F. (2015). Analytic drain current model for III–V cylindrical nanowire transistors. *Journal of Applied Physics, 118*(4), 044502. doi:10.1063/1.4927041

Ohno, T., Shiraishi, K., & Ogawa, T. (1992). Intrinsic origin of visible light emission from silicon quantum wires: Electronic structure and geometrically restricted exciton. *Physical Review Letters, 69*(16), 2400–2403. doi:10.1103/PhysRevLett.69.2400 PMID:10046475

Ouyang, H., Striemer, C. C., & Fauchet, P. M. (2006). Quantitative analysis of the sensitivity of porous silicon optical biosensors. *Applied Physics Letters*, *88*(16), 163108. doi:10.1063/1.2196069

Park, J. T., Colinge, J. P., & Diaz, C. H. (2001). Pi-gate soi mosfet. *IEEE Electron Device Letters*, *22*(8), 405–406. doi:10.1109/55.936358

Park, T. S., Yoon, E., & Lee, J. H. (2003). A 40nm body-tied FinFET (OMEGA MOSFET) using bulk Si wafer. *Physica E, Low-Dimensional Systems and Nanostructures*, *19*(1-2), 6–12. doi:10.1016/S1386-9477(03)00285-6

Roy, K., Mahmoodi, H., Mukhopadhyay, S., Ananthan, H., Bansal, A., & Cakici, T. (2005, November). Double-gate SOI devices for low-power and high-performance applications. In *ICCAD-2005. IEEE/ACM International Conference on Computer-Computer Aided Design*, *2005*, 217–224.

Sakaki, H. (1980). Scattering suppression and high-mobility effect of size-quantized electrons in ultrafine semiconductor wire structures. *Japanese Journal of Applied Physics*, *19*(12), L735–L738. doi:10.1143/JJAP.19.L735

Salfi, J., Philipose, U., De Sousa, C. F., Aouba, S., & Ruda, H. E. (2006). Electrical properties of Ohmic contacts to ZnSe nanowires and their application to nanowire-based photodetection. *Applied Physics Letters*, *89*(26), 261112. doi:10.1063/1.2424653

Sharma, S., Raj, B., & Khosla, M. (2016). Comparative Analysis of MOSFET CNTFET and NWFET for High Performance VLSI Circuit Design. *Journal of VLSI Design Tools & Technology*, *6*(2), 19–32.

Sharma, S. K., Raj, B., & Khosla, M. (2015, October). Performance enhancement of junctionless nanowire FET with laterally graded channel doping and high-K spacers. In *2015 IEEE 4th Global Conference on Consumer Electronics (GCCE)* (pp. 556-559). IEEE. 10.1109/GCCE.2015.7398608

Sharma, S. K., Raj, B., & Khosla, M. (2017). Subthreshold Performance of In 1–x Ga x As Based Dual Metal with Gate Stack Cylindrical/Surrounding Gate Nanowire MOSFET for Low Power Analog Applications. *Journal of Nanoelectronics and Optoelectronics*, *12*(2), 171–176. doi:10.1166/jno.2017.1961

Sharma, V. K., Pattanaik, M., & Raj, B. (2014). PVT variations aware low leakage INDEP approach for nanoscale CMOS circuits. *Microelectronics and Reliability*, *54*(1), 90–99. doi:10.1016/j.microrel.2013.09.018

Shaygan, M., Davami, K., Shaygan, M., Davami, K., Kheirabi, N., Baek, C. K., Cuniberti, G., Meyyappan, M., & Lee, J. S. (2014). Single-crystalline CdTe nanowire field effect transistors as nanowire-based photodetector. *Physical Chemistry Chemical Physics*, *16*(41), 22687–22693. doi:10.1039/C4CP03322A PMID:25230829

Simpson, M. L., Ericson, M. N., Jellison, G. E., Dress, W. B., Wintenberg, A. L., & Bobrek, M. (1999). Application specific spectral response with CMOS compatible photodiodes. *IEEE Transactions on Electron Devices*, *46*(5), 905–913. doi:10.1109/16.760396

Soci, C., Zhang, A., Bao, X. Y., Kim, H., Lo, Y., & Wang, D. (2010). Nanowire photodetectors. *Journal of Nanoscience and Nanotechnology*, *10*(3), 1430–1449. doi:10.1166/jnn.2010.2157 PMID:20355534

SOPRA infobase. (n.d.). http://refractiveindex.info

Takato, H., Sunouchi, K., Okabe, N., Nitayama, A., Hieda, K., Horiguchi, F., & Masuoka, F. (1991). Impact of surrounding gate transistor (SGT) for ultra-high-density LSI's. *IEEE Transactions on Electron Devices*, *38*(3), 573–578. doi:10.1109/16.75168

Tanaka, J., Toyabe, T., Ihara, S., Kimura, S. I., Noda, H., & Itoh, K. (1993). Simulation of sub-0.1-mu m MOSFETs with completely suppressed short-channel effect. *IEEE Electron Device Letters*, *14*(8), 396–399. doi:10.1109/55.225591

Vega, R. A., & Liu, T. J. K. (2009). Three-dimensional FinFET source/drain and contact design optimization study. *IEEE Transactions on Electron Devices*, *56*(7), 1483–1492. doi:10.1109/TED.2009.2021439

Vishvakarma, S. K., Agrawal, V., Raj, B., Dasgupta, S., & Saxena, A. K. (2007). Two dimensional analytical potential modeling of nanoscale symmetric double gate (SDG) MOSFET with ultra thin body (UTB). *Journal of Computational and Theoretical Nanoscience*, *4*(6), 1144–1148. doi:10.1166/jctn.2007.2390

Xie, X., Kwok, S. Y., Lu, Z., Liu, Y., Cao, Y., Luo, L., Zapien, J. A., Bello, I., Lee, C.-S., Lee, S.-T., & Zhang, W. (2012). Visible–NIR photodetectors based on CdTe nanoribbons. *Nanoscale*, *4*(9), 2914–2919. doi:10.1039/c2nr30277b PMID:22499254

Xie, X., Kwok, S. Y., Lu, Z., Liu, Y., Cao, Y., Luo, L., Zapien, J. A., Bello, I., Lee, C.-S., Lee, S.-T., & Zhang, W. (2012). Visible–NIR photodetectors based on CdTe nanoribbons. *Nanoscale*, *4*(9), 2914–2919. doi:10.1039/c2nr30277b PMID:22499254

Young, K. K. (1989). Short-channel effect in fully depleted SOI MOSFETs. *IEEE Transactions on Electron Devices*, *36*(2), 399–402. doi:10.1109/16.19942

Yuan, Y., Yu, B., Song, J., & Taur, Y. (2009). An analytic model for threshold voltage shift due to quantum confinement in surrounding gate MOSFETs with anisotropic effective mass. *Solid-State Electronics*, *53*(2), 140–144. doi:10.1016/j.sse.2008.10.010

Chapter 7
Brace of Nanowire FETs in the Advancements and Miniaturizations of Recent Integrated Circuits Design

Debapriya Chakraborty
National Institute of Technology, Sikkim, India

Jeetendra Singh
National Institute of Technology, Sikkim, India

Shashi Bala
Chandigarh Engineering College, Landran, India

ABSTRACT

This chapter encompasses the gradual requirements, basic working principle, inbuilt physics, structural and functional characteristics, and applications of nanowires, especially that of semiconductor nanowires in depth. Today, research and development in material science and electronics going hand in hand have opened up numerous directions for the exploration and utilization of several unique semiconducting materials in the design of novel field-effect-transistors (FETs) in the nano-scale architecture. The performance results of the basic NWFETs structures and hetero-structures along with methods to organize nanowires in the form of arrays to fulfill the requirement of integration of devices and circuits are described in detail. This chapter would be beneficial for students of undergraduate and postgraduate, researchers, and the industrial peoples as well who are working in the regime of the advancement of semiconductor technology because every aspect of nanowire and NWFETs is discussed here deeply in a single platform.

DOI: 10.4018/978-1-7998-6467-7.ch007

INTRODUCTION TO NANOWIRES

The timeline of the 1950s witnessed a breakthrough invention of integrated circuits and it is since then that the gradual growth of technology has focused on developing transistors and techniques to enhance energy efficiency along with speed and area efficiency. However, difficulties faced during integration and miniaturization of devices because of the different short channel effects in a metal-oxide-semiconductor field-effect-transistor (MOSFET), have culminated to an increase in off-state leakage current and also high power dissipation to create a major concern that is inevitably supposed to slow down MOSFET scaling as it approaches atomic dimension. Hence the urgent need to strike a balance between costs, speed, density, reliability, power dissipation in designing modern devices has made researchers realize the potential of nanoelectronics that led to the advent of nanowires.

Nanowire

A nanowire is an anisotropic cylindrical structure that looks like a normal electrical wire whose diameter can vary between 1nm to 100 nm and lengths of about 10 to 100 microns that is normally yielded from materials available sufficiently in nature including metals, semiconductors, and insulators built using silver, copper, gold, silicon, germanium and carbon nanotubes.

Evolution of Nanowires from a Bulk

CMOS technology has been abiding by Moore's law to continuously scale the dimensions of transistors. Over the past few decades, the planar bulk structure-based device architecture has been improvised by decreasing the length of MOSFET gates with every new production technology (Bondyopadhyay, 1998, pp. 78-81). Optimization techniques to modify junctions, to exemplify, the incorporation of a lightly-doped-drain (LDD), building graded channel in laterally doped profile, reduced depth of junction, and many other methods have been made use of. In addition to this, the mobility factor has been enhanced a lot by strain engineering. To enhance the gate controllability, minimize the gate leakage current, and to achieve the wide adjustability of the least amount of voltage to turn the device on, i.e. the threshold value, high K value materials are nowadays in extensive use as gate dielectric material. As a consequence of gate length shortening, the gradual degradation of the grip of the gate over the channel slowly decays down with overpowering short-channel effects. Hence planning for options of the maximally-scaled CMOS structure has led to the advent of optimized MOSFET structures, like the ultrathin-body (UTB) based FET having one gate, silicon-on-insulator (SOI), and the multiple-gate FET

(Mug-FET) with time (Choi et al., 1999, 2001), the Mug-FET has taken the shapes of FinFET and the gate-all-around (GAA) FET (Sharma, 2017, Jain, 2018). Thus nanowires being used to shape up the new Mug-FET bodies have come into the picture of the present as well as a future study. Fig. 1 shows the gradual developments of the MOSFET technology from the single gate MOS to the nanowire FETs.

Figure 1. A gradual evolution of the bulk-based planar MOSFET to a present GAA-MOSFET

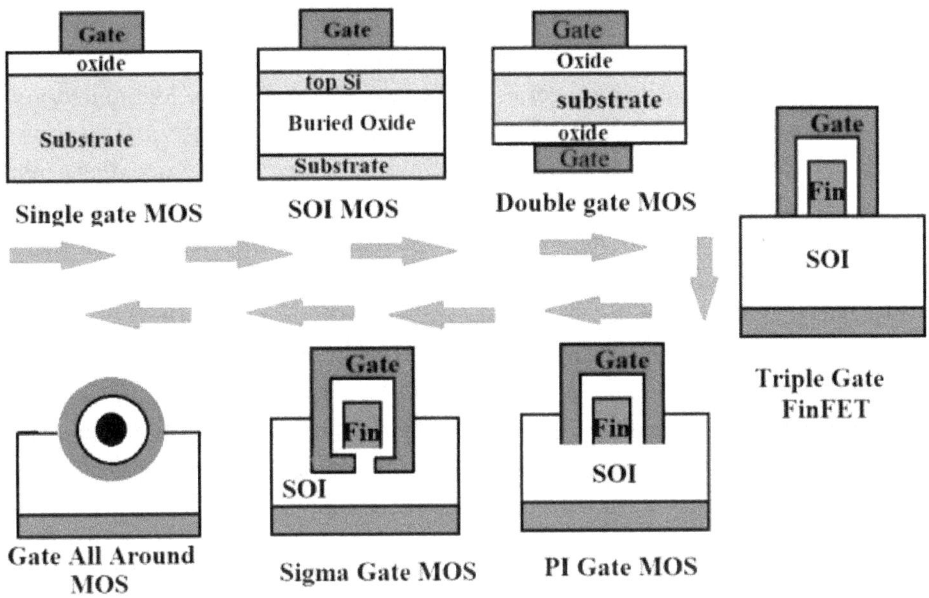

Special Features of Nanowires Compared to that of Bulk Materials Based Planar Devices

A massive ratio of length to diameter leading to a huge surface area is the most attractive feature of a nanowire which is responsible for making nanowires to be considered as 1-dimensional materials that have fewer defect sites and can exhibit unique quantum confinement effects of nanoscale geometry as compared to bulk materials. The increased surface-to-volume ratio renders nanowires highly sensitive to surface chemistry leading to their efficient utilization, concerning building sensors that are superior in quality as compared to those built using planar devices in bulk.

Again, to discuss the GAA structure which can characterize a nanowire FET, we find that it allows electrostatic gate control from all directions over the nanowire channel. Fin-FET has been the precursor to GAA NWFET and the horizontal NWFET structure resembles a lot to that of the Fin-FET. Process nodes of 10 nm or even 7 nm have seen a wide use of Fin-FET and are still much dependent on the same for complex circuit-level applications. Now, considering sub-7nm scaling, GAA FETs are being realized to have an extremely promising future implementation with a complexity level that is much lower as compared to other alternative techniques to enhance scalability (Mertens et al., 2016). The particularly small diameter of nanowires allows their easy entry into living entities causing no harm, while their length supports them to pass bi-directional electrochemical signals within a cell (Dasguta et al., 2014). These unique features of nanowires help them to be effectively used as battery electrodes fast charging time (Shao et al., 2010). A nanowire also has a diameter that is shorter than the wavelength characterized with visible light, hence making it permissible to operate in the wave optics domain. Manipulation of geometrical dimensions of a nanowire can successfully tune reflectivity and absorption of light energy resulting in the selection of definite optical resonant modes that cannot be done using bulk materials (Tian et al., 2009). Similarly, nanowire dimensions allow curtailing thermal conductivity as compared to that in bulk materials due to their being smaller than the phonon-mean free path (Dasguta et al., 2014).

Enhancement of flexibility of nanowires is a direct advantage of strain dissipation on their huge surface area making them quite different in mechanical characteristics when compared to the bulk materials (Dasguta et al., 2014). Proper utilization of all the above-mentioned properties demanded reliability and controllability of nanowire synthetic processes, which differ according to how the generation of highly anisotropic structures are initiated. Classification of synthesis techniques orients mainly along the two broad directions followed: top-down method and bottom-up process. Fabrication in the top-down method is a conventional subtractive process that uses chemicals to get a nanoscale level of control on patterning or selective removal of material producing nanowires. It makes use of many prevalent techniques that are commonly seen in the semiconductor industry, to cite, lithography using dry or wet chemical etching, converting a wafer made of bulk material into nanowire forms. Usually, the top-down method engages bulky, precision-oriented, and costly instrumentation expected to be found in some cleanrooms or nanofabrication facilities. The other picture is of bottom-up synthetic processes that differ from the one discussed above as these assemble molecules one-by-one and scale up to form materials possessing dimensions that are hundred times(and even more) bigger than the size of a molecule on an average. Just like a tree grows from a seed, bottom-up additive processes can also be thought to proceed in that way using chemistry for structure controlling. The benefits and issues with the two kinds of synthesis approaches will be studied along with the different synthesis techniques that are covered in later part of this chapter after a quick view of the inbuilt physics of nanowires followed by a brief section on the study of materials used in producing semiconducting nanowires. After synthetic procedures, the structure effects on device parameters, i.e. cross-section shape variations, channel orientation variations size of nanowire variations, the physical properties including mechanical, transport, optical and other properties of nanowire FETs (NWFETs), an introduction to nanowire heterostructures (NWHSFETs), methods to organize nanowires necessary for fabricating arrays of devices, NWFETs circuit applications, and a conclusive summary will follow one by one.

IN-BUILT PHYSICS OF SEMICONDUCTOR NANOWIRES

The functioning of NWFET (Si-NWFET) can be well understood with the basic reference of the operational logic of a MOSFET. Here we would take a look at how the n-channel Si-NWFETs work. The main focus will be on the I-V characteristics of intrinsic Si-NWFET. In the case of this type of FETs, it is noteworthy to know that both n-type and p-type doped channels can be fabricated within a single nanowire, as shown in Fig. 2. Again, Fig. 3 (a) and Fig 3(b) depict how the drain current varies with the change in drain-to-source as well as gate-to-source voltages when the channel has an approximate length of 100nm. It can be noticed that the I–V behavior in Si-NWFET resembles that of a MOSFET in general, indicating that the operational basics are essentially alike.

Figure 2. Si-NWFET cross-section showing n-type (top) and p-type (bottom) doped poly-Si structures.

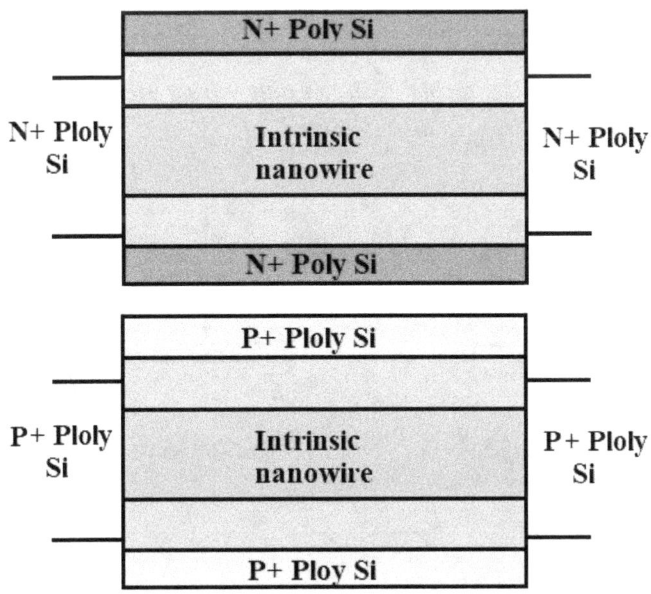

Figure 3. I–V curve and transfer characteristics of Si-NWFET having a 100nm long channel (a) for various gate voltage (b) for Various Drain voltage.

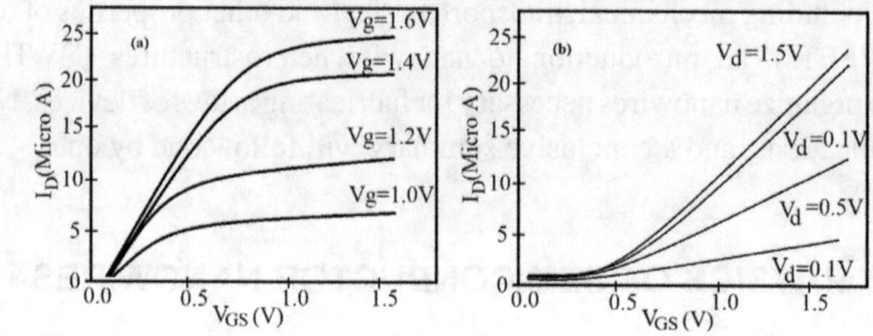

The replacement of the p-type substrate of n-type MOSFET by the intrinsic layer is the only change in the case of a nanowire construction keeping the analogy of the n+ ploy-Si gate-SiO_2-intrinsic contact at equilibrium condition in nanowire with that of an NMOS intact (Pan et al., 2001), as shown in Fig. 4. Electrons that leave the gate electrode and travel into the nanowire do so due to the change in potential from one Fermi-level to the next. The electrons residing in a distributed manner inside the sublevels hence cannot get the chance to accumulate at the proximity of an oxide interface like what we find for a typical NMOS. The underlying reason for this phenomenon is the wave-particle dual nature exhibited by electrons, which can be studied in detail under quantum mechanics and is currently not taken under the scope of this chapter. A cross-sectional study of nanowire reveals an approximately uniform probability density of electrons in each sublevel across it. Carrier transport

Figure 4. n^+ poly-Si–SiO_2-intrinsic silicon nanowire system, The Fermi levels, as well as the affinity factors, are also depicted.

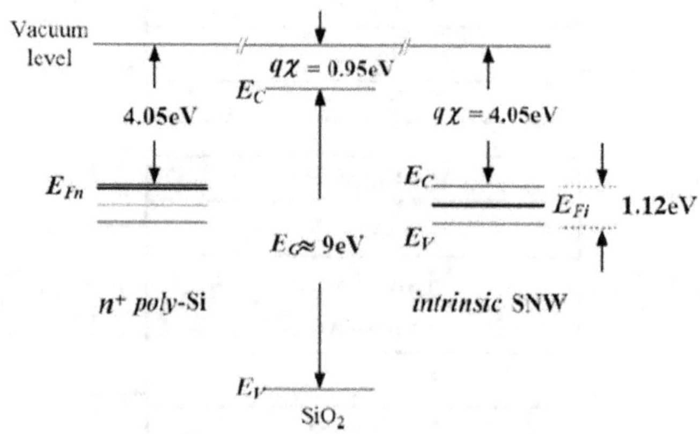

takes place mainly by the drift-diffusion mechanism in case of a long channel, while short channel experiences mainly ballistic transport. Hence, FETs experience carrier transport by a combination of drift-diffusion and ballistic mechanisms.

An important aspect of the I–V characterization is the number of charges present on the surface, Q_n that get induced by the gate voltage (V_G). Since Q_n rises almost linearly with the gate overdrive, it is possible to write,

$$Q_n=C_{eff}[V_G-V_T-V] \tag{2.1}$$

Here, V is the induced voltage across the channel, and C_{eff} the effective capacitance throughout the nanowire. The capacitance, C_{eff} has a value that is to some extent smaller than C_{OX} above V_T for the following reasons. It is a well-known fact that C_{eff} comprises of the series-connected oxide (C_{OX}) and surface (C_S) capacitances i.e., $1/\ C_{eff} = 1/\ (C_{OX} + C_S)$. It is practically found that the electrons gather at the silicon dioxide interface in case of an NMOS. As a result, C_S is much greater than C_{OX}, and hence C_T, the total capacitance, can be taken to be the same as C_{OX} above V_T. As mentioned earlier, the picture of a nanowire FET is quite different where the electrons distribute themselves almost uniformly across the cross-section region, as a consequence of the wave-like property of electrons. Therefore, in the case of a nanowire FET, the ratio of C_S concerning C_{OX} is much smaller in comparison to that in an NMOS, also making C_{eff} somewhat reduced in value than C_{OX}. Nevertheless, the $C_S\ /\ C_{OX}$ value is still maintained large enough to yield a constant C_{eff} value that is measured to be at a slightly lower level than C_{OX}.

In short channel FETs, a prevalence of ballistic transport can be observed. For example, in an ultra-scaled MOSFET, the movement of carriers occurs by the ballistic component by as much as 50%. The detail of ballistic transport is outside the scope of this chapter and readers can study quantum mechanics to have an in-depth knowledge of the same.

DIFFERENT TYPES OF NANOWIRE MATERIALS

The variety of nanowire materials highly depends upon the methods used in fabricating the materials. For example, template-based synthesis results in the fabrication of numerous inorganic materials like Au, Ag, Pt, TiO_2, MnO_2, ZnO, SnO_2, electronically conductive polymers polypyrrole, poly(3-methyl thiophene), and polyaniline, and carbon nanotubules (Huczko, 2002, pp. 617-638). A topotactic reaction has successfully yielded Ag_2Se nanowires of crystalline nature having a single crystal structure (Gates et al., 2001). Apart from the elemental semiconductors such as Si and Ge, binary compounds including III-V semiconductors (e.g. GaN, GaAs and

InAs), oxides (e.g. ZnO, SiO_2 and MgO), carbides(SiC, B_4C), nitrides(Si_3N_4) and chalcogenides (e.g. Bi_2Se_3, In_2Se_3, and CdS) and simple II-VI semiconductors(ZnS, ZnSe, CdSe) have also undergone a Vapor-Liquid-Solid (VLS) fabrication process for morphological study of nanowires (Trentler et al., 1995).

In addition to all these, aromatic compounds with planar structure as shown in Fig. 5 provides a good research area since their faces being stacked together can help in increasing mobility in FET-like devices. 2-D extended aromatic structures are more likely to exhibit this sort of stacking leading to strong Π-Π bonds to form 1-dimensional nanostructures (Sharma, 2016, Hoeben, 2016). Some examples of such 1-D nanowires are those obtained from hexabenzocoronene derivatives (Xiao, 2006, Briseno, 2007) by self-assembly, perylene tetracarboxylic diimide(PTCDI) derivatives (Wurthner, 2004, Che, 2007, Singh, 2016), pentacene derivatives (Lee et al., 2007), metal phthalocyanines and porphyrins (Luca et al., 2007), polythiophene polymers (Briseno et al., 2007) and several other organic semiconductors (Kim, 2007, Li, 2005). Researches on modern synthetic methods and molecular design have enabled chemists to come across intermolecular interactions with an increased degree of anisotropy, giving rise to new organic semiconductors that show face-to-face packing, and are capable of self-assembling into nanowires.

Figure 5. Various polymer semiconductors that can Π-stack to produce 1-D organic FETs synthesized from a single crystal

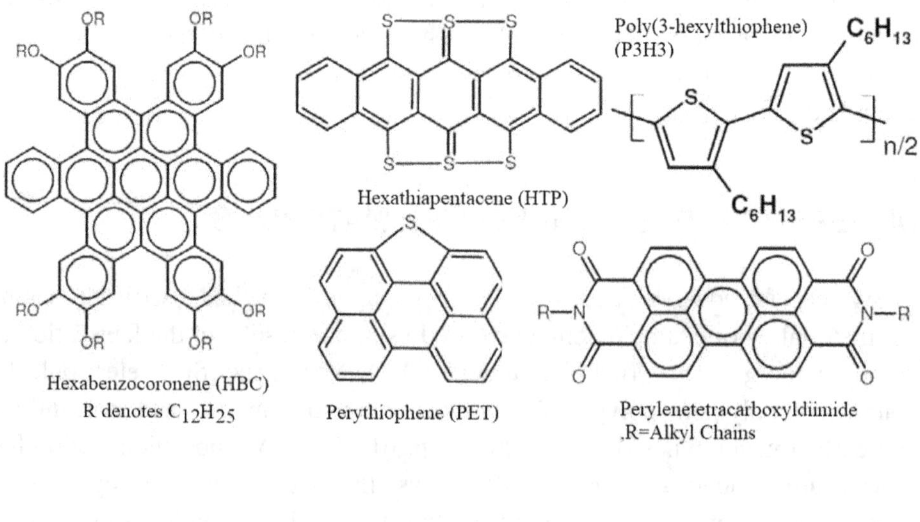

SYNTHETIC METHODOLOGIES OF NANOWIRES

As already been introduced, the synthesis of nanowires can follow two basic ways namely top-down and bottom-up. The different important methods of nanowire synthesis are described below along with the benefits and issues of the top-down and bottom-up techniques.

Lithography and its Types (Top-Down Fabrication Process)

Over the years, lithography is extensively used in microelectronics and can be counted as one of the best top-down processes for fabricating devices at nano-order. It is essentially a subtractive mechanism and it involves steps like chemical film deposition, photolithography, and etching. The definition of patterns to be masked on a coat of photoresist by photolithography marks the beginning of this method. The patterns are then designed on a substrate by an etching process that follows. Be it an etching process using reactive ions (RIE) or one using inductively-coupled-plasma (ICP), both are equally popular in the industry.

Optical lithography, also termed as photolithography, is based on allowing light to fall on a light-sensitive photoresist through a photo-mask to copy a region or node defining pattern from the latter to a substrate. The projection range of a clear image of a small feature onto a wafer is limited by the wavelength of the light. The minimum feature size that a projection system can define is specified by the Rayleigh criterion (Rai & Chaudhary, 1997).

$$CD = K_1\lambda / NA \quad (1)$$

Here, CD stands for critical dimension or the minimum feature size, K_1 the coefficient that reflects process-dependent attributes, λ the wavelength of the incident light used, and NA the numerical aperture of the lens as observed from the wafer. It can be understood that the critical dimension may be reduced by a shorter wavelength and a higher value of the numerical aperture. To state in simple words, the above equation 1 points to the statement that the minimum feature size is found to be restricted by the wavelength and the projection system.

The tools used in optical lithography engage deep ultraviolet (DUV) rays having beam-length of 248 nm from a krypton fluoride laser (KrF) and 193 nm from an argon fluoride (ArF) and can extend up to a feature size as low as 50 nm and lower than that, employing a 193 nm ArF excimer laser with the support of immersion lithography (Saib et al., 2015). Ion beam lithography, electron beam (e-beam) lithography, extreme ultraviolet (EUV) lithography, X-ray lithography are regarded as some of the potential techniques for sub-20 nm patterning. Even though EUV lithography using a 13.5 nm wave and x-rays using a sub-1 nm wave have got a huge prospect in achieving good results in nanowire patterning, these methods are yet to be fully developed to serve on a large scale, mostly due to the challenges they face in handling proper source power, mask, and photoresist.

Benefits and Issues of Top-Down Techniques

In respect of benefits, top-down nanowire fabrication being a subtractive method can provide much simpler ways to fabricate nanowire arrays. It helps in the establishment of electrical contacts to the nanowires for integrating them for device-level implementation. Moreover, a good number of the said methods exhibit high compatibility with already established techniques of the microelectronics industry which are extensively used in the well-known CMOS technology.

Bottom-up Methods of Fabrication of Nanowires

Over the years, many bottom-up methodologies have been introduced by researchers to synthesize 1-D nanostructures. All these methodologies are divided into two major streams depending upon the medium in which the reaction occurs while preparing nanowires: solution and gaseous-medium based methods.

Solution-Based Synthesis for Nanowire Fabrication

Template-Oriented Synthesis of One-Dimensional Nanostructures

This convenient and versatile technique helps to synthesize different kinds of materials that use the template simply as a shaping agent having morphology similar to the one required by the materials to grow to perform their desired functions. These templates generally are made inside mesoporous materials or polycarbonate membranes and alumina having pores, forming nanochannels as shown in Fig. 6. Filling up the nanoscale channels with a solution like sol-gel results in the generation of 1-D nanostructures. Finally, selectively peeling off the host matrix can release the required nanowires from those templates.

Another way of obtaining an ensemble of nanostructures resembling the appearance of the bristles of a brush is by a templating methodology that has been found to have pretty good efficiency in respect of controlling morphological growth of the material yielding nanowires and nanorods. The difficulty that this method faces is regarding the synthesis of nanowire materials that can possess single crystallinity. For example, leaving some single-crystalline nanowires of bismuth that have been obtained by templating bismuth melt against alumina membranes (Zhang, 2000),

Figure 6. Schematic view of nanoscale template-based synthesis of nanowires

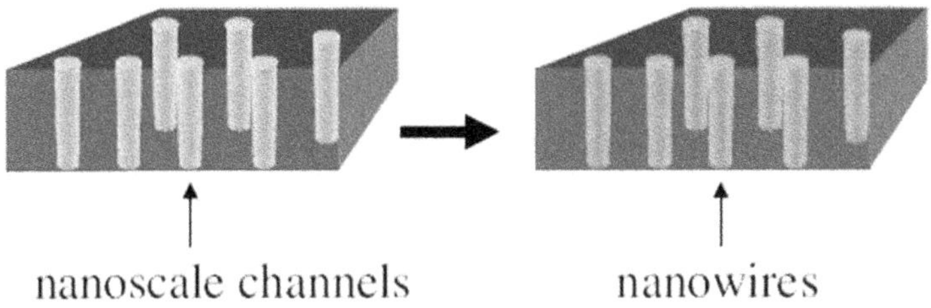

nanowires prepared are polycrystalline in the majority, thus potentially barring extensive utilization in field applications as well as studies linked with electronic transport. In addition to the benefits that can be derived due to large surface area and uniform pores present in polymer membranes and porous alumina, mesoporous silica materials have also been considered effective in hosting growth of catalysts, polymers, metal and semiconductor nanoparticles which are of prime importance in catalytic, environmental and optoelectronic applications. Summing up, it can be said that, the solution-based templating approach is quite a powerful technique, though it is likely to yield polycrystalline products most of the time.

Solution-Liquid-Solid (SLS) Technique

This approach yields semiconductor nanowires with high crystallinity at fairly low temperatures (approximately 203°C). Fiber like or near-single-crystal whiskers are synthesized that have widths ranging from 10 to 150 nm and lengths up to several micrometers. The growth of InAs, GaAs, and InP nanowhiskers follow this method[13].

Solvothermal Process of Nanowire Synthesis Using Chemicals

Semiconductor nanowire and nanorod synthesis using solvothermal methodology have been lately introduced in the microelectronics industry. These processes involve mixing a solvent with some kind of metals and possibly a crystal growth regulator like amine. The solution thus prepared is then placed within an autoclave for carrying out crystal growth followed by an assembly process. The inherent complexity of a solvothermal process, however, leads to products having a good amount of impurities, as well as the monodispersity of the sample, deviates much from the ideal case. As a whole, it can be said that these simple solution-centric techniques have yielded nanowires of different compositions to date. Once the problems of not having the desired extent of crystallinity and monodispersity are solved by building the better concept of crystal nucleation associated with growth techniques in solution, it can be expected to produce well-defined morphologies of crystalline nanowires having highly controllable aspect ratios.

Growth of Nanowires in the Gaseous Medium

Vapor-Liquid-Solid (VLS) Mechanism of Preparing Nanowires

A frequently used technique of growing nanowires using reactions in the gaseous medium is the so-called VLS process proposed by Wagner in the 1960s while he was doing his studies on the growth of nano-whiskers having the single-crystalline structure (Levitt, 1970). VLS method in the chemical vapor deposition(CVD) is a mechanism that finds extensive use for the yield of silicon nanowires. Fig. 7 depicts how a nanoparticle catalyst molecule is deposited to serve as a "seed" material on the silicon wafer. A nanoparticle catalyst in a liquid phase adsorbs a vapor-phase precursor and reaches a super-saturation level to mark the beginning of one-dimensional crystalline nanowire growth from the nucleation site at the liquid-solid interface. Processing the diameter of silicon nanowires that usually ranges between 3–500 nm is possible by altering the diameter of the nanoparticle catalyst (Morales & Lieber 1998). Although Au can be used very commonly as a catalyst for growing nanowires in the VLS method, Au has a compatibility issue with the widely used CMOS process due to its diffusivity being intolerably high accompanied by the creation of deep level traps in the bandgap of silicon leading to the formation of recombination centers as performance effectors. Au is also a highly stable chemical for which a fine cleaning process becomes difficult. Also, Au etchants containing alkali ions degrade the device performance.

For the above-mentioned reasons, many alternative processes have been devised to produce catalysts that can suitably be applied in CMOS based architecture or to invent growth techniques, entirely free from catalyst (Molnar, 2010). It is possible to grow silicon nanowires in epitaxy involving no use of a catalyst when we opt for molecular beam epitaxy (MBE). MBE can generate silicon atoms in the form of a directionally aligned beam having long mean-free paths. The evaporated particles from a highly purified silicon source get deposited over the Au seed on the wafer following its crystal orientation. This process of MBE makes the evaporation source switch and provides good control on the beam of gas hence producing layers of different dopants gradually developing a heterogeneous body. Now, the approximate limit of the possible diameter of the silicon nanowire is 40 nm as an effect of weak super-saturation that can be noticed in an MBE process (Werner, 2006). The heterogeneous structure will be discussed in the latter part of this chapter.

Figure 7. Vapor-liquid-solid (VLS) method of Si nanowire growth: (a) nanodroplet is exposed to a silane (SiCl) and H gas mixture; H reduces SiCl to form Si; Au forms an alloy with Si and liquefies; (b) Precipitation of Si begins when the liquid is supersaturated, producing crystalline nanowires; (c) Increase in length of the nanowire results from further condensation and dissolution of Si

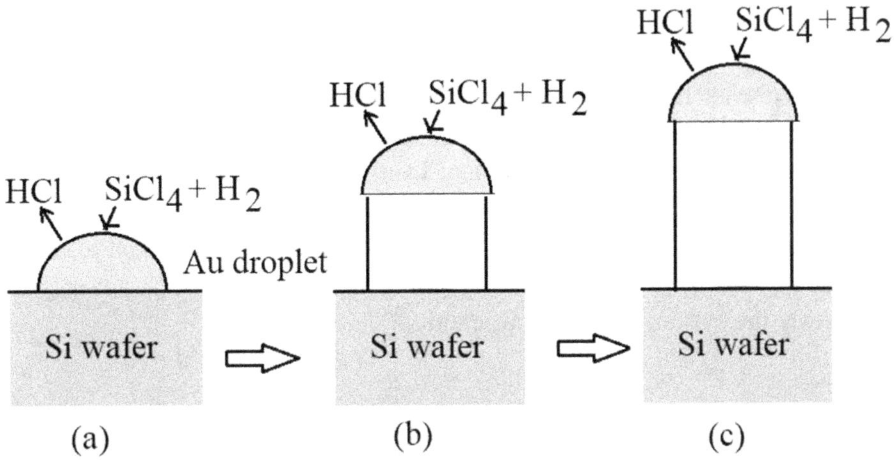

It can be said that the VLS method is the most exploited one among all the bottom-up methods to produce 1-100 micron diameter 1-D nanowires.

Oxide-Dependent Growth of Nanowire

The beginning of oxide-dependent synthesis was marked by treating a mixture of GaAs and Ga_2O_3 to yield GaAs nanowires under laser ablation technique and then this method was extended to produce Si nanowires. The key advantage offered here is that it requires none of a metal catalyst, or a metal template hence making it simple to purify the nanowires for their next applications. Si_xO (x>1), either thermally evaporated or after undergoing a laser ablation, produces the vapor phase that is the primary need of this method to proceed. The multiple decompositions of silicon oxide as given in equations (2) and (3), lead to the nucleation of nanoparticles on the substrate.

$$Si_xO \rightarrow Si_{x-1}(s) + SiO(s)\ (x>1) \tag{2}$$

$$2SiO(s) \rightarrow Si(s) + SiO_2(s) \tag{3}$$

Fig. 8 shows the schematic representation of the oxide-assisted growth of Si nanowires. Precipitation of silicon nanoparticles occurs in the deposited bed of silicon oxide that in turn gets formed by the decomposition of Si oxide described in the above equations (2) and (3). Few nanoparticles group together on the matrix surface Fig. 8(a). When the growth directs normally to the substrate, the nanowire nuclei can undergo rapid growth Fig. 8(b). The four main factors on which the Si growth rate depends are suggested by researchers as (1) the catalytic effect of the SiOx (x>1) layer on nanowire tips. Atomic absorption, diffusion, and deposition get enhanced when the materials at the Si nanowire tips remain in or almost in their molten state. (2) The pace with which the nanowires grow laterally gets retarded by the formation of the SiO_2 component from the decomposition of SiO. (3) Si nanowires grow faster as a result of the dislocation defects present at the tip regions. (4) Nucleation and growth of Si nanowires occurring along the axes parallel to the {111} surfaces having the minimum surface energy among all the Si surfaces, help to minimize the system energy significantly.

Figure 8. Schematics of oxide-dependent nucleation and growth of nanowires from Si. {112} orientation is depicted using parallel lines. (a) Deposited silicon oxide vapor creates a matrix for Si nanoparticles to get precipitated in. (b) The fast growth of nanoparticles in the orientation of preference results in forming nanowires. Nanoparticles accumulated in non-preferred orientations may take the shape of chains of nanoparticles

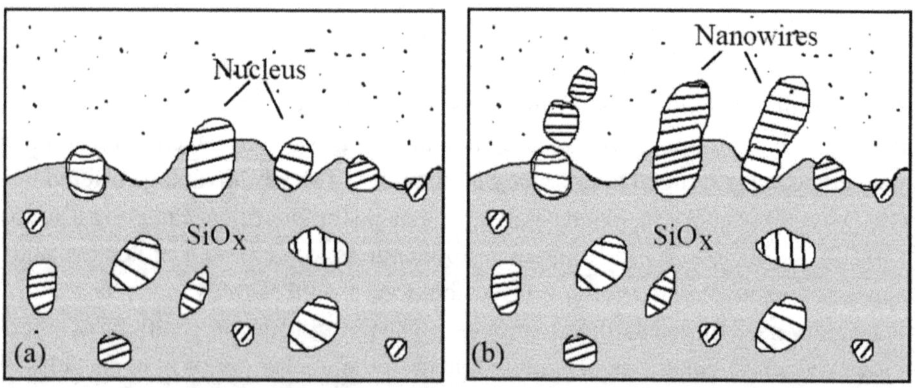

With the advancement of research, few other methods of nanowire fabrication using bottom-up techniques like electro-spinning mechanism and others have come up which are not been currently covered here.

Benefits and Issues of Bottom-Up Methods

Silicon nanowires prepared by different bottom-up methods possess unique properties like their small diameters, their heterogeneous composition, the possibility of in situ doping, and fewer surface dangling bonds. It is difficult to achieve such properties using top-down methods. Bottom-up synthesis techniques can easily produce superlattice structures with complex features, at a reliable cost and with great control over the growth of the composition of nanowires. Selective etching of these superlattice structures can build structures having a wide range of shapes and sizes that can, in turn, find various applications in photonics, electronics, and construction of memory elements.

How to have rational control over the nanoparticle assembly so that the 2-D and 3-D superstructures of the nanowires can have dimensions and interfaces tailor-made for their supposed purpose of the application, is the main concern that appears when bottom-up mechanisms are employed. Researchers continuously focus on developing new organizational procedures for nanowires so that their interfacing can be at par with all conventional manufacturing steps of the microelectronics industry. The implementation of these techniques forms the basis for mapping laboratory-generated nanowires to the factory.

PHYSICAL PROPERTIES OF NANOWIRES

Compared with bulk materials used in general MOSFETs, the nanoscale materials with reduced dimensions that find use in NWFETs, offer the benefits of their increased surface area and possible quantum confinement effect, thereby inducing much distinctive electric, optical, chemical and thermal properties some of which are discussed below.

Thermal Stability

How exactly the semiconductor nanowires respond to the change in temperature, is a very important aspect of study regarding their potential of being used in devices operated by nanoelectronics. It has been observed by researchers that by varying formation and annealing temperatures, different morphological nanostructures of silicon are produced (Pan et al., 2001). The thermal evaporation of Si/SiO_2 mixture in an alumina tube can lead to such synthesis. It has been also found that apart from silicon nanowires, multiple pin-like, chain-like, tadpole-like, and octopus-like nanostructures were also formed. It is demonstrated that control over these formation and annealing temperatures can precisely control these morphologies.

The significant lowering of the melting temperature of these nanowires reveals that the annealing temperature needed for making good-quality nanowires with defect levels almost equal to nil goes down to a fraction of that required in case of bulk annealing. The potential of cutting, linking, and welding nanowires at comparatively low temperatures has therefore opened up the research area of integrating the 1-D nanomaterials into electronic devices and circuitry. However, nanoscale devices are expected to experience a decreased chemical as well as thermal stability, thus demanding special attention.

Optical Properties

Nanowires witness some special and unique optical properties resulting from quantum confinement effects which occur due to their size entering a range comparable to the wavelength of electrons. Studies have revealed that in case of Si nanowires, the nanowires that orient themselves in <110> direction, show molecular-type transitions while those which are <100> oriented, show a gradually increasing phonon-supported optical transition. The intrinsic anisotropy of nanowires induces polarization-sensitivity to nanoscale photodetectors which find numerous applications in integrated photonic circuits, high-resolution detectors, interconnects, optical switches and near-field imaging.

Electronic Properties

The electronic properties of silicon nanowires become interesting because of the fact of their having direct bandgap in most of the directions of growth. It happens such that k-space can get a certain point where the alignment of the highest level of the valence band with the lowest level of the conduction band becomes possible, making the nanowires potentially useful in fields of optoelectronics and photonics (Canham, 1990, pp. 1046-1048). The nanowire growth direction, area of cross-section, and diameter determine the electronic properties. Interestingly, the bandgap of 1nm nanowire is found to remain almost constant and varying just less than 0.09eV for cross-sections that are dissimilar in shape but maintain an equal value of surface-to-volume ratio (Yao, 2008). The quasi-ballistic transport of electrons in nanowires due to their 1-D structure is also responsible for their electronic behavior and can be utilized in NWFETs design (Gnani, 2008).

NANOWIRE HETEROSTRUCTURES (NWHSFETS)

As discussed earlier, a very popular process of producing nanowires is the VLS growth process that defines nucleation and subsequent nanowire growth from a nanocluster. Chemical vapor deposition (CVD) most of the time introduces the source material in the vapor phase as it offers a great amount of control on the concentration of dopant molecules as well as the electrical properties of SiNWs during their growth. Fabricating metal electrodes to form source and drain suggests that NWFETs contain Schottky barriers differing from degenerately doped semiconductor source/drain contacts in usual MOSFETs. One way out to improve the contacts by producing apparent ohmic behavior is annealing, for example, an effective way only for highly doped materials is by establishing NiSi/Si contacts (Wu et al., 2004). Despite the use of post-growth treatment (Cohen et al., 2007) as well as different in-situ doping methods (Yang et al., 2005), several challenges of integrating semiconductor having a high concentration of dopants embedded in source or drain contacts, with nanowire channels (i.e. MOSFET type devices) having a much lower concentration of dopants show up from time to time. Today, bandgap engineering has brought heterostructure nanowire devices in the limelight as these are capable of establishing transparent ohmic contacts characterized by a much small value of contact resistance to the intrinsic channel or channel having a very less number of dopants. The growth of two or more materials on top of each other gives rise to a heterostructure. The three commonly formed heterostructures are depicted in Fig. 9.

Figure 9. Nanowire heterostructures drawn schematically with the dark grey and light grey colors representing two different materials. (a) Separate substrate-based heterostructure (b) heterostructure having axial segments of different materials (c) heterostructure having diverse materials placed radially. The substrate is only shown in (a), where it is a component of the heterostructure.

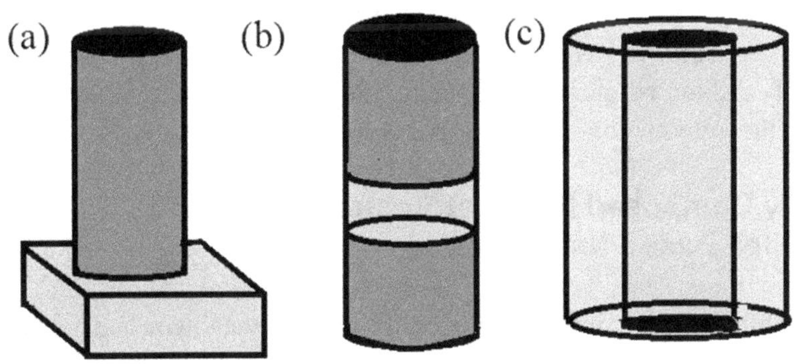

Substrate-Variant Nanowire Heterostructures

The easiest way to form a heterostructure is by taking different materials for nanowire growth and the substrate on which it will be grown. The fuelling factor to drive researches in this path is mainly the possibility of combining III–V materials with silicon. On achieving this, monolithic integration of highly specified optoelectronic and high-speed III–V devices can become possible only using the standardized CMOS technology.

Axial Heterostructures

Axial heterostructure (Fig. 9a) exploits the major benefit that can be drawn from a nanowire geometry i.e. the ability to relax strain laterally at an angle of 90° to the nanowire axis, along the smallest dimension. Unlike the limited capability of bulk systems, the nanowire dimensions being typically below 100 nm, can very well relax strain without causing any defect due to lattice mismatch. Defect-free axial heterostructures in the Si/SiGe (Wu et al., 2002), InAs/InP (Bjork et al., 2002), and GaAs/GaP (Gudiksen et al., 2002) systems were achieved as early as 2002 and all of these can have lattice mismatch of about 3%. The importance of axial heterostructures lies in their applications in electron transport devices where enough thin plus sharply-defined layers of materials having multiple variants of band structures can serve like tunnel barriers counting single-electron transistors (Thelander et al., 2003), resonant tunneling diodes (Kumar & Raj, 2016), and memory components (Thelander et al., 2005). Developing even and unidirectional growth, interfaces with sharp borders, and proper selectivity along the axis have been always very tough to achieve. Failing to optimize the growth conditions can lead to what is called a kinking effect. A detailed model of aspects related to kinking in seeded systems (Schwarz, 2009, 2011) can be studied to understand the necessary conditions for straight axial growth. It has been demonstrated that heterostructure nanowires comprising of some definite proportions of materials typically mature in a straight manner only in one interface direction, but a kink in any other (Dick, 2007). The problem of kinking can be most effectively tackled by gradually changing the material across the interface hence avoiding the introduction of any sharpness at the interface.

Radially Composed Heterostructure or Core/Shell Based Nanowires

The possibilities of passivation the sidewalls of a core nanowire and the fabricating radially oriented quantum wells and superlattices are the triggering factors of designing radial, or core/shell, heterostructures (Fig. 9c) (Van et al., 2010). It is important to take care of the growth parameters for maintaining a maximum possible ratio of radial and axial growth to ensure the desired growth rate of radial heterostructures (Laihon et al., 2004). In most of the cases, a higher temperature is needed for the desired growth, though other means of making the vapor phase more concentrated with reactant molecules can also direct radial growth (Skold, 2005, Lauhon, 2002). In certain cases of growth equipment, pure axial or pure radial growth becomes a genuine challenge rendering the system of materials impossible to be used as radial nanowires (Borgstrom, 2010).

A short note about the requirement of various heterostructures in developing NWFETs has been covered above leaving aside the experimental results from the enormous number of research papers that need a deep understanding of the characterization techniques related to transmission electron microscopy (TEM), scanning electron microscopy (SEM) and others.

ORGANIZATION METHODS FOR DEVICE ARRAY FABRICATION

Nanowires are grown on different substrates essentially require to be transferred to the actual device substrates for accomplishing the purpose of their use in the respective devices. Several techniques are tried and tested to arrange a bunch of nanowires into the desired integrated circuits. The two basic points that need to be considered in organizing arrays of nanowires are: first, there should be full control on orientation and spatial placement of the nanowires in a method that is applied; second, actions must be taken to build nanowire device that can be equipped for application on different scales of length ensuring unambiguous interconnect mapping between all nano level, micro-level, and macro-level subsystems.

Assembly processes that commonly address the first point mentioned above, take the help of a force from outside for desirable alignment of nanowires along a specific direction. For example, an electrostatic force is applied in an electric field directed assembly (Daun et al., 2001) and shear force promotes the fluid-assisted assembly methods (Sharma, 2018, Haung, 2001). In the latter, the nanowire surface coverage, on an average, is controlled by the time for which the fluid flows while improving the alignment is dependent on the increased shear force that can enhance the rate of flow. The allowable size of fluidic channels ranges up to a few hundreds of micrometers and in turn, determines the array size of aligned nanowires.

The Langmuir–Blodgett (LB) technique at the centimeter level (Whang, 2003, Kim, 20001) and the blown-bubble method at the wafer level (Jain, 2016, Yu, 2008) can be much interesting to study. In the LB process, a single layer of nanowire-surfactant undergoes a compression against some kind of aqueous sub-phase creating aligned nanowires that can be controllably spaced (Ulman, 2013). A single next step transfers the compressed layer yielding nanowires to cover the entire planar surface of the substrate (Whang et al., 2003). The formation of nanowires with variable density can be possible by finely performing space generating close-packed structures (Singh, 2018, Whang 2003).

Mostly similar to the fluidic assembly, the blown bubble film (BBF) approach [61, 62] uses the shear stress built up during the expansion of a film, to align nanowires at a wafer scale. In this technique, a homogeneous solution is prepared to disperse nanowires or nanotubes in it and a bubble from the solution is expanded with a monitored speed and direction for its subsequent transfer to target device substrates. As much as 10^6 cm^{-2} dense nanowire films have been demonstrated on 8-inch wafers producing an angular spread of less than 10° over the total surface of the wafer. Better optimization techniques concerning surface chemistry can produce even denser nanowire arrays that are likely to give rise to large-scale nanowire-based electronics at a reasonable cost.

Any material from crystalline silicon to flexible plastics can act as substrates to undergo a repetitive process of nanowire alignment/transfer yielding crossed and more complex nanowire. On an ending note of these common mechanisms of organizing nanowire arrays, it will be good to mention a recent optimized growth technique of designing nanowires using dry deposition strategy on substrates (Javey, 2007, Singh, 2019) and then directly transferring the structures from the substrate in which they are grown initially to one on which the device is built, using contact printing.

NWFETs CIRCUIT APPLICATIONS

NWFETs have got a huge number of applications in different fields of science and technology. A quick view of some of the significant applications of NWFETs is given in what follows.

Nanowire Bio-FET for Bio-Sensing

A biosensor is a device that can detect and/or can determine what amount of a biochemical species is contained inside a sample. Numerous application areas of biosensors in present days include diagnostics and healthcare appliances, water and food quality monitoring, environmental monitoring, and many more (Daniels, 2007, Singh, 2019, Lee, 2009). An exclusive type of FET used for sensing biomolecules in a fluid system is termed as bio-FET. It is a well- known fact that the conventional unipolar current carried either by holes or by electrons due to a voltage difference from source to drain terminals of a FET, is regulated by the third terminal of the device, the gate electrode. In the case of a Bio-FET, a gate dielectric, an electrolyte, and a reference electrode together act as an alternative system of the gate (Bergveld, 1970, Singh 2019). The reference electrode that is kept dipped into the electrolytic solution provides electrical contact and modulates resistance of a channel by generating an electrostatic field (Schoning, 2002, Singh, 2019). Commonly used dielectric materials that cover the channel include Al_2O_3, Si_3N_4, SiO_2, Ta_2O_5 (Jin et al., 2004). Channel potential across the electrolyte varies according to the concentration level of the ions present inside the solution. Biosensors contain a probe molecule in immobilized condition, placed on the dielectric surface such that it can choose and bind with the target molecule present in the electrolyte. As a result, the channel potential gets altered and reflects itself by changing the drain current. Unfortunately, the full-fledged utilization of silicon NWFET is restricted by several environmental and technical factors and hence it is yet to provide practical benefits in areas of life science and development of medical equipment. The sensing ability faces a challenge to identify neutral molecules or biomolecules with low electrostatic charge levels. Also, the frequently encountered complex samples of serum, blood, urine, etc. possess an ionic strength high enough to hamper the device sensitivity. Despite these challenges, there awaits a bright future for silicon NWFET based biosensors that have the potential to implement an entire testing system on a biochip making it possible to design a platform for personalized medical treatment.

NWFETs in Optoelectronics

Nowadays, multiple types of devices that operate following the physics of optoelectronics, like photodetectors and photovoltaic cells use homogeneously doped semiconductor nanowires as building blocks. Studies on NWFETs have revealed that they can be well distinguished from the older planar devices built from bulk materials based on numerous applicability related parameters. For example, demonstration of results obtained from NWFETs fabricated using boron, phosphine doped silicon nanowires can exhibit performance somewhat better than the best observed in case of planar devices built using the same materials (Zheng et al., 2004). A notably large value of carrier mobility of epitaxial InAs NWFETs has been observed when a wrap-around gate structure is used. So naturally, the switching speed and the rapidity of conversion of light energy to electrical energy get enhanced in NWFETs satisfying a crucial aspect of improving the performance of any kind of photodetector. Researches are going on to build NWFETs using different hybrid organic materials with potentially high photosensitivity to exploit the structural and optical property-based benefits of nanowires to extend their applications beyond what is possible in traditional electronic and optoelectronic systems that use single crystal wafers.

NWFETs in Flash-Like Non-Volatile Memory Design

Today, speed, storage, and portability form the basic three demands of the semiconductor industry. Solid-state mass storage devices are the main attract a huge market because of their mostly non-volatile nature, ability to easily gel with CMOS scaling technology, and endurance in harsh environmental conditions. These secondary storage areas do not require high operational speed or high integration density as these are meant for permanent storage of data which need not be accessed very often. Now, memory on-chip that is to say volatile storage that accounts for the cache memory in the central processing unit (CPU), comprises mainly of static random access memory (SRAM). Over the years, the size of cache memory has been doubled to store more data but being volatile makes it occupy a large chip floor space make it lose on the net data throughput. Thus, the replacement of SRAM by a non-volatile on-chip memory has a huge prospect in the microelectronics research world.

Flash memory has occupied a special place in the electronics market due to many of its advantages like fast read access times, longer retention time as well as fine reliability and compatibility with the CMOS fabrication process (Singh, 2018, Lai, 2008). A flash memory device stores the trapped electrons in the floating gate until they are eliminated by applying a separate electric potential as shown in Fig. 10. Representation of logical "0" and "1" states is done by the presence or absence of

Figure 10. Schematic representation of a floating-gate flash memory

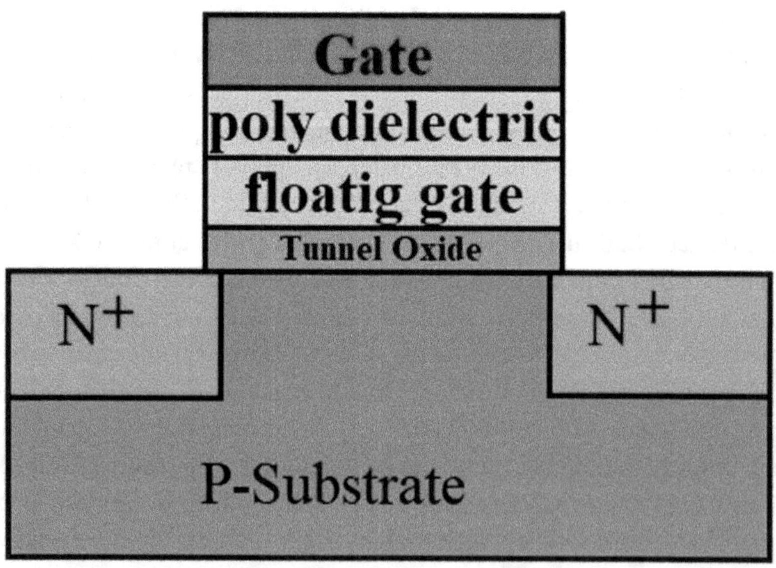

electrons trapped in the floating gate. The floating gate, being sandwiched between the two oxide layers separating it from the controlling gate on top and the conducting channel on the bottom, can store the data for many years.

However, current flash memory also exhibits disadvantages such as the relatively low speed of write/erase operations and medium endurance power that make it fail to meet the on-chip memory application standards. The necessity of data retention hinders the scaling of tunneling oxide to barely reach a value of less than 7nm. Added to this, is the issue of extreme stress-induced leakage current (SILC) resulting from ultra-thin tunneling oxide. Other concerns are the high voltages required to program the device and endurance limitation of up to 10^5 operation cycles only in the case of conventional poly-Si and oxy-nitride materials.

To fight the problems of conventional flash memory, a novel approach using poly-oxide-nitride-oxide-silicon (SONOS) compositional charge-trapping non-volatile memory is found as a promising solution to researchers since they offer some exciting features lower power consumption, better reliability, and simpler structure and fabrication process (Wu et al., 2005). The fact that SONOS memory is incompatible with dimensional scaling thus making it tough to strike a balance between speed and retention has opened up the horizon for non-planar structural modifications as noticed in nanowires and nanotubes that can potentially induce a stronger control of gate over the channel. The Si nanowire FET based dielectric charge-trapping flash memory is prepared by following the self-alignment method (Zhu et al., 2014). Typically, a 300 nm SiO_2 is first grown on Si substrate by dry oxidation. Then a thin film of Au catalyst (1–3 nm) is deposited on patterned area pre-defined by photolithography. Then the Si nanowires are grown from the catalyst following the VLS mechanism and are oxidized at 750°C for 30 minutes to form a ≈ 3 nm SiO_2 which functions as the surface for tunneling. Patterning of 3/100 (in nanometer-scale) Ti/Pt source and drain regions is achieved utilizing photolithography which was followed by deposition using electron beam evaporation. The source and drain electrodes get implanted at the two ends of a 5 μm long channel. In the next step, multiple layers that can trap charges can be incorporated (Zhu et al., 2014). A 25nm thick Al_2O_3 layer is used as blocking oxide in these memory devices. So, nanowires have a good scope of application in the construction of flash memory to support future advancement of speed, density, and endurance of non-volatile memories.

CONCLUSION

An overview of almost all the perspectives related to the present day researches going on related to nanowire FETs is being covered here in this entire chapter. The understanding of the very basic topics like the evolution of nanowire FETs, the fundamental working principle of NWFETs, a compact overview of the different commonly used materials in nanowires, the important synthetic techniques of nanowires, the contrast picture of top-down and bottom-up mechanisms and the corresponding benefits and challenges, a brief discussion on the structural effects on device performance of NWFETs, the different physical properties of NWFETs, an introduction to nanowire heterostructures, a glimpse of the methods to integrate arrays of nanowires for fabrication and some of the different bright areas of applications of NWFETs in the fields of optoelectronics, biosensors, and non-volatile flash memory construction are being compiled in this chapter. Despite the several challenges of assembling nanowires and integrating them into application-specific device components, NWFET seems to be one of the most promising areas of research and development in microelectronics domain that can keep up the pace of scaling and miniaturization to follow the trend well set by CMOS technology over the past decades. The future is expected to see tremendous growth of new devices employing NWFETs, highly optimized in performance to power consumption, cost-effectiveness, and area efficiency.

REFERENCES

Bergveld, P. (1970). Development of an ion-sensitive solid-state device for neurophysiological measurements. *IEEE Transactions on Biomedical Engineering*, *BME-17*(1), 70–71. doi:10.1109/TBME.1970.4502688 PMID:5441220

Björk, M. T., Ohlsson, B. J., Sass, T., Persson, A. I., Thelander, C., Magnusson, M. H., Deppert, K., Wallenberg, L. R., & Samuelson, L. (2002). One-dimensional steeplechase for electrons realized. *Nano Letters*, *2*(2), 87–89. doi:10.1021/nl010099n

Bondyopadhyay, P. K. (1998). Moore's law governs the silicon revolution. *Proceedings of the IEEE*, *86*(1), 78–81. doi:10.1109/5.658761

Borgström, M. T., Wallentin, J., Trägårdh, J., Ramvall, P., Ek, M., Wallenberg, L. R., Samuelson, L., & Deppert, K. (2010). In situ etching for total control over axial and radial nanowire growth. *Nano Research*, *3*(4), 264–270. doi:10.100712274-010-1029-x

Briseno, A. L., Mannsfeld, S. C., Lu, X., Xiong, Y., Jenekhe, S. A., Bao, Z., & Xia, Y. (2007). Fabrication of field-effect transistors from hexathiapentacene single-crystal nanowires. *Nano Letters*, *7*(3), 668–675. doi:10.1021/nl0627036 PMID:17288487

Briseno, A. L., Mannsfeld, S. C., Reese, C., Hancock, J. M., Xiong, Y., Jenekhe, S. A., Bao, Z., & Xia, Y. (2007). Perylenediimide nanowires and their use in fabricating field-effect transistors and complementary inverters. *Nano Letters*, *7*(9), 2847–2853. doi:10.1021/nl071495u PMID:17696562

Canham, L. T. (1990). Silicon quantum wire array fabrication by electrochemical and chemical dissolution of wafers. *Applied Physics Letters*, *57*(10), 1046–1048. doi:10.1063/1.103561

Che, Y., Datar, A., Balakrishnan, K., & Zang, L. (2007). Ultralong nanobelts self-assembled from an asymmetric perylene tetracarboxylic diimide. *Journal of the American Chemical Society*, *129*(23), 7234–7235. doi:10.1021/ja071903w PMID:17506565

Choi, Y. K., Asano, K., Lindert, N., Subramanian, V., King, T. J., Bokor, J., & Hu, C. (1999, December). Ultra-thin body SOI MOSFET for deep-sub-tenth micron era. In *International Electron Devices Meeting 1999. Technical Digest* (Cat. No. 99CH36318) (pp. 919-921). IEEE.

Choi, Y. K., Ha, D., King, T. J., & Hu, C. (2001, June). Ultra-thin body PMOSFETs with selectively deposited Ge source/drain. In *2001 Symposium on VLSI Technology. Digest of Technical Papers* (IEEE Cat. No. 01 CH37184) (pp. 19-20). IEEE.

Cohen, G. M., Rooks, M. J., Chu, J. O., Laux, S. E., Solomon, P. M., Ott, J. A., Miller, R. J., & Haensch, W. (2007). Nanowire metal-oxide-semiconductor field effect transistor with doped epitaxial contacts for source and drain. *Applied Physics Letters*, *90*(23), 233110. doi:10.1063/1.2746946

Daniels, J. S., & Pourmand, N. (2007). Label-free impedance biosensors: Opportunities and challenges. Electroanalysis. *An International Journal Devoted to Fundamental and Practical Aspects of Electroanalysis*, *19*(12), 1239–1257. PMID:18176631

Dasgupta, N. P., Sun, J., Liu, C., Brittman, S., Andrews, S. C., Lim, J., Gao, H., Yan, R., & Yang, P. (2014). 25th anniversary article: Semiconductor nanowires–synthesis, characterization, and applications. *Advanced Materials*, *26*(14), 2137–2184. doi:10.1002/adma.201305929 PMID:24604701

De Luca, G., Liscio, A., Maccagnani, P., Nolde, F., Palermo, V., Muellen, K., & Samori, P. (2007). Nucleation-Governed Reversible Self-Assembly of an Organic Semiconductor at Surfaces: Long-Range Mass Transport Forming Giant Functional Fibers. *Advanced Functional Materials*, *17*(18), 3791–3798. doi:10.1002/adfm.200700549

Dick, K. A., Kodambaka, S., Reuter, M. C., Deppert, K., Samuelson, L., Seifert, W., Wallenberg, L. R., & Ross, F. M. (2007). The morphology of axial and branched nanowire heterostructures. *Nano Letters*, *7*(6), 1817–1822. doi:10.1021/nl0705900 PMID:17489640

Duan, X., Huang, Y., Cui, Y., Wang, J., & Lieber, C. M. (2001). Indium phosphide nanowires as building blocks for nanoscale electronic and optoelectronic devices. *Nature*, *409*(6816), 66–69. doi:10.1038/35051047 PMID:11343112

Gates, B., Wu, Y., Yin, Y., Yang, P., & Xia, Y. (2001). Single-crystalline nanowires of Ag2Se can be synthesized by templating against nanowires of trigonal Se. *Journal of the American Chemical Society*, *123*(46), 11500–11501. doi:10.1021/ja0166895 PMID:11707137

Gnani, E., Gnudi, A., Reggiani, S., & Baccarani, G. (2008). Quasi-ballistic transport in nanowire field-effect transistors. *IEEE Transactions on Electron Devices*, *55*(11), 2918–2930. doi:10.1109/TED.2008.2005178

Gudiksen, M. S., Lauhon, L. J., Wang, J., Smith, D. C., & Lieber, C. M. (2002). Growth of nanowire superlattice structures for nanoscale photonics and electronics. *Nature*, *415*(6872), 617–620. doi:10.1038/415617a PMID:11832939

Hoeben, F. J., Jonkheijm, P., Meijer, E. W., & Schenning, A. P. (2005). About supramolecular assemblies of π-conjugated systems. *Chemical Reviews*, *105*(4), 1491–1546. doi:10.1021/cr030070z PMID:15826018

Huang, Y., Duan, X., Wei, Q., & Lieber, C. M. (2001). Directed assembly of one-dimensional nanostructures into functional networks. *Science*, *291*(5504), 630–633. doi:10.1126cience.291.5504.630 PMID:11158671

Huczko, A. (2002). Synthesis of aligned carbon nanotubes. *Applied Physics. A, Materials Science & Processing*, *74*(5), 617–638. doi:10.1007003390100929

Jain, A., Sharma, S. K., & Raj, B. (2016). Design and analysis of high sensitivity photosensor using Cylindrical Surrounding Gate MOSFET for low power applications. *Engineering Science and Technology, an International Journal, 19*(4), 1864-1870.

Jain, N., & Raj, B. (2018). Analysis and performance exploration of high performance (HfO2) SOI FinFETs over the conventional (Si3N4) SOI FinFET towards analog/ RF design. *Journal of Semiconductors*, *39*(12), 124002. doi:10.1088/1674-4926/39/12/124002

Javey, A., Nam, S., Friedman, R. S., Yan, H., & Lieber, C. M. (2007). Layer-by-layer assembly of nanowires for three-dimensional, multifunctional electronics. *Nano Letters*, *7*(3), 773–777. doi:10.1021/nl063056l PMID:17266383

Jin, S., Whang, D., McAlpine, M. C., Friedman, R. S., Wu, Y., & Lieber, C. M. (2004). Scalable interconnection and integration of nanowire devices without registration. *Nano Letters*, *4*(5), 915–919. doi:10.1021/nl049659j

Kim, D. H., Lee, D. Y., Lee, H. S., Lee, W. H., Kim, Y. H., Han, J. I., & Cho, K. (2007). High-mobility organic transistors based on single-crystalline microribbons of triisopropylsilylethynyl pentacene via solution-phase self-assembly. *Advanced Materials*, *19*(5), 678–682. doi:10.1002/adma.200601259

Kim, F., Kwan, S., Akana, J., & Yang, P. (2001). Langmuir– Blodgett nanorod assembly. *Journal of the American Chemical Society*, *123*(18), 4360–4361. doi:10.1021/ja0059138 PMID:11457213

Kumar, S., & Raj, B. (2016). Analysis of I ON and Ambipolar current for dual-material gate-drain overlapped DG-TFET. *Journal of Nanoelectronics and Optoelectronics*, *11*(3), 323–333. doi:10.1166/jno.2016.1902

Lai, S. K. (2008). Flash memories: Successes and challenges. *IBM Journal of Research and Development, 52*(4.5), 529-535.

Lauhon, L. J., Gudiksen, M. S., & Lieber, C. M. (2004). Semiconductor nanowire heterostructures. *Philosophical Transactions of the Royal Society of London. Series A: Mathematical, Physical and Engineering Sciences, 362*(1819), 1247-1260.

Lauhon, L. J., Gudiksen, M. S., Wang, D., & Lieber, C. M. (2002). Epitaxial core–shell and core–multishell nanowire heterostructures. *Nature*, *420*(6911), 57–61. doi:10.1038/nature01141 PMID:12422212

Lee, C. S., Kim, S. K., & Kim, M. (2009). Ion-sensitive field-effect transistor for biological sensing. *Sensors (Basel)*, *9*(9), 7111–7131. doi:10.339090907111 PMID:22423205

Lee, W. H., Kim, D. H., Jang, Y., Cho, J. H., Hwang, M., Park, Y. D., Kim, Y. H., Han, J. I., & Cho, K. (2007). Solution-processable pentacene microcrystal arrays for high performance organic field-effect transistors. *Applied Physics Letters*, *90*(13), 132106. doi:10.1063/1.2717087

Levitt, A. P. (1970). *Whisker technology*. John Wiley & Sons.

Li, Q., Kumar, V., Li, Y., Zhang, H., Marks, T. J., & Chang, R. P. (2005). Fabrication of ZnO nanorods and nanotubes in aqueous solutions. *Chemistry of Materials*, *17*(5), 1001–1006. doi:10.1021/cm048144q

Mertens, H., Ritzenthaler, R., Chasin, A., Schram, T., Kunnen, E., Hikavyy, A., & Devriendt, K. (2016, December). Vertically stacked gate-all-around Si nanowire CMOS transistors with dual work function metal gates. In *2016 IEEE International Electron Devices Meeting (IEDM)* (pp. 19-7). IEEE. 10.1109/IEDM.2016.7838456

Molnar, W., Lugstein, A., Pongratz, P., Auner, N., Bauch, C., & Bertagnolli, E. (2010). Subeutectic synthesis of epitaxial Si-NWs with diverse catalysts using a novel Si precursor. *Nano Letters*, *10*(10), 3957–3961. doi:10.1021/nl101744q PMID:20843058

Morales, A. M., & Lieber, C. M. (1998). A laser ablation method for the synthesis of crystalline semiconductor nanowires. *Science*, *279*(5348), 208–211. doi:10.1126cience.279.5348.208 PMID:9422689

Pan, Z. W., Dai, Z. R., Xu, L., Lee, S. T., & Wang, Z. L. (2001). Temperature-controlled growth of silicon-based nanostructures by thermal evaporation of SiO powders. *The Journal of Physical Chemistry B*, *105*(13), 2507–2514. doi:10.1021/jp004253q

Pavan, P., Bez, R., Olivo, P., & Zanoni, E. (1997). Flash memory cells-an overview. *Proceedings of the IEEE*, *85*(8), 1248–1271. doi:10.1109/5.622505

Rai-Choudhury, P. (Ed.). (1997). *Handbook of microlithography, micromachining, and microfabrication. 2. Micromachining and microfabrication*. SPIE Optical Engineering Press. doi:10.1117/3.2265071

Saib, S. S., Yadav, S., Rahul, J., Srivastava, A., & Raj, B. (2015). Impact of HfO2 in Graded Channel Dual Insulator Double Gate MOSFET. *Journal of Computational and Theoretical Nanoscience*, *12*(6), 950–953. doi:10.1166/jctn.2015.3833

Schöning, M. J., & Poghossian, A. (2002). Recent advances in biologically sensitive field-effect transistors (BioFETs). *Analyst (London)*, *127*(9), 1137–1151. doi:10.1039/B204444G PMID:12375833

Schwarz, K. W., & Tersoff, J. (2009). From droplets to nanowires: Dynamics of vapor-liquid-solid growth. *Physical Review Letters, 102*(20), 206101. doi:10.1103/PhysRevLett.102.206101 PMID:19519043

Schwarz, K. W., & Tersoff, J. (2011). Elementary processes in nanowire growth. *Nano Letters, 11*(2), 316–320. doi:10.1021/nl1027815 PMID:21188963

Shao, M., Ma, D. D. D., & Lee, S. T. (2010). Silicon nanowires–synthesis, properties, and applications. *European Journal of Inorganic Chemistry, 2010*(27), 4264–4278. doi:10.1002/ejic.201000634

Sharma, S., Raj, B., & Khosla, M. (2016). Comparative Analysis of MOSFET CNTFET and NWFET for High Performance VLSI Circuit Design. *Journal of VLSI Design Tools & Technology, 6*(2), 19–32.

Sharma, S. K., Raj, B., & Khosla, M. (2017). Subthreshold Performance of In 1–x Ga x As Based Dual Metal with Gate Stack Cylindrical/Surrounding Gate Nanowire MOSFET for Low Power Analog Applications. *Journal of Nanoelectronics and Optoelectronics, 12*(2), 171–176. doi:10.1166/jno.2017.1961

Sharma, S. K., Singh, J., Raj, B., & Khosla, M. (2018). Analysis of Barrier Layer Thickness on Performance of In1–x Ga x As Based Gate Stack Cylindrical Gate Nanowire MOSFET. *Journal of Nanoelectronics and Optoelectronics, 13*(10), 1473–1477. doi:10.1166/jno.2018.2374

Singh, A., Khosla, M., & Raj, B. (2016). Circuit compatible model for electrostatic doped Schottky barrier CNTFET. *Journal of Electronic Materials, 45*(10), 5381–5390. doi:10.100711664-016-4743-7

Singh, J., & Raj, B. (2018). Modeling of mean barrier height levying various image forces of metal–insulator–metal structure to enhance the performance of conductive filament based memristor model. *IEEE Transactions on Nanotechnology, 17*(2), 268–275. doi:10.1109/TNANO.2018.2793953

Singh, J., & Raj, B. (2019). An accurate and generic window function for nonlinear memristor models. *Journal of Computational Electronics, 18*(2), 640–647. doi:10.100710825-019-01306-6

Singh, J., & Raj, B. (2019). Design and Investigation of 7T2M-NVSRAM With Enhanced Stability and Temperature Impact on Store/Restore Energy. *IEEE Transactions on Very Large Scale Integration (VLSI) Systems, 27*(6), 1322–1328.

Singh, J., & Raj, B. (2019). Enhanced Nonlinear Memristor Model Encapsulating Stochastic Dopant Drift. *Journal of Nanoelectronics and Optoelectronics*, *14*(7), 958–963. doi:10.1166/jno.2019.2570

Singh, J., & Raj, B. (2019). Tunnel current model of asymmetric MIM structure levying various image forces to analyze the characteristics of filamentary memristor. *Applied Physics. A, Materials Science & Processing*, *125*(3), 203. doi:10.100700339-019-2482-3

Singh, J., Raj, B., & Khosla, M. (2018). Design and performance analysis of nano-scale memristor-based nonvolatile static random access memory. *Sensor Letters*, *16*(10), 798–805. doi:10.1166l.2018.4021

Sköld, N., Karlsson, L. S., Larsson, M. W., Pistol, M. E., Seifert, W., Trägårdh, J., & Samuelson, L. (2005). Growth and Optical Properties of Strained GaAs– Ga x In1-x P Core– Shell Nanowires. *Nano Letters*, *5*(10), 1943–1947. doi:10.1021/nl051304s PMID:16218714

Thelander, C., Mårtensson, T., Björk, M. T., Ohlsson, B. J., Larsson, M. W., Wallenberg, L. R., & Samuelson, L. (2003). Single-electron transistors in heterostructure nanowires. *Applied Physics Letters*, *83*(10), 2052–2054. doi:10.1063/1.1606889

Thelander, C., Nilsson, H. A., Jensen, L. E., & Samuelson, L. (2005). Nanowire single-electron memory. *Nano Letters*, *5*(4), 635–638. doi:10.1021/nl050006s PMID:15826100

Tian, B., Kempa, T. J., & Lieber, C. M. (2009). Single nanowire photovoltaics. *Chemical Society Reviews*, *38*(1), 16–24. doi:10.1039/B718703N PMID:19088961

Trentler, T. J., Hickman, K. M., Goel, S. C., Viano, A. M., Gibbons, P. C., & Buhro, W. E. (1995). Solution-liquid-solid growth of crystalline III-V semiconductors: An analogy to vapor-liquid-solid growth. *Science*, *270*(5243), 1791–1794. doi:10.112 6cience.270.5243.1791

Ulman, A. (2013). *An Introduction to Ultrathin Organic Films: From Langmuir—Blodgett to Self—Assembly*. Academic press.

Van Tilburg, J. W. W., Algra, R. E., Immink, W. G. G., Verheijen, M., Bakkers, E. P. A. M., & Kouwenhoven, L. P. (2010). Surface passivated InAs/InP core/shell nanowires. *Semiconductor Science and Technology*, *25*(2), 024011. doi:10.1088/0268-1242/25/2/024011

Werner, P., Zakharov, N. D., Gerth, G., Schubert, L., & Gösele, U. (2006). On the formation of Si nanowires by molecular beam epitaxy: Dedicated to Professor Dr. Knut Urban on the occasion of his 65th birthday. *Zeitschrift fur Metallkunde*, *97*(7), 1008–1015.

Whang, D., Jin, S., Wu, Y., & Lieber, C. M. (2003). Large-scale hierarchical organization of nanowire arrays for integrated nanosystems. *Nano Letters*, *3*(9), 1255–1259. doi:10.1021/nl0345062

Wu, K. H., Chien, H. C., Chan, C. C., Chen, T. S., & Kao, C. H. (2005). SONOS device with tapered bandgap nitride layer. *IEEE Transactions on Electron Devices*, *52*(5), 987–992. doi:10.1109/TED.2005.846347

Wu, Y., Fan, R., & Yang, P. (2002). Block-by-block growth of single-crystalline Si/SiGe superlattice nanowires. *Nano Letters*, *2*(2), 83–86. doi:10.1021/nl0156888

Wu, Y., Xiang, J., Yang, C., Lu, W., & Lieber, C. M. (2004). Single-crystal metallic nanowires and metal/semiconductor nanowire heterostructures. *Nature*, *430*(6995), 61–65. doi:10.1038/nature02674 PMID:15229596

Würthner, F. (2004). Perylene bisimide dyes as versatile building blocks for functional supramolecular architectures. *Chemical Communications*, (14), 1564–1579. doi:10.1039/B401630K PMID:15263926

Xiao, S., Tang, J., Beetz, T., Guo, X., Tremblay, N., Siegrist, T., Zhu, Y., Steigerwald, M., & Nuckolls, C. (2006). Transferring self-assembled, nanoscale cables into electrical devices. *Journal of the American Chemical Society*, *128*(33), 10700–10701. doi:10.1021/ja0642360 PMID:16910663

Yang, C., Zhong, Z., & Lieber, C. M. (2005). Encoding electronic properties by synthesis of axial modulation-doped silicon nanowires. *Science*, *310*(5752), 1304–1307. doi:10.1126cience.1118798 PMID:16311329

Yao, D., Zhang, G., & Li, B. (2008). A universal expression of band gap for silicon nanowires of different cross-section geometries. *Nano Letters*, *8*(12), 4557–4561. doi:10.1021/nl802807t PMID:19368006

Yu, G., Li, X., Lieber, C. M., & Cao, A. (2008). Nanomaterial-incorporated blown bubble films for large-area, aligned nanostructures. *Journal of Materials Chemistry*, *18*(7), 728–734. doi:10.1039/b713697h

Zhang, Z., Sun, X., Dresselhaus, M. S., Ying, J. Y., & Heremans, J. (2000). Electronic transport properties of single-crystal bismuth nanowire arrays. *Physical Review B: Condensed Matter and Materials Physics*, *61*(7), 4850–4861. doi:10.1103/PhysRevB.61.4850

Zheng, G., Lu, W., Jin, S., & Lieber, C. M. (2004). Synthesis and fabrication of high-performance n-type silicon nanowire transistors. *Advanced Materials*, *16*(21), 1890–1893. doi:10.1002/adma.200400472

Zhu, H., Bonevich, J. E., Li, H., Richter, C. A., Yuan, H., Kirillov, O., & Li, Q. (2014). Discrete charge states in nanowire flash memory with multiple Ta2O5 charge-trapping stacks. *Applied Physics Letters*, *104*(23), 233504. doi:10.1063/1.4883717

Chapter 8
Soft Error-Resilient RHBD16T SRAM Cell in 32nm Technology

Gaurav Kaushal
Atal Bihari Vajpayee Indian Institute of Information Technology and Management, Gwalior, India

Balamurugan Murgan
Qualcomm, India

Manisha Pattanaik
Atal Bihari Vajpayee Indian Institute of Information Technology and Management, Gwalior, India

Chinnapurapu Naga Raghuram
Department of ECE, National Institute of Technology, Patna, India

Surendra Singh Rathod
Sardar Patel Institute of Technology, India

ABSTRACT

Radiation environment generates high soft error rates in conventional SRAM. To overcome this issue, several radiation hardened by design SRAM circuits (12TRHBD, 13TRHBD, DICE, etc.) have been developed. Although many of the radiation hardened SRAM cells are there, all the circuits mainly concern a single node upset only. In this chapter, 16T radiation hardened static random-access memory bit cell is designed and verified for a single node and multi-node upset. RHBD 16T bit cell is designed with SAED-PDK 32nm technology and compared with recently reported RHBD 12T and has a 99% improvement in recovery rate. Simulation results show that RHBD 16T is more resilient to a single node and multi-node upset. This shows that the proposed RHBD16T cell is highly tolerant against radiation strikes.

DOI: 10.4018/978-1-7998-6467-7.ch008

INTRODUCTION

Radiation hardened SRAM cell is well-known in the field of latch design for radiation environments such as aerospace and military. In a modern technological era, scaling down transistor size is carried out for reducing power consumption, and to increase the speed of a circuit. Scaling the transistor length to nanometer regimes reduces the nodal capacitance of VLSI circuits, which therefore increases the susceptibility of the circuits towards radiation strikes(Chang, Chiu, Lai, & Hwang, 2011; P. E. Dodd et al., 1996; Paul E. Dodd & Massengill, 2003; McLain, Barnaby, Esqueda, Oder, & Vermeire, 2009a; Zheng et al., 2018). Moreover, nowadays, VLSI circuits in terrestrial atmospheres are also affected by alpha particles or cosmic ray traces(Baumann, 2005; Bohr & Young, 2017; Ch, Manohar Reddy, Kishore Kumar, & Kaushal, 2019; Ferlet-Cavrois, Massengill, & Gouker, 2013; Gaurav Kaushal, Surendra S. Rathod, Ch Naga Raghuram, & Sudeb Dasgupta, 2019; Narasimham et al., 2018), which confirmed the need for the design of the radiation-hardened SRAM cell in non-military memory devices. Literature categorized soft errors into three levels, such as Single-event single-node upset (SEU), Single-event multi-node upset (SEMU), Multi-event multi-node upset (MEMU). Most of the circuits are vulnerable to a multi-node upset. Error detection and Error Correction Codes (ECC)(Gracia-Moran, Saiz-Adalid, Gil-Tomas, & Gil-Vicente, 2018; Guo, Xiao, & Mao, 2014; Raghuram & Member, 2020) are employed for soft error mitigation in a register file as a system-level architectural method. However, the ECC technique needs bit interleaving architecture for easy implementation, complex internal circuits, significant area overhead, and performance overhead (ECC method needs separate timing control circuits for reducing performance overhead to nanosecond level delay). Other than circuit-level techniques, layout-level techniques like Shallow Trench Isolation (STI) and annular gate are proposed in (Chao Peng et al., 2018). However, STI does not give full protection from soft error and difficult to implement due to demanding design rules in a nanometer technology. Triple Modular Redundancy (TMR)(Prokhorov, Tyurin, & Chudinov, 2018) also one of the primary circuit level techniques in radiation-hardened memory cells. TMR design means employing the same circuit three times. If data flips due to radiation strike at one node, two nodes which maintained their original data would bring back the flipped node to its value. The main disadvantage of a TMR technique is its higher area and power overhead. Due to these disadvantages, TMR is more unlikely preferable by a designer as a circuit-level method. The authors in (Calin, Nicolaidis, & Velazco, 1996) proposed the Dual Interlock storage CEll (DICE), which is SEU immune, but not SEDU immune. 8T SRAM proposed in (Prasad & Sandya, 2014) used four storage nodes for storing the data, is not SEU immune, and its read operation occurs in a stringent condition. RHD11T(McLain, Barnaby, Esqueda, Oder, & Vermeire,

2009b) and RHD12T(Guo et al., 2017) used a clocked pass-gate technique. That is a pass transistor between the inverters of the latch. Operating pass transistors need a separate refresh and time controlling circuits. Also, the critical charge of the circuit is minimum when pass-gates are in ON state. In (Jung, Kim, & Lombardi, 2012), PS-10T and NS-10T are designed with a pMOS and nMOS stacked 10T model for soft error robustness. However, PS-10T provides only 1 to 0 robustness, and NS-10T provides only 0 to 1 robustness. In (Atias, Teman, Giterman, Meinerzhagen, & Fish, 2016), 13T SRAM cell used the dual feedback mechanism to achieve high robust against SEU, but it is not SEDU tolerant. RHBD12T in (Guo et al., 2017) is SEU immune and SEDU tolerant.

MODELLING OF SOFT ERROR

A soft error is due to a radiation strike that alters the data in an SRAM cell. The charge deposited value at a junction exceeds the critical charge (Q_c) value of that junction, the stored data in the SRAM will flip. The Qc is the lower limit of the charge at a junction to alter the data of the SRAM(Chatterjee et al., 2014)(Chunyu Peng et al., 2019). To replicate the radiation strike, a double exponential current pulse as shown in (1) is introduced at a storage junction which is given by (Guo et al., 2017)

$$I_{inj}(t)=\frac{Q_c}{T_f-T_r}\left(e^{-\frac{t}{T_f}}-e^{-\frac{t}{T_r}}\right) \tag{1}$$

Where

I_{inj} is the Injected current pulse
T_f is fall time constant
T_r is rise time constant
Qc is the Charge Value

T_r is the time needed for ion track formation. Tr is assumed to be 50psec in (1) and is negligible compared to the value of T_f (Calin et al., 1996).

$$\frac{1}{T_f}=\frac{K\epsilon_0}{q\mu N_D} \tag{2}$$

Where N_D doping concentration, which is a technology-based variable.

RHBD 16T RAM BIT CELL

The circuit is made of a totally 16 devices, Six NMOS and ten PMOS. The schematic diagram of the proposed RHBD 16T SRAM is shown in Fig. 1. The proposed SRAM has four storage junctions. QN and Q are the two primary junctions, and S0, and S1 are the extraneous junctions where S1 and Q have the same storage value and vice versa. The PMOS devices P6 and P5 are access devices, and aids to eliminate the occurrence of negative transient pulse at the storage junctions. This concept is depending on the consideration that when radiation strikes an junction of the NMOS device, it produces nothing but negative current pulses, while radiation strikes on the junction of a PMOS device produces nothing but positive current pulses(Jung et al., 2012). As a matter of the fact that access transistor produces an only positive transient pulse (0 to 1)[8]. Considering the stored values of storage junctionss as Q=1, QN=0, S0=0, S1=1.

Figure 1. RHBD16T SRAM cell

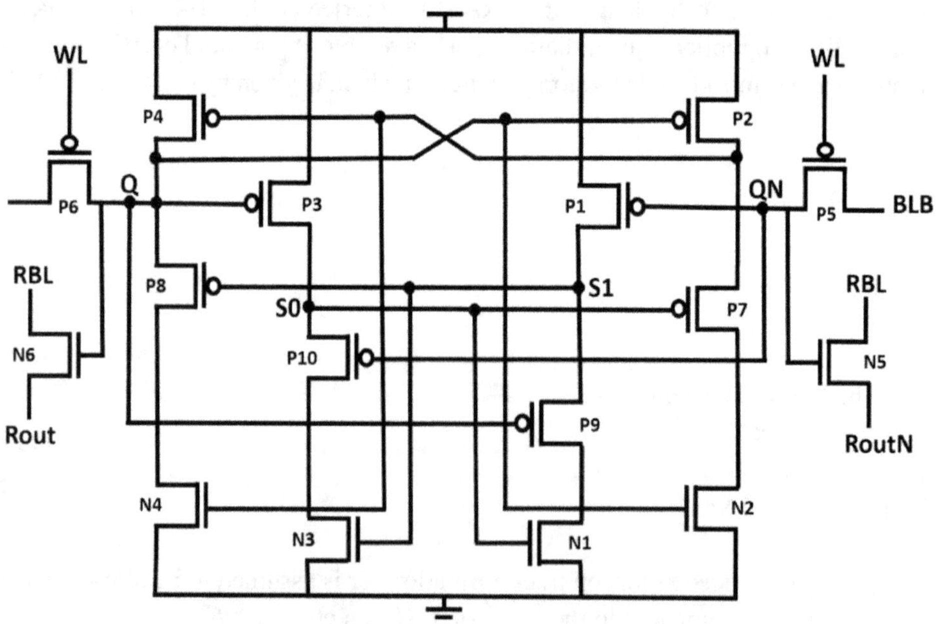

Figure 2. Write Operation of RHBD16T

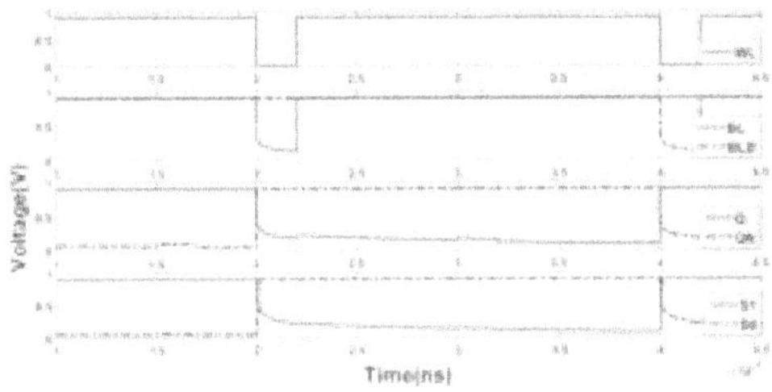

Figure 3. Read Operation of RHBD16T

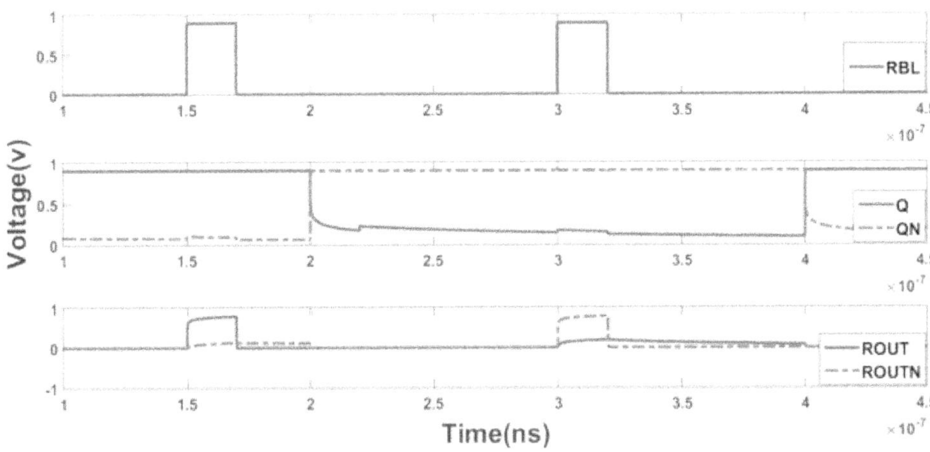

Write Operation

For writing value as Q=0, BL and BLB have values of 0 and 1 respectively; WL is low to switch ON access transistors of P6 and P5. When access transistors are ON Q value will be to pullled down by BL and it will ON the P3 and P2 transistor. Due to this junction value of S0 and QN start pull up to VDD. If once Q < QN is attained P1, P4, P7, P10, N3, N2 becomes ON and P2, P3, P8, P9, N4, N1 becomes OFF. When WL is pulled back to high the Q value becomes 0. And the write operation is shown in Fig (2).

Read Operation

Fig. (3) displays the timing diagram of the read operation. Two NMOS devices N6 and N5 are dedicated to read the data from primary junctions of QN and Q. QN and Q are attached to gate input of N5 and N6 respectively and read bit line is connected to both the NMOS drain region. During read operation RBL is high and N6 starts to conduct RBL but on the other side N5 is OFF due to low gate input and remains low.

Fig. 4 displays the layout of the proposed RHBD 16T SRAM cell.

Figure 4. Layout of 16T RHBD

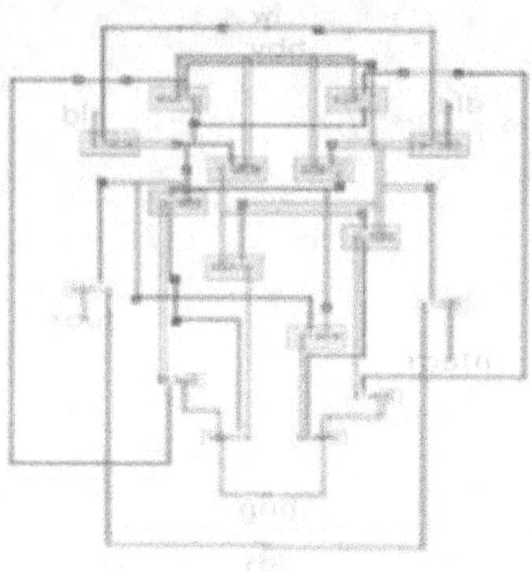

Soft Error Analysis

Single node and multi-node upset tolerance of the proposed RHBD 16T SRAM are discussed as follows. Two primary and secondary junctions are covered with PMOS devices. Therefore, only positive transient current will produce. So, (0 to 1) upset will happen but not 1 to 0. Q is the primary junction where 1 is stored and if particle strike happens only (1 to 1) upset will happen. And RHBD16T SRAM contains four storage junctions. Eight combinations of multi-node upset are possible.

QN (0 to 1): When junction QN is upset by a radiation strike it will alter to 1, then devices P4 and P1 will move into OFF state. But on the other side, junctions S1 and Q remain the same because P9 and P8 is OFF. So, N3 will be in ON state. P7 and N2 are also ON at the same Time. QN is pulled back to original value 0. Fig. 5 displays the recovery of data when a radiation strike is introduced at junction **QN (0 to 1).**

S0 (0 to 1): When junction S0 is affected by a radiation strike, it will switch ON the devices N1 and switch OFF the P7 transistor. However, P9 is OFF, and the S1 value will not change. P10 and N3 are ON so, S0 will pull back to original value of 0. Fig. 6 displays the recovery of data when a radiation strike is introduced at junction QN.

Figure 5. Fault injection simulation waveform of node QN is upset

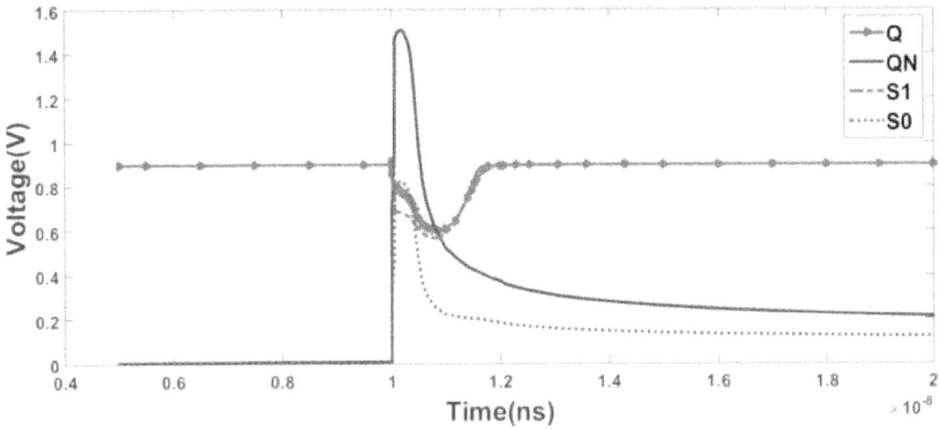

Figure 6. Fault injection simulation waveform of junction S0 is upset

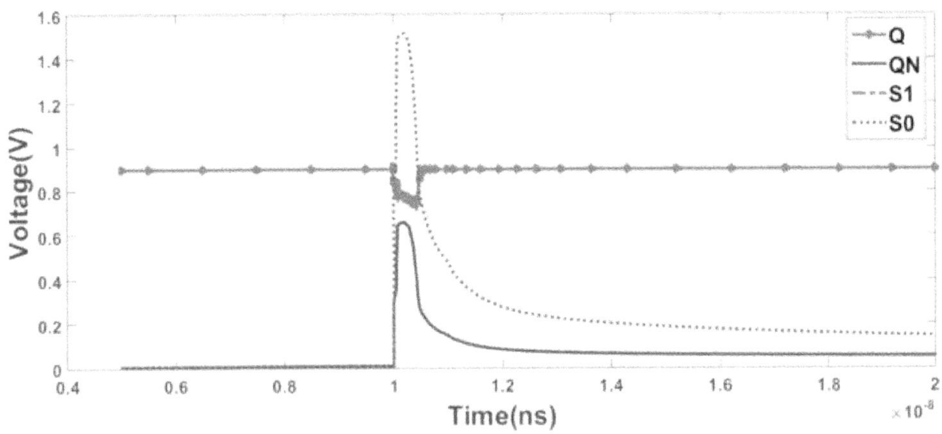

QN (0 to 1), S1 (1 to 0): When junctions QN and S1 are affected by radiation particles (note that (1 to 0) upset will not happen directly to junction S1. So, we model the strike in between P9 and N1. Since P9 is in OFF condition, it will affect the S1 junction. And, as explained previously for QN single node upset. QN junction will pull back to the original value of 0. Fig. 7 displays the recovery of data when a radiation strike is introduced at junctions QN and S1.

S0 (0 to 1), S1 (1 to 0): When S0 and S1 are affected by particle strike, S0 strikes will switch ON N1 and switch OFF P7 transistor. But Q and QN are in its original value P9, and P10 doesn't change its state. Therefore, S1 and S0 pull back to the original state. Fig. 8 displays the recovery of data when a radiation strike is introduced at junctions S0 and S1.

Finally, also for the maximum combination of multi-node up-sets, RHBD 16T can recover the stored data from the corrupted one. When junction pair QN-S0 is upset, the stored data can flip because pMOS is used between all four inverters, and it directly takes gate inputs from the storage junctions.

EVALUATION

Soft Error Verification

A double exponential current pulse is introduced into the storage junction to mimic the soft error effect. For (1 to 0) effect double current pulse is connected to the drain junction of the NMOS to resemble the negative transient pulse. Similarly, for (0 to 1) effect, a double current pulse is introduced to the drain junction of the PMOS to resemble the positive transient pulse. The simulation was taken out in Synopsys tool at SAED-PDK 32nm technology with a power supply of 0.9v at room temperature. Fig. 9 shows the critical charge comparison of RHBD 12T and RHBD 16T. The SEU and SEMU critical charges of the proposed SRAM are enhanced and hence the robustness of the proposed SRAM is greater than the previously proposed radiation hardened SRAM cells.

Figure 7. Fault injection simulation waveform of node-pair QN & S1 is upset

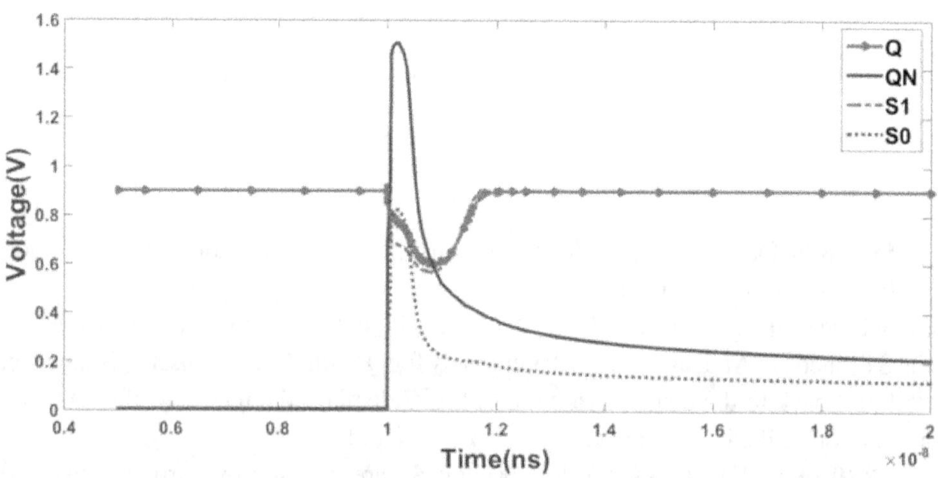

Figure 8. Fault injection simulation waveform of node-pair S0 & S1 is upset

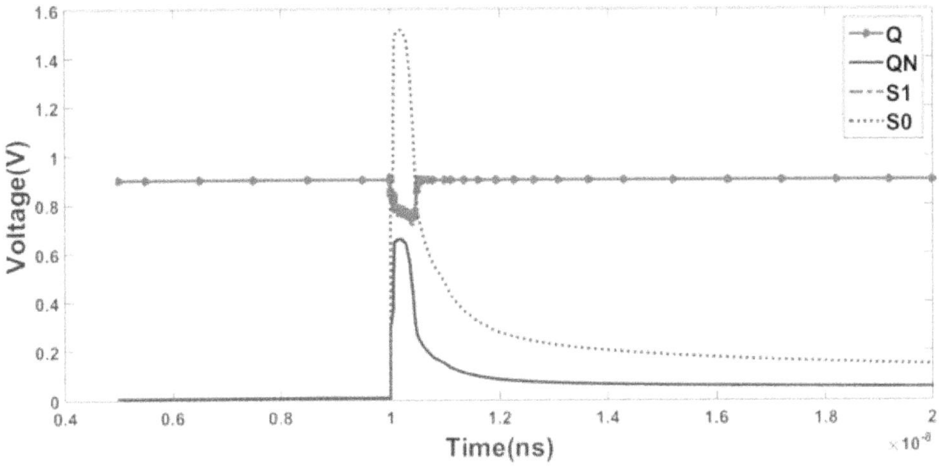

Figure 9. Critical charge comparison between 12T & 16T

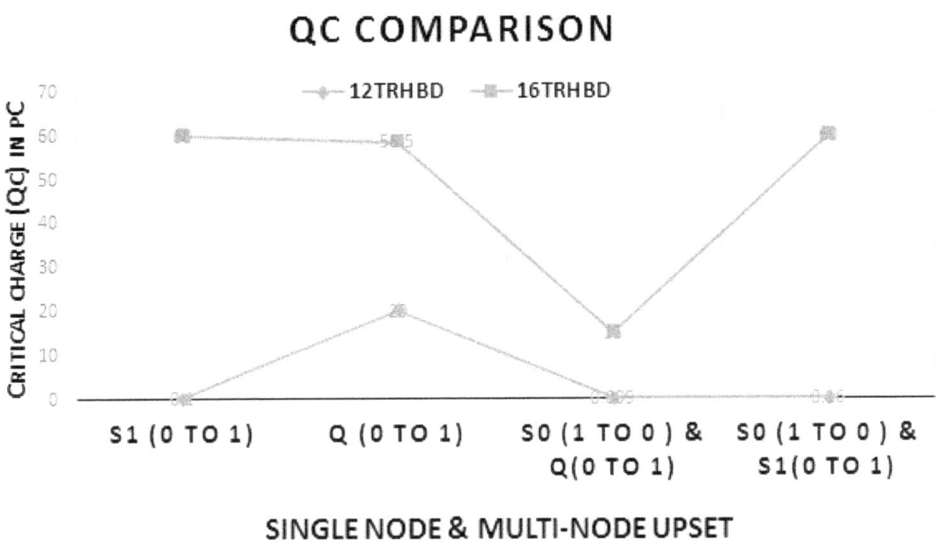

Process Variation

Process variation is the variation in attributes of a device due to the imperfect foundry process(Dang, Kim, & Chang, 2017). Its effects are predicted by using the Monte Carlo Simulation. In this paper, 1000MC simulation (5 Gaussian distribution) was carried out to evaluate the multi-node upset robustness of the memory cell. The authors in

Table 1. Failure In Rate in SRAM cells in 32nm technology

Cell	Failure In Rate
6T	40%
DICE	50%
8T SRAM	50%
11TSRAM	100%
13TSRAM	70%
NS-10T	60%
PS-10T	70%
12TRHBD	90%
16TRHBD	1%

(Guo et al., 2017) explained the multi-node verification, according to Monte Carlo simulation as per that 30fC charge is introduced in primary and secondary affected junctions. Monte Carlo simulation results of Table. 1 displays that proposed RHBD 16T memory cell has a 99% recovery rate, and RHBD12T cell has a 10% of recovery rate. Failure-In-Rate(RAJAEI, ASGARI, TABANDEH, & FAZELI, 2017)(Guo et al., 2018)(Khandelwal, Raj, & Gupta, 2015) is calculated as (3).

$$Failure\ inRate = \frac{No\ of\ flips}{Total\ No\ of\ Simulation} \tag{3}$$

Write Access Time

The Time taken for writing the data in a memory cell is defined as write access time. Write access time is calculated by giving 0 and 1 as an initial condition to the storage nodes, and opposite 1 and 0 is given as an initial condition to respective bit line. Where 1 storage node bit line will not discharge, but another side of the bit line will discharge because of the storage node is in '0' conditions. The total Time taken for '0' storing nodes into vdd/2 is calculated as write access time. And Fig. 10 shows the result of the write access time of the different RHBD circuits. In that 16TRHBD circuit is having access time, which is slightly larger than that 12TRHBD because of an increased number of transistors need a strong feedback path for writing data.

Read Access Time

Read access time is defined as the Time taken for reading the data in a memory cell. Read access time is calculated by giving the 0 and 1 initial condition to the storage nodes and one as an initial condition to both of the bit line. And once read operation begins, 0 storage nodes start to discharge the bit line. When bit line attains the vdd/2 value, consider this as a read access time(Bellaouar, Elmasry, Bellaouar, & Elmasry, 1995). Fig. 11 shows the operation of read access time for the different RHBD circuits. 16RHBD having read access, which is less as compared to that of 11T and 12TRHBD cell. A separate nMOS transistor is employed for reading data in which the gate terminal of that nMOS is connected to the storage nodes.

Figure 10. Write access time comparison between RHBD cells

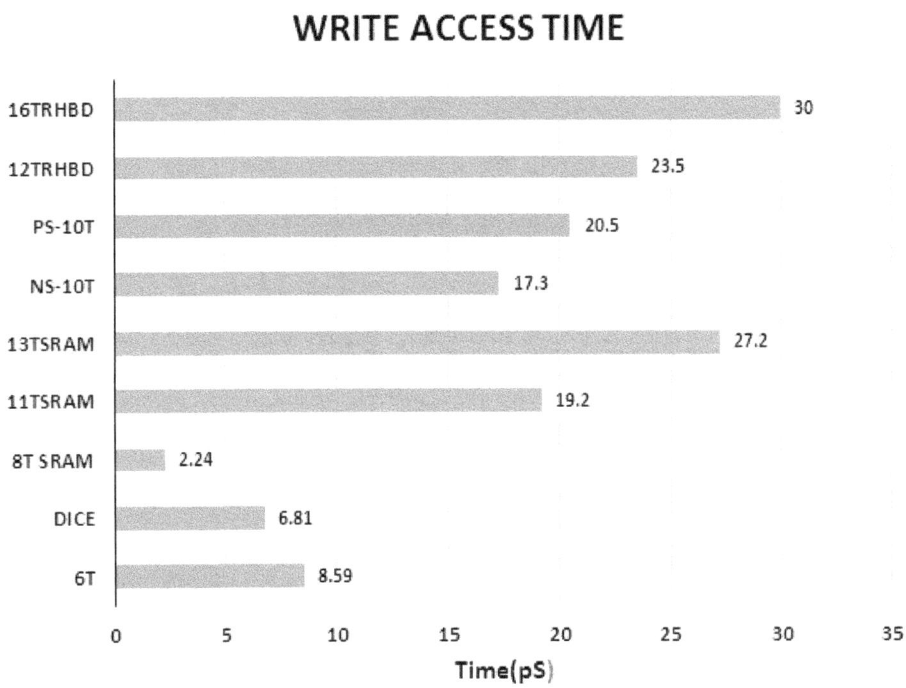

Power Comparison

The Static power consumption of the circuit is measured by giving initial condition values to the storage nodes. And only VDD is connected with the circuit. Now the

Figure 11. Read access time comparison between RHBD cells

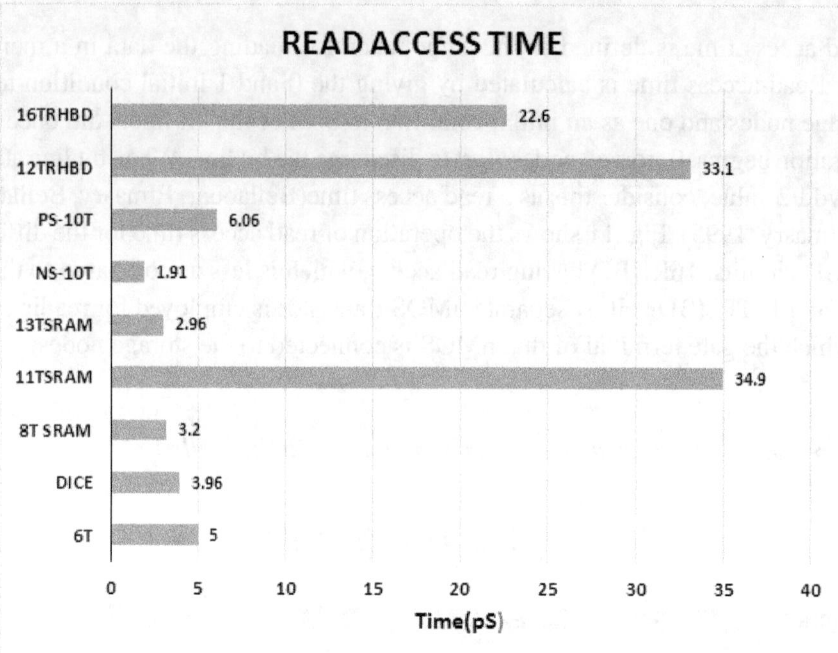

Figure 12. Power comparison between RHBD cells

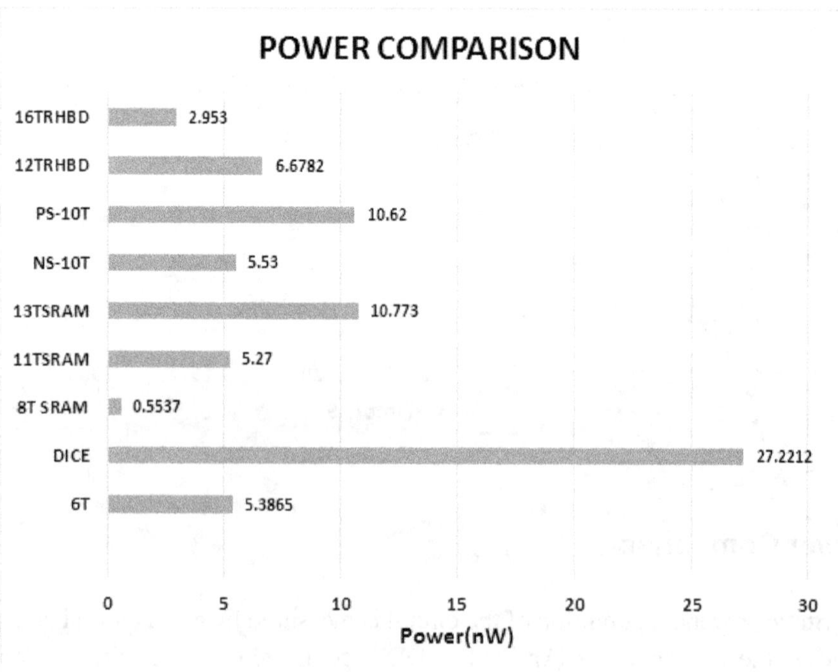

average power of the circuit is calculated. Fig. 12 shows the static power comparison between the RHBD circuits. When compared to all other RHBD cells, 16TRHBD consumes less static power because all storage junctions are separated by a pMOS device between the inverters. So, it consumes less current.

Critical Charge Comparison

Double Node Upset is checked by injecting a Double Exponential Current pulse in both primary and secondary junctions of the SRAM storage cells(Rajaei, Asgari, Tabandeh, & Fazeli, 2015). And start by injecting the same current pulse value for both the nodes. In which current pulse, the circuit starts to flip that current value is taken for a critical charge calculation of a circuit. Fig. 13, Fig. 14, Fig. 15, shows that the (Double Node Upset) Critical Charge comparison between the all significant radiation-hardened SRAM cells using technological model cards of 45nm, 65nm and 90nm. 16TRHBD have the highest QC value as compare to any other circuits, as shown in the graph. The storage nodes are protected by PMOS due to this, there are no possibilities of 1 to 0 as well as 0 to 1 SEU attack. By this radiation transient event eventually gets slow down due to a long feedback path.

Figure 13. Critical charge comparison between RHBD cells(45nm)

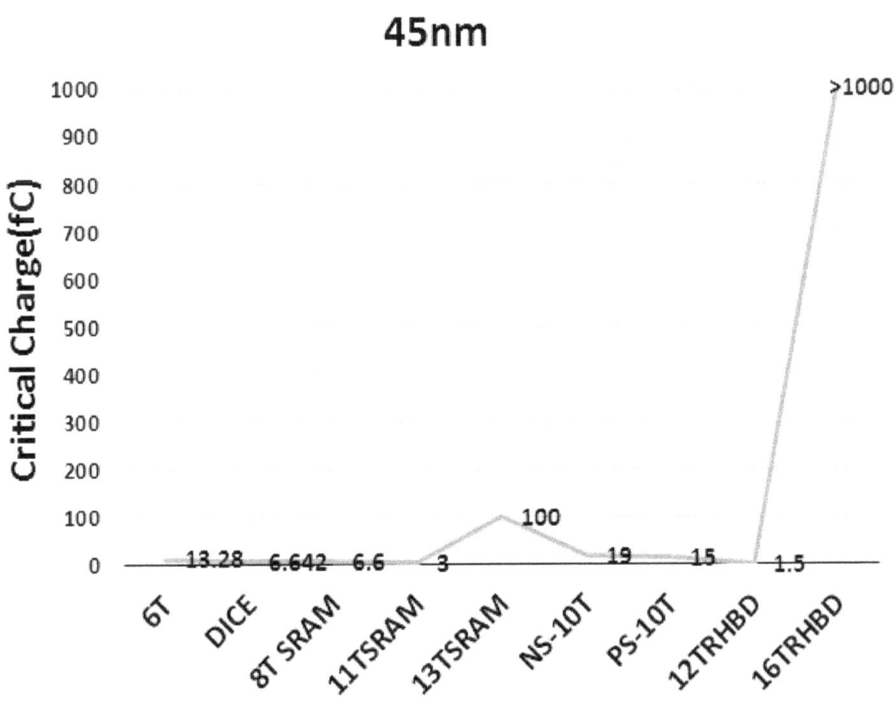

Figure 14. Critical charge comparison between RHBD cells(65nm)

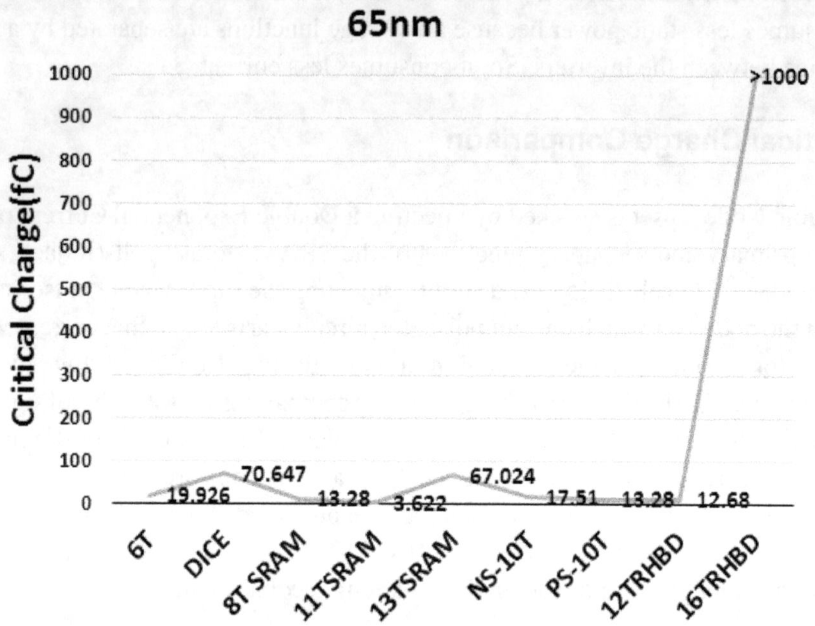

Figure 15. Critical charge comparison between RHBD cells(90nm)

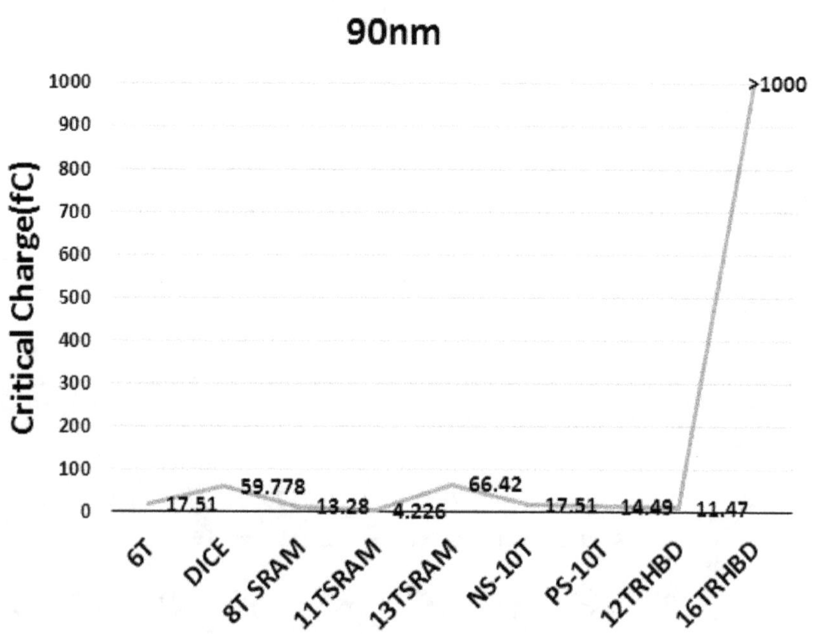

A comparison between the critical charge value of various RHBD circuits in different technological model cards (45nm,65nm,90nm,130nm) was carried out. After this comparison, it is concluded that by changing the model cards, Qc value increased due to the increase in technological nodes. During this comparison, it is observed that the Qc values of some of the RHBD circuits are saturating due to the increased capacitance of the storage node, which deposits the more particle strike charge by this it creates a disturbance in the circuits.

CONCLUSION

In this paper, RHBD 16T memory cell is designed for mitigating, a soft error in 32nm CMOS technology is proposed. And the result shows that the proposed circuit restores stored value for any single node upset and seven combinations of multi-node upset, and its critical charge value is above 10pc. The shortcoming of the proposed RHBD16T memory cell is sensitive to one combination (node pair S0-QN is upset to (0 to 1). As compared to the previously reported RHBD12T, our proposed design is immune against three more combinations of multi-node upset and single node upset. Hence RHBD16T memory cell is robust against soft error compared to the state-of-the-art hardened SRAM cells.

REFERENCES

Atias, L., Teman, A., Giterman, R., Meinerzhagen, P., & Fish, A. (2016). *A Low-Voltage Radiation-Hardened 13T SRAM Bitcell for Ultralow Power Space Applications. IEEE Transactions on Very Large Scale Integration (VLSI) Systems*. doi:10.1109/TVLSI.2016.2518220

Baumann, R. C. (2005). Radiation-induced soft errors in advanced semiconductor technologies. *IEEE Transactions on Device and Materials Reliability*, *5*(3), 305–316. Advance online publication. doi:10.1109/TDMR.2005.853449

Bellaouar, A., Elmasry, M. I., Bellaouar, A., & Elmasry, M. I. (1995). Low-Power CMOS Random Access Memory Circuits. In Low-Power Digital VLSI Design (pp. 313–407). doi:10.1007/978-1-4615-2355-0_6

Bohr, M. T., & Young, I. A. (2017). CMOS Scaling Trends and beyond. *IEEE Micro*, *37*(6), 20–29. Advance online publication. doi:10.1109/MM.2017.4241347

Calin, T., Nicolaidis, M., & Velazco, R. (1996). Upset hardened memory design for submicron CMOS technology. *IEEE Transactions on Nuclear Science, 43*(6), 2874–2878. doi:10.1109/23.556880

Ch, N. R., Manohar Reddy, D., Kishore Kumar, P., & Kaushal, G. (2019). Robust SRAM Cell Development for Single-Event Multiple Effects. In Communications in Computer and Information Science (Vol. 892). doi:10.1007/978-981-13-5950-7_29

Chang, M.-H., Chiu, Y.-T., Lai, S.-L., & Hwang, W. (2011). A 1kb 9T subthreshold SRAM with bit-interleaving scheme in 65nm CMOS. *IEEE/ACM International Symposium on Low Power Electronics and Design*, 291–296. 10.1109/ISLPED.2011.5993652

Chatterjee, I., Narasimham, B., Mahatme, N. N., Bhuva, B. L., Reed, R. A., & Schrimpf, R. D. … Monzel, C. (2014). Impact of technology scaling on SRAM soft error rates. *IEEE Transactions on Nuclear Science*. Advance online publication. doi:10.1109/TNS.2014.2365546

Dang, L. D. T., Kim, J. S., & Chang, I. J. (2017). We-Quatro: Radiation-Hardened SRAM Cell With Parametric Process Variation Tolerance. *IEEE Transactions on Nuclear Science*. Advance online publication. doi:10.1109/TNS.2017.2728180

Dodd, P. E., & Massengill, L. W. (2003). Basic mechanisms and modeling of single-event upset in digital microelectronics. *IEEE Transactions on Nuclear Science, 50*(3), 583–602. doi:10.1109/TNS.2003.813129

Dodd, P. E., Sexton, F. W., Hash, G. L., Shaneyfelt, M. R., Draper, B. L., Farino, A. J., & Flores, R. S. (1996). Impact of technology trends on SEJJ in CMOS SRAMs. *IEEE Transactions on Nuclear Science, 43*(6 PART 1), 2797–2804. doi:10.1109/23.556869

Ferlet-Cavrois, V., Massengill, L. W., & Gouker, P. (2013). Single event transients in digital CMOS - A review. *IEEE Transactions on Nuclear Science, 60*(3), 1767–1790. Advance online publication. doi:10.1109/TNS.2013.2255624

Gracia-Moran, J., Saiz-Adalid, L. J., Gil-Tomas, D., & Gil-Vicente, P. J. (2018). *Improving error correction codes for multiple-cell upsets in space applications. IEEE Transactions on Very Large Scale Integration (VLSI) Systems*. doi:10.1109/TVLSI.2018.2837220

Guo, J., Xiao, L., & Mao, Z. (2014). Novel Low-Power and Highly Reliable Radiation Hardened Memory Cell for 65 nm CMOS Technology. *IEEE Transactions on Circuits and Systems. I, Regular Papers, 61*(7), 1994–2001. doi:10.1109/TCSI.2014.2304658

Guo, J., Zhu, L., Liu, W., Huang, H., Liu, S., Wang, T., … Mao, Z. (2017). Novel Radiation-Hardened-by-Design (RHBD) 12T Memory Cell for Aerospace Applications in Nanoscale CMOS Technology. *IEEE Transactions on Very Large Scale Integration (VLSI) Systems*. doi:10.1109/TVLSI.2016.2645282

Guo, J., Zhu, L., Sun, Y., Cao, H., Huang, H., Wang, T., … Mao, Z. (2018). Design of Area-Efficient and Highly Reliable RHBD 10T Memory Cell for Aerospace Applications. *IEEE Transactions on Very Large Scale Integration (VLSI) Systems*. doi:10.1109/TVLSI.2017.2788439

Jung, I. S., & Kim, Y. B., & Lombardi, F. (2012). A novel sort error hardened 10T SRAM cells for low voltage operation. *Midwest Symposium on Circuits and Systems*. 10.1109/MWSCAS.2012.6292120

Kaushal, G., Rathod, S. S., Raghuram, C. N., & Dasgupta, S. (2019). Radiation hard circuit design: flip-flop and SRAM. In VLSI and Post-CMOS Electronics. Volume 2: Devices, circuits and interconnects (Vol. 2, pp. 249–278). doi:10.1049/pbcs073g_ch12

Khandelwal, S., Raj, B., & Gupta, R. D. (2015). FinFET Based 6T SRAM cell design: Analysis of performance metric, process variation and temperature effect. *Journal of Computational and Theoretical Nanoscience*, *12*(9), 2500–2506. Advance online publication. doi:10.1166/jctn.2015.4055

McLain, M. L., Barnaby, H. J., Esqueda, I. S., Oder, J., & Vermeire, B. (2009a). Reliability of high performance standard two-edge and radiation hardened by design enclosed geometry transistors. *IEEE International Reliability Physics Symposium Proceedings*. 10.1109/IRPS.2009.5173247

McLain, M. L., Barnaby, H. J., Esqueda, I. S., Oder, J., & Vermeire, B. (2009b). Reliability of high performance standard two-edge and radiation hardened by design enclosed geometry transistors. *2009 IEEE International Reliability Physics Symposium*, 174–179. 10.1109/IRPS.2009.5173247

Narasimham, B., Gupta, S., Reed, D., Wang, J. K., Hendrickson, N., & Taufique, H. (2018). Scaling trends and bias dependence of the soft error rate of 16 nm and 7 nm FinFET SRAMs. *IEEE International Reliability Physics Symposium Proceedings*. 10.1109/IRPS.2018.8353583

Peng, C., Hu, Z., En, Y., Chen, Y., Lei, Z., Zhang, Z., … Li, B. (2018). Radiation hardening by the modification of shallow trench isolation process in partially depleted SOI MOSFETs. IEEE Transactions on Nuclear Science. doi:10.1109/TNS.2018.2798295

Peng, C., Huang, J., Liu, C., Zhao, Q., Xiao, S., Wu, X., … Zeng, X. (2019). Radiation-hardened 14T SRAM bitcell with speed and power optimized for space application. *IEEE Transactions on Very Large Scale Integration (VLSI) Systems*. doi:10.1109/TVLSI.2018.2879341

Prasad, V., & Sandya, S. (2014). Single event transient tolerant high speed phase frequency detector for PLL based frequency synthesizer. *Proceedings of International Conference on Circuits, Communication, Control and Computing, I4C 2014*. 10.1109/CIMCA.2014.7057761

Prokhorov, A. S., Tyurin, S. F., & Chudinov, M. A. (2018). A radiation resistant voting circuit. *Proceedings of the 2018 IEEE Conference of Russian Young Researchers in Electrical and Electronic Engineering, ElConRus 2018*. 10.1109/EIConRus.2018.8317068

Raghuram, C. H. N., & Member, S. (2020). *Double Node Upset Tolerant RHBD15T SRAM Cell Design for Space Applications, 4388*(c). doi:10.1109/TDMR.2020.2970089

Rajaei, R., Asgari, B., Tabandeh, M., & Fazeli, M. (2015). Design of Robust SRAM Cells Against Single-Event Multiple Effects for Nanometer Technologies. *IEEE Transactions on Device and Materials Reliability*, *15*(3), 429–436. doi:10.1109/TDMR.2015.2456832

Rajaei, R., Asgari, B., Tabandeh, M., & Fazeli, M.RAJAEI. (2017). Single event multiple upset-tolerant SRAM cell designs for nano-scale CMOS technology. *Turkish Journal of Electrical Engineering and Computer Sciences*, *25*, 1035–1047. doi:10.3906/elk-1502-124

Zheng, Q., Cui, J., Lu, W., Guo, H., Liu, J., & Yu, X. (2018). The Increased Single-Event Upset Sensitivity of 65-nm DICE SRAM Induced by Total Ionizing Dose. *IEEE Transactions on Nuclear Science*. Advance online publication. doi:10.1109/TNS.2018.2816583

Section 4
Future of Nanowire Technology

Chapter 9
Safe and Secure Home Automation Through IoT Applications:
A Sensor and IC-Based Implementation for Digital Transformation

Rohit Rastogi
https://orcid.org/0000-0002-6402-7638
Dayalbagh Educational Institute, India & ABES Engineering College, Ghaziabad, India

Puru Jain
ABES Engineering College, Ghaziabad, India

Rishabh Jain
ABES Engineering College, Ghaziabad, India

ABSTRACT

In current conditions, robotization has changed into the fundamental piece of our lives. Everybody is completely subject to mechanization whether it is an extraordinary bundling or home robotization. So as to bring home automation into thought, everybody now needs a heterogeneous state security, and in our task on residential robotization, such high security highlights are completely on the best possible consumption for this reason. In light of the structure of the interruption zone, there are some fundamental interests in it. Piezoelectric sensors are compelling for sharpening appropriated wellbeing checking and structures. An intrusion detection system (IDS) is a structure that screen for suspicious movement and issues alarms when such advancement is found. While impossible to miss worthiness and presentation is, some obstruction divulgence structures are fit to take practice when poisonous improvement or peculiar action is perceived.

DOI: 10.4018/978-1-7998-6467-7.ch009

INTRODUCTION

The Manuscript presents a detailed study wherein the Introduction section, the basic concepts of Big data, and CPS and IoT are explained. In the Literature Survey, the recent work of a few researchers has been explained. In the next section, the structure and functioning of the components used in this experiment like Arduino Uno, Piezoelectric-Transducer Sensor, RFID, LCD, GSM, and fingerprint sensor have been presented. Their working procedure has been explained in detail. At last, the Application aspect for home security using these components is elaborated. Then the future research direction, novelty, limitations, and conclusion are reflected.

An Intrusion Detection System (IDS) is a structure working on the principle of the Cyber-Physical System principle that screen for suspicious movement and issues alarms when such advancement is found. While impossible to miss worthiness and presentation is as far as possible, some obstruction divulgence structures are fit to take practice when poisonous improvement or peculiar action is perceived.

Security is an important issue nowadays, as the possibilities of intrusion are increasing day by day. The Cyber-Physical System is a network of physical objects devices, vehicles, buildings, and other items embedded with electronics software sensors, and network connectivity that enables these objects to collect and exchange data.

Big Data Analysis

It is the technology that is used to handle big data. Data science and predictive analytics can help you to achieve your business goals. Learn the process and benefits of implementing big data into your business. Big data analytics is the often complex process of examining large and varied data sets, or big data, to uncover information -- such as hidden patterns, unknown correlations, market trends, and customer preferences -- that can help organizations make informed business decisions.

Cyber-Physical System

It is a system in which a mechanism is controlled or monitored by computer-based algorithms. It includes autonomous automobile systems, robotics, automatic pilot avionics, and many more. In light of the structure of the interruption zone and there are some fundamental interests in it. The cyber-physical system involves enhancing the network to proficiently collect and analyze the data from various sensors and actuators then sends the data to the mobile phone or a personal computer over a wireless connection. Cyber-Physical Systems have progressed essentially in the last couple of years since it has created a new era in the world of information and communication technologies.

The Cyber-Physical System allows objects to be sensed and controlled remotely across existing network infrastructure, creating opportunities for more direct integration of the physical world into computer-based systems, and resulting in improved efficiency, accuracy, and economic benefit; when it is augmented with sensors and actuators, the technology becomes an instance of the more general class of cyber-physical systems, which also encompasses technologies such as smart environment grids, smart homes, intelligent transportation, and smart cities.

IoT and Smart Automation

It is a technology that has made the non-connectivity appliance a connectivity appliance. The appliances that contain technology that helps us to communicate with human and technology. Smart Automation is a technology that is to automate or automatic something by giving a single command. This technology reduces human efforts in the completion of work [Hua Li. et al.,2017] and [DivyanshGoel. et al., 2016].

LITERATURE SURVEY

The important goal of the Cyber-Physical System is to build a human interactive system using an RF controlled remote. Now technology is accelerating so homes are also getting smarter. Modern homes are deliberately relocating from current l switches to the centralized control system, containing RF controlled switches. Today traditional wall switches situated in various parts of the home make it laborious t for the end-user to go near them to control and operate (Li, H. et al., 2017).

Even further it turns into more problematic for the old persons or physically handicapped people to do so. Cyber-Physical Systems using remote implements an easier solution with Wireless sensor networks. To accomplish this, an RF remote is combined to the microcontroller on the transmitter side that sends ON/OFF signals to the receiver where devices are connected. By operating the stated remote switch on the transmitter, the loads can be turned ON/OFF globally using wireless technology (Asensio, J.A., . et et al.,2019).

EXPERIMENTAL DESCRIPTION

Our project contains six major components which are as follows:

1. ARDUINO (UNO)
2. Piezoelectric-Transducer Sensor
3. RFID
4. LCD
5. GSM
6. Buzzer
7. Fingerprint Sensor

Arduino- Uno

The microcontroller board depends upon ATmega328P (datasheet). 14 are moved data/yield pins (6 of which can be used as PWM yields), 6 direct information sources, a 16 MHz quartz gem, a USB affiliation, a power jack, an ICSP header, and a reset gets. This consolidates everything expected to help the microcontroller; add it to a PC with a USB affiliation or power it with an AC-to-DC connector or battery to start it. Arduino is an open-source stage used to make Gadgets Ventures. Arduino uses a physical programmable circuit board (routinely perceived as a microcontroller) and uses a touch of programming, or IDE (fused improvement condition), which continues running on your PC, which is used to make and move PC code. The Arduino stage has still ended up being critical with individuals, which is just beginning with the hardware, and in light of the current conditions. Despite the most recent programmable circuit sheets, Arduino does not require a free (planner) contraption to stack the new code on the board - you can generally use a USB interface. Also, Arduino uses the streamlined kind of IDE C ++, which makes the program more straightforward to learn. For a long time, Arduino gives a standard structure factor that breaks the segments of the microcontroller into a progressively open gathering (Asensio, J.A., et al.,2019) and (Arora, S., et al., 2016)(As per Fig. 1).

Uno Board and Form 1.0 Arduino Software (IDE) were the reference groupings of Arduino, which are by and by made for a snappy release. Arduino is the most preferred microcontroller which is used in Cyber-Physical security systems. UNO is the best board in the first place rigging and coding. UNO is the most used and recording driving social event of the whole Arduino family (Li, H. et al.,2017) and (Goel, D. et al., 2016].

The above-stated Table gives the whole specifications about ARDUINO UNO and which is retrieved from its Datasheet (As per Table 1).

Figure 1. Arduino Uno Rev3 (ATmega328P)(Agarwal A. et al.,2018)

Table 1. Technical speculations of Arduino Uno

Component	Specifications
Microcontroller	ATMEGA328P
Voltage	5V
Input	7-12V
Digital pins	14
PWM pins	6
Analog pins	6
DC per i/o pins	20 mA
Dc current	50 mA
Flash memory	32 KB
Sram	2 KB
Eeprom	1 KB (ATmega328P)
Clock speed	16 MHz
Led_builtin	13
Length	68.6 mm
Width	53.4 mm
Weight	25 gm

Arduino Functionality

Power USB: Arduino board is powered by using USB 2.0 which is connected from the computer or power bank for providing program and power.

Barrel Jack: Arduino can also be powered directly from the AC mains power by connecting this jack.

Voltage Regulator: It controls the voltage given by the USB port and Barrel Jack and distributes it into the processor and other elements.

Crystal Oscillator: It deals with time calculation. It has a frequency of 16 MHz.

Reset: This button is used to reset or refresh the Arduino board. By doing this the functioning starts from the first line of the program written in the board

Microcontroller: Arduino has many microcontroller series but in this project, we are using ATMEGA328P.

Led Indicator: It glows up when the power is provided to the Arduino. If it does not turn on there is something wrong with the connection of power.

Digital I/O: It has digital I/O pins that are used to read the logic or in providing the output to the different modules such as LEDs, Relays, Bluetooth, etc.[Nicole D. Karpinskya.et al.,2018]and[Agarwal A. et al.,2018].

Piezoelectric Transducer Sensor

A device that uses the piezoelectric effect, to measure changes in pressure, acceleration, temperature, strain, or force by converting them to an electrical charge. A piezoelectric sensor is generally used in Cyber-Physical security. Piezoelectric sensors are versatile tools for the measurement of various processes like:-

1. Sensing Touch
2. The aroma in the Air
3. Measuring Pressure/ Force
4. Measuring Speed
5. Measuring the impact
6. Realizing Direction & position (Goel, D. et al., 2016) and (Hofmann, P., et al., 2019).

Working of a Piezoelectric Sensor

The physical sums normally evaluated by the piezoelectric sensor are accelerating and weight. Both weight and accelerating sensors manage a comparable rule of piezoelectricity, yet the crucial qualification between them is that the way wherein the power is associated with their recognizing part.

Figure 3. Connection of the piezoelectric sensor(Gupta, R., and Rastogi, R2015)

Figure 2. Piezoelectric Sensor(Agarwal A. et al.,2018)

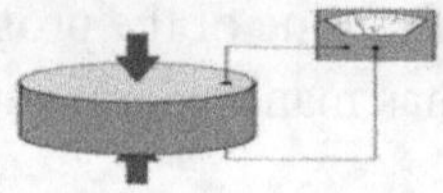

In this sensor, a wobbly film is put on an enormous scale to move the power associated with the piezoelectric part. It is generally a Cyber-Physical technology that works on change with pressure in the surrounding physical surfaces. On weight on this slight film, the piezoelectric material loads and starts making electric voltages. The yield voltage is concerning the proportion of associated weight(As per Fig. 2).

Representation of a simple Piezoelectric Transducer is displayed in Fig. 3.

Working Principle

The force applied with a neutral axis (y) displaces the charge with direction, is perpendicular to the line of force (Niua, J., et al., 2018) and (Aggarwal, S., et al., 2015).

Piezoelectric Coefficient

A piezoelectric transducer has a great deal of DC yield impedance and it is displayed as a relative voltage source and channel organ. The voltage at the source is valid relative to the power, weight, or stress associated. The yield signal is then identified with this mechanical strength as if it has passed a similar circuit(As per Fig. 4 and 5).

For use in the form of sensors, the repetition reaction area of the level of the plot is generally used between the high reaction pass-plot and the magnificent corpuscle. Burden and Spitz protest must be bigger for the fact that it does not need to lose less frequent intrigue. A disorderly similar circuit model can be used in the form, in which Cs speaks to the surface capacitance of the plane, which is guided by a standard recipe for the capacitance of parallel plates. Similarly, it can be added as an active source parallel to the source capacitance, the accent is connected to the direct charge (Kaul, U. et al., 2015) and [Zheng, S., et al., 2011) (As per Fig. 6 and 7).

Radio Frequency Identification (RFID)

RFID uses electromagnetic fields to typically recognize and seek after the name that is fused into the things. Names have electronic-remote difficult to reach data. In Cyber-Physical System, wireless network protocols are used in communication

Figure 4. Piezoelectric sensors a voltage source in series with the sensor's capacitance (Nigam, A., et al., 2015)

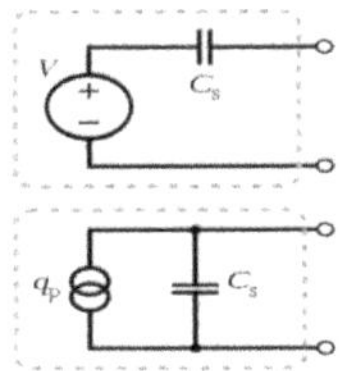

Figure 5. A piezo sensor outer specifications(Nigam, A., et al.,2015)

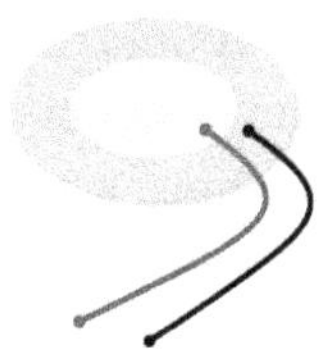

Figure 6. Piezoplate description for a piezoelectric sensor(Gupta, R. et al., 2015)

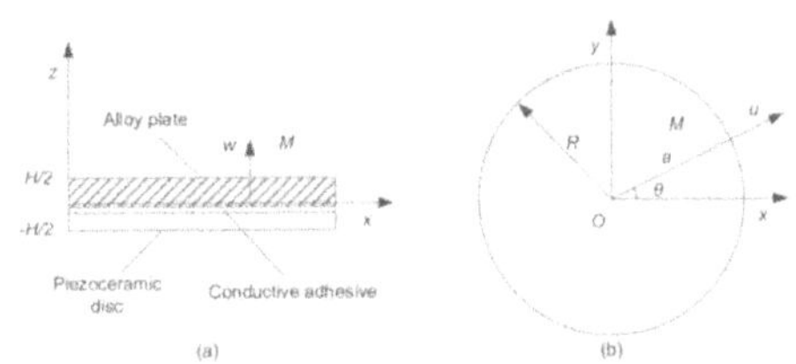

Figure 7. The dimension of a piezoelectric sensor (Gupta, R., et al., 2015)

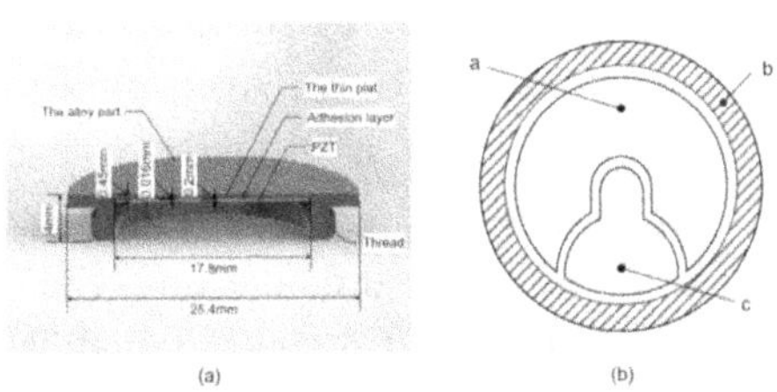

Figure 8. RFID module used in our experimental work(Gupta, R., et al.,2015)

between the microcontroller and the human. In this project, we are using the Radio Frequency-Identification technology is being used so that communication can become easy and also safe. It unconstrained imprints assemble essentialness from bordering RFIDs to customer's grilling radio waves. There is a region control source (battery) in the dynamic name and a couple of meters for every customer can work from RFID. Industry 4.0 brought new age benchmarks and among them, splendid preparing plant thought is surely the most huge (Zheng, S. et al., 2011).

The Smart Factory thought relies upon the blend of the Manufacturing Execution System (MES) with the shop-floor level on one side, and on the contrary side with the Enterprise Resource Planning (ERP) level. The outstanding response for assembling things data is the RFID (Radio-repeat Identification) and Cyber-Physical Systems (CPS) structure. In this letter, the RFID-engaged Smart Factory has been arranged and facilitated inside the Learning Factory, and its show has been attempted. As for the plan of the system, it was made as fundamental as possible to enable understudies to manage it in the creation line condition to fathom and learn it. Dissimilar for an organized tag, the imprint should not be inside the customer's recognizable way, so it will, in general, be presented in the going with article. RFID is an advancement for customized ID and data get (AIDC) (Gluhak et al., 2011) (As per Fig. 8).

These stresses accomplished the standard delineation of progress for prosperity and security issues. Use ISO/IEC 18000 and ISO/IEC 29167 on-chip cryptography methods to perceive the United Nations character, tag, and per customer confirmation and over-the-air protection.

Specification of RFID MFRC522

Module MFRC-522 fuses; RC-522 RFID RF-IC Card Sensor module for sending the S50 Fudan card and keychain for RC50. According to the MFRC522 client 13.56MHz contactless correspondence is related to "excellent three-table" of an excellent schematic chip card, a low voltage, unimportant effort, non-contact according to the customer chip, and "instrumentation" NXP convenient "handicraft contraception. The Mutual Fund uses RC522 Propelled Twik and Demodulation

Idea to a wide extent of 13.56MHz in exceptional systems and shows without contact. 14443 A Strong Transponder Signal. The computerized part is surrounded and handles ISO14443A and Blonder affirmation. Also, for the support of Bolster QUICK CRYPTO-1 encryption probe, MIFARE cheese (Gluhak et al., 2011).

MFRC522 Bolster Mifare Game Plan Quickly form non-contact correspondence, formed in Table 2.2, as 424kbit or two-way information change standard. Customers of chips have been shown by the new assembling, MF RC522 MF RC500, and MF RC530 as an extraordinarily sorted 13.56 MHz card, and have different comparable properties, yet there are various highlights and irregularities are high. It is between host correspondence using SPI mode, reduces the connection, reduces the size and cost of the PCB board. MF522-A module receives exceptional Philips MFRC522 according to the client's circuit chip plot, easy to use, unimportant effort, sensible for gear changes, according to customer customers, the progress of edge applications, RF card terminal structure or customer's need to produce.

This module can be stacked in the illustration according to the client's position. The module uses 3.3V voltage through the SPI interface, the pair of straight lines can be exceptionally united with the client. Any CPU board correspondence module can ensure reliability and reliable work. Specifications of RFID are displayed in Table 2(Sharma, S. et al., 2015) and (Sharma, P. et al., 2014).

Table 2. Specifications and components of RFID

Component	Specifications
CURRENT	13-26mA
INPUT VOLTAGE	DC 3.3V
SLEEP CURRENT	<80uA
PEAK CURRENT	<30mA
FREQUENCY	13.56MHz
SUPPORTED CARDS	MIFARE1 S50,MIFARE1 S70, MIFARE ULTRALIGHT,MIFARE PRO, MIFARE DESFIRE PRODUCT
DIMENSIONS	40mm - 60mm
ENVIRONMENTAL TEMP	-20 to 80 C
RELATIVE HUMIDITY	5-95%
DATA TRANSFER RATE	10MBPS

Buzzer

For a chime or beeper, for example, a sound hilling connotation, which can be mechanical, electromechanical, or snoozes electric, as shown in Fig. 2.5, signature and beepers include the usual work-ready device, timekeeper, and confirmation of customer data, for example. A mouse snap or keystroke.

A Piezoelectric component is a gem or artistic that turns to some extent when connected to the voltage. Many types of Buzzers of different voice frequencies are being used in Cyber-Physical Systems (CPS) so that it can give the red signal before any tragedy happens. So if you supply the AC voltage on a two kHz, then at that point it turns further and reverses with the same speed as the AC signal, and gives a melodious voice. A similar effect works backward.

The most productive piece of the sound transducer is a metering plate of the piezoelectric, which is attached to the small metal stomach. When a voltage is connected to the plate of dirt, the cycle is connected incorrectly, causing the metal stomach to bend. At the point when a discontinuous voltage is connected to the point, the wired / metal-tilted section vibrates on the recurrence of the related voltage, and it fits to enable the hearing sound. In this occasion there is the mechanism of the mechanical resurgence of dirt/metal segment and the redistribution of related power banner, the vibration will be the most accurate, and sound construction will be the largest (Reverse redundancy of the lamp oil segment alone is exceptionally high, the option to listen to the sound is quite high, later it is a requirement for the metal stomach)(As per Fig. 9) (Aggarwal, S. et al., 2015) and (Shaquib, S.T. et al., 2013).

Figure 9. Piezoelectric Buzzer (5v)(Sharma, R. et al.,2013)

Several characteristics need to be taken a mind of are:-

1. **Dimensions (Not Including Pins):** 12mm diameter, 9.7mm tall.
2. **Pins Length:** 6mm long and 0.3” apart.
3. **Crystal Oscillation:** 2 kHz
4. **Input Voltage:** 5v to 9v
5. **Weight:** 18.1gm
6. **Principle Operation:** Reverse Piezoelectric Effect

7. **Material Used:** Piezoceramic (Saquib et al., 2013) and (Shekhar, S. et al.,2015)

Working of Piezoelectric Buzzer

Pizzeria buzzers use inverted piezoelectric theory to create ceramic disks for producing sound waves. A built-in oscillation circuit is included in the buzzer. Piezabuzer works on a wide temperature range and creates noise from soft and gentle to loud and aggressive (Marie, A. et al., 2015).

When piezoelectric material is under pressure, the pressure causes the change in the surface of the material, as a result of these pressure differences, there are a surface and tension with the other. As a result, the positive charge collects on one side of the material, and the negative charge collects on the opposite side. This generator effect changes mechanical energy into electricity. In reverse piezoelectric effect, it is used in buzzer, by applying an electric field, the length of the surface changes and transforms electrical energy into mechanical energy, which creates sound waves, capable of detecting human ears. is.

Jacques and Pierre Curie discovered the piezoelectric theory in 1880, which was kept under a fixed crystal when they were under pressure, they used to produce electricity. When he continued to research this theory, he found that by applying electric fields in the piezoelectric crystal, he got accelerated. The discovery of ceramic materials made with barium titanate displays similar changes, which allow the use of both piezoelectric and inverse piezoelectric principles, including the development of Piezo buzzer (Saquib et al., 2013).

GSM 900A GSM / GPRS Modem

RS232 is working with dual-band GSM / GPRS engine - SIM900A, it works on repeat 900/1800 MHz. The modem is going with the RS232 interface, which gives you the right to stay connected with the microcontroller with the PC in the form of an RS232 chip (MAX232). BOD Rate is configurable from 9600-115200 through the AT Course. GSM is one of the most commonly used electronic devices in Cyber-Physical Systems so that communication can be possible between owner and machine. GSM is used for communication purposes as it provides the feature of calling, SMS, and also works at a different frequency. It just needs a SIM of any company so that it can communicate as instructed by the user. This is a GSM / GPRS-error-free quad-band remote, which works on the repetition of 850/900/1800/1900MHz and can be used not only on the Internet but also for oral correspondence.

In this case that it is related to the increase and a small derogatory speaker and for SMS. Remotely, undoubtedly a tremendous group (0.94-inch x 0.94-inch x 0.12 inch) with four L-shaped contacts on both sides so that they can be secured on different sides and floors. Inside, the module is guided by the AMR926EJ-S processor, which controls phone correspondence, Cyber Security, data correspondence (methods for an organized TCP / IP stack), and circuit correspondence (via UART and TTL interference interfaces) (Saquib et al., 2013) and (Qun, C., et al., 2018).

The processor is also in charge of a SIM card (3 or 1,8 V) that should be related to the external mass of the module. Also, the GSM900 device integrates a basic interface, A / D converter, an RTC, an SPI transport, an IOC, and a PWM module. Radio part GSM phase 2/2 + is great and Class 4 (2W) is 850/900 MHz or Class 1 (1W) at 1800/1900 MHz. GSM / GPRS modem has an inward TCP / IP stack that enables you to communicate with the internet through GPRS. This M2M interface is suitable for SMS, the voice is similar to data-driven applications. Given the wide range of power supply, it gives the right to interfere with the supply of unregistered power. Using this modem, you can call and send via the voice call, SMS, read SMS, original AT course, and visit the web (Rungta, S. et al., 2014).

Features

- Dual-Band 900/1800MHz.
- Configured baud rate.
- Holds SIM Card.
- Network Status LED.
- Inbuilt powerful TCP/IP protocol stack for internet data transfer over GPRS.

Applications

- Access control devices
- Cyber-Physical System technology using devices
- Supply chain management

Working

In contrast to cell phones, a GSM modem does not have a keypad and show. It just acknowledges a few directions through a sequential interface and acknowledges those. These directions are called AT directions. The modem is a rundown of AT directions to teach you to do your errands. Each direction begins with "AT". That is the reason they are called AT directions. AT represents reflection(As per Fig. 10).

In our straightforward venture, the program trusts that the portable PIN will be entered through the console. At the point when a ten-digit portable PIN is given, the program teaches the modem to send an instant message utilizing the succession of ATM directions(As per Fig. 11).

Figure 10. GSM 900A kit and its components (Agarwal, A. et al., 2018)

Figure 11. The graph between frequency and magnitude (Gupta, R., et al., 2015)

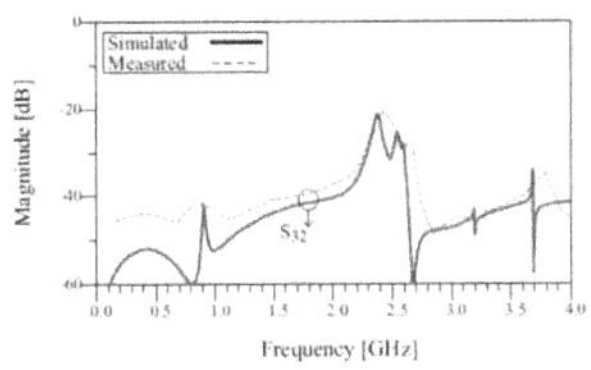

LCD (Liquid Crystal Display)

A dimension board is visible or another electronically bent optical device that uses light-directed properties of liquid valued stones. Liquid pearls typically do not emit light, although they use foundation knowledge or reflector to create images in cinematography or monochrome. Cyber-Physical Systems uses LCD so that messages can be displayed on the screen and can be shown to the owner of the machine so that works get completed early. LCDs are open to indicate intangible images, (for example, usually in a valuable PC show) or fixed images with twisted objects that can be seen safe, for example, similar to a mechanical clock, presets Displays the words, digits, and seven-section. The LCD (liquid precious stone presentation) screen is an electronic showcase module and identifies a wide scope of usage. A 16x2 LCD show is an extraordinary fundamental module and it is usually used in various gadgets and circuits. These modules are favored by more than seven pieces and other multi-part LEDs. Due to being: LCDs are appropriate; easily programmable; There is no restriction for showing extraordinary and even custom characters (instead of seven pieces), movements, etc.

TheCyber-Physical System also uses 16x2 LCD so that message of long length can be displayed and also it features 2 line display. It implies that it can show 16 characters for each line and there are 2 lines of this type. In this LCD, each character is shown in a 5x7 pixel lattice. There are two registers in this LCD, for example, instructions and information (Sharma, R. et al., 2013).

The order register stores the guidelines given to the LCD. A guide is given to the LCD that it is used to run a predefined assignment, for example, it's beginning, its screen cleaned, controlling the cursor, controlling the showcase, and so on. The information store stores information shown on the LCD. The information is the ASCII estimate of the character shown on the LCD. Snap to know about the incoming structure of an LCD (As per Fig. 12).

They use the same fundamental progress, at that point, to be told the truth, there are innumerable pixels in alternative images, whereas different grandstands have huge sections.

As shown in below the use of LCD is a 16x2 alphanumeric display with a yellow monochromatic backlight. It is commonly used on 8051, AVR, Arduino, PIC, ARM and so on.

Figure 12. LCD (16x2 alphanumeric display) (Kaul, U. et al., 2015)

There are certain Specifications in regards to these LCD's:-

1. **LCD Type:** 16x2 Alphanumeric Displays
2. **Size:** 20x4 (2 Rows and 16 characters per Row)
3. Built-in industry standard HD44780 equivalent LCD controller
4. **Operating Voltage**: 5v to 9v DC
5. **Backlight Color:** Yellow
6. **Weight:** 41gm
7. **Chipset:** HD44780 parallel interface chipset.
8. **I/O Pins:** 16 pins
9. **Mode of Operation:** 8-bit Mode and 4-bit Mode.
10. **Pixel:** Each character is built by a 5×8 pixel box (Agarwal, S. et al.,2014).

Fingerprint Sensor

A fingerprint is the unique identification of a person who cannot be the same as the two people. Cyber-Physical System uses this technology so that security in terms of

Data, personal information, files and many more can be secure. Also, fingerprint can be used in starting or stopping, authentication of the identity so that a false Id cannot disturb the physical system. The fingerprint is recovered from the crime through which guilt person is found and punished. The impression of the fingerprint can be taken or copied by using moisture, grease, ink, and other substance on surfaces such as glass, plastic sheet, paper, and glass. Fingerprint through unique but keeps on changing due to age factor.

In India, Aadhaar Card is in trending these days due to its uniqueness as it has stored our two unique print which is Fingerprint and Retina which cannot be copied and is used in security checks whether the person is genuine or not. Data recorded in Cyber-Physical Systems so that it can be checked at the time of the function. It has recorded our all fingers including thumbs(As per Fig. 13).

Yes, we can use this unique Biometric identification in making a security system so that there is a minimum number of loopholes present in it. In this project, we are using a biometric fingerprint reader R305 module which can be easily interfaced with a microcontroller. The user can store around 1000 number of fingerprint data in this module and can configure it in 1:1 or 1: N mode for identifying the person (Nigam, A. et al., 2015).

Figure 13. Fingerprint module[Kaul, U. et al.,2015]

Working Principle

Fingerprint processing includes two parts, fingerprint enrollment and fingerprint matching (the matching can be 1:1 or 1:N). When enrolling, the user needs to enter the finger two times. The system will process the two-time finger images, generate a template of the finger based on processing results, and store the template. When matching, the user enters the finger through the optical sensor, and the system will generate a template of the finger and compare it with templates of the finger library.

The system will compare the live finger with a specific template designated in the Module; for 1: N matching, or searching, the system will search the whole finger library for the matching finger. In both circumstances, the system will return the matching result, success, or failure (Zheng, S. et al., 2011).

Security with biometrics can be done with the help of the R305 Fingerprint Module. This fingerprint sensor module will make adding fingerprint detection and verification super simple. These modules are typically used in safes – there's a high powered DSP chip that does the image rendering, calculation, feature-finding, and searching. Connect to any microcontroller or system with TTL serial, and send packets of data to take photos, detect prints, hash, and search. You can also enroll new fingers directly- up to 162 fingerprints can be stored in the onboard FLASH memory. There's a red LED in the lens that lights up during a photo so you know it's working(As per Fig. 14 and Table 3).

Fingerprint processing includes two parts: fingerprint enrolment and fingerprint matching (the matching can be 1:1 or 1: N). When enrolling, a user needs to enter the finger two times. The system will process the two -time finger images, generate

Table 3. Specifications of the fingerprint sensor

Component	Specifications
Model	R307
Type	Optical Fingerprint Module
Interface	USB, UART(RS232 TTL)
Resolution	508 DPI
Voltage	DC 4.2-6V
Fingerprint Capacity	1000
Sensing array	300000 pixel
Fingerprint module size	52*20*22 (mm)
Effective collection area	11 * 15 (mm)
Scanning Speed	< 0.2 second
Verifying Speed	< 0.3 second
Matching Method	1:1; 1:N
FR	£0.1%
FAR	£0.001%
Temperature	-20C ---50C
Humidity	10-85%
Anti-static capacity	15KV
Abrasive resistance intensity	1 million times
Communications baud rate (UART):	(9600 × N) bps where N = 1 ~ 12(default N = 6, ie 57600bps)

a template of the finger based on processing results, and store the template[Zhang L.et al.,2019]. When matching, the user enters the finger through an optical sensor, and the system will generate a template of the finger and compare it with templates of the finger library. For 1:1 matching, the system will compare the live finger with a specific template designated in the Module; for 1: N matching, or searching, the system will search the whole finger library for the matching finger. In both circumstances, the system will return the matching result, success, or failure (Arora, S. et al., 2016).

Figure 14. Fingerprint sensor circuit

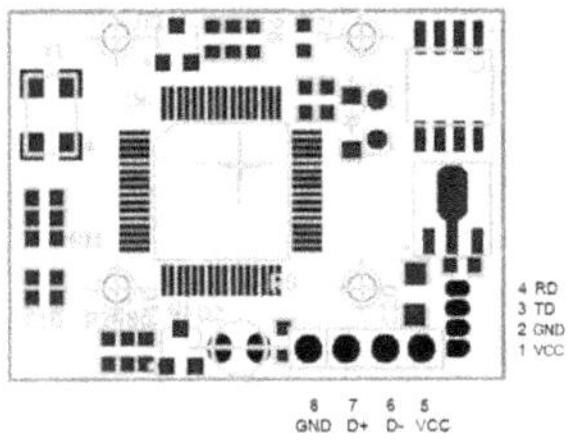

Pin no.	Name	Function
1	VCC	Power input
2	GND	Signal ground
3	TD	Data output TTL logic
4	RD	Data input TTL logic
5	VCC	+5 VDC
6	D-	Data -
7	D+	Data +
8	GND	Ground

PROJECT IMPLEMENTATION

The Project Implementation of IDS can be explored in basic two sections:

- Hardware Design and Circuitry.
- Software Design and Programming

HARDWARE DESIGN AND CIRCUITRY

The IDS Module in our Project can be divided into 4main Circuit Connections:

1. Arduino To Piezoelectric Sensor Array
2. Arduino To RFID
3. Arduino To LCD
4. Arduino To GSM

5. Arduino To Fingerprint Sensor (Gluhak et al., 2011) and (Aggarwal, S. et al., 2015)

Arduino to Piezoelectric Sensor Array

As per our requirement to get the input of intrusion from the piezoelectric sensor, by sensing the pressure-induced from the foot of the intruder, we have connected the Piezo sensors in a calculated fashion as shown in Figure and thus making an array of Sensors.

Figure 15. Parallel connection of piezoelectric sensor (Shekhar, S. et al.,2015)

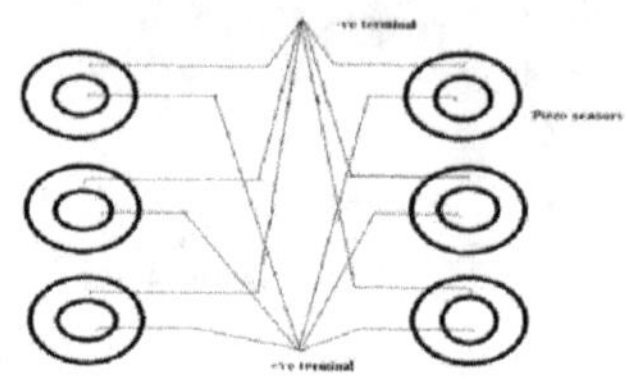

Parallel Connection of Piezoelectric Sensor

As per Figure 15, thus 6-8 piezoelectric sensors are connected in parallel to increase the current.

Voltage of 1 sensor = 2.14v (normal pressure)
Voltage of Array = 2.14v
Current of 1 sensor = 0.5 mA

Figure 16. Arduino To RFID connection (Shekhar, S. et al.,2015)

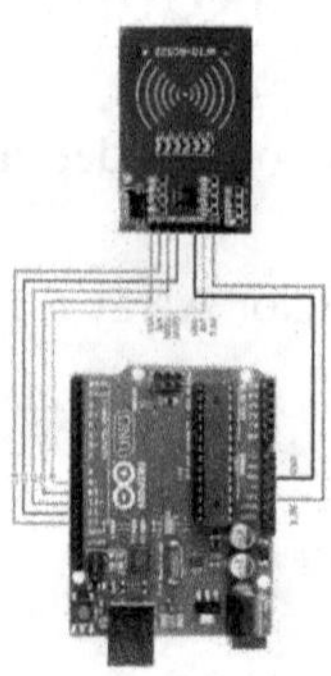

The current of Array = 4mA (All Pressed Properly)

Arduino RFID

As we needed to tackle the problem of false triggering of buzzer due to foot pressure induced by the authorized person, we deployed an RFID for the same purpose as in Fig. 16.

Circuit Connection of Arduino with RFID

The Connection made in above Fig. 15 is as follows tabulated in Table 4;

Pin Connections for RFID

Table 4. RFID pin connections specifications

Components	Specifications
SDA	Digital 10
SCK	Digital 13
MOSI	Digital 11
MISO	Digital 12
IRQ	Unconnected
GND	GND
RST	Digital 9
3.3V	3.3V

ARDUINO to LCD

As in any research experiment or any device, the most important thing which is needed to be there is its User interfacing and its user-friendly nature. Thus LCD is doing the same purpose here as shown in Figure 3.3; it is not only giving directions to the user but also making it easy for the user to use the IDS more efficiently. On the programmers end it makes it easy to debug and check the proper functioning of the Module. Pin Connections are as follows in Table 5;

LCD to ARDUINO

Table 5. LCD to ARDUINO connection

LCD PIN	ARDUINO PIN
RS	PIN 2
ENABLE	PIN 3
D4	PIN 4
D5	PIN 5
D6	PIN 6
D7	PIN 7
R/W	GROUND

LCD TO PERIPHERALS AND POWER SYSTEM

Table 6. LCD to peripherals and power system

Specifications	Components
VSS PIN	GROUND
PIN 16	GROUND
VCC	5V
PIN 15	5V0
10K RESISTOR	+5V
10K RESISTOR	GROUND
PIN 3	WIPER

Figure 17. Arduino To LCD connection (Gupta, R. et al., 2015)

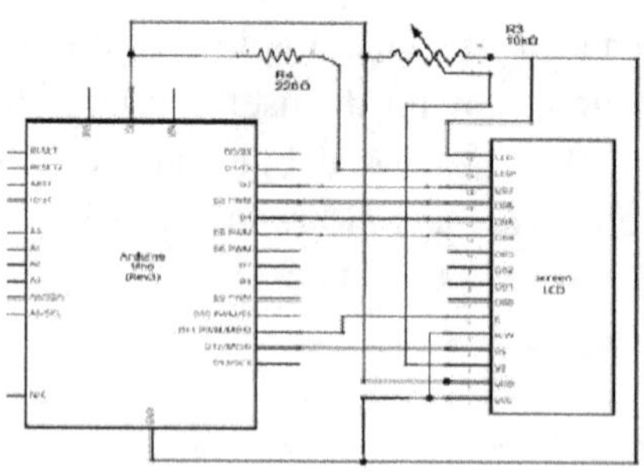

ARDUINO TO GSM

Connections of GSM to Arduino(As per Fig. 18, Table 7)

Table 7. GSM vs ARDUINO module

GSM MODULE	ARDUINO UNO
TX PIN	RX PIN
RX PIN	TX PIN
GROUND PIN	GROUND PIN

Figure 18. Arduino to GSM (Aggarwal, S. et al., 2015)

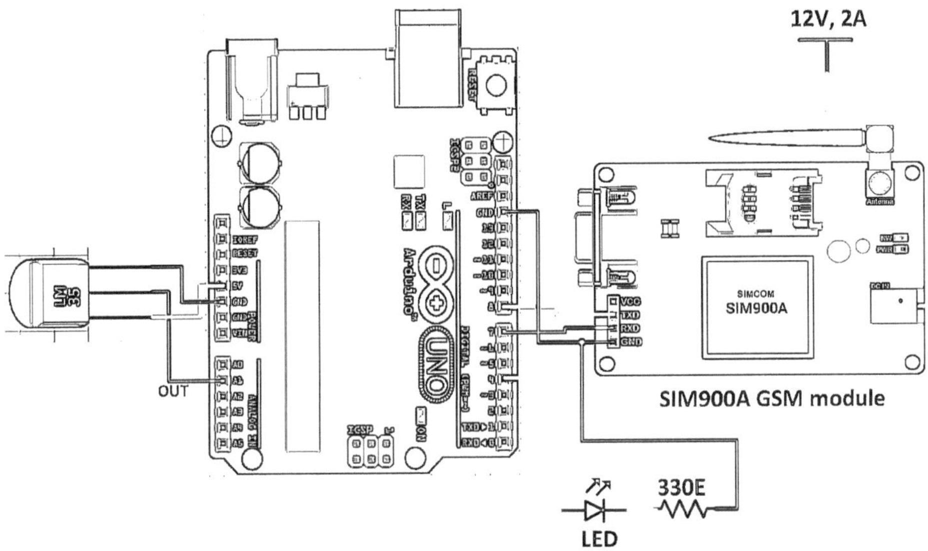

Arduino to Fingerprint Sensor

Table 8. Fingerprint vs ARDUINO module

FINGERPRINT MODULE	ARDUINO
VCC	5V
GND	GND
CS	PIN 10
RST	PIN 9
A0	PIN 8
SDA	PIN 11
SCK	PIN 13
LED	3.3V

SOFTWARE DESIGN AND PROGRAMMING

For accomplishing the need to integrate LCD, RFID, Buzzer &Arduino through Software, we required 4 different libraries.

1. RFID Read Library
2. RFID Function Library
3. Piezo Knock Sensor Library
4. LCD Read/Write Library (Sharma, P. et al., 2014).

Applications of RFID using Arduino to GSM can be understood by several appliances which can be designed by a simple integration of components. Refer the Fig. 19 to understand the application of this module in different day to day perspectives.

PROCESS

- The Piezo-Electric Transducer-sensor detects intrusion and triggers an alarm as soon as a person applies pressure on the piezo – plate.
- Here the pressure energy gets changed into voltage needed to trigger the Arduino and thus making alarm to go on.

- To avoid a false Alarm due to intrusion of a legal person we will be using RF-ID security to stop any of the false triggerings so that our ID system will produce alarm only at intrusions.
- A message will be sent to the owner about the authorization access.

Figure 19. Applications of RFID using Arduino to GSM (Srivastava, S. et al., 2014)

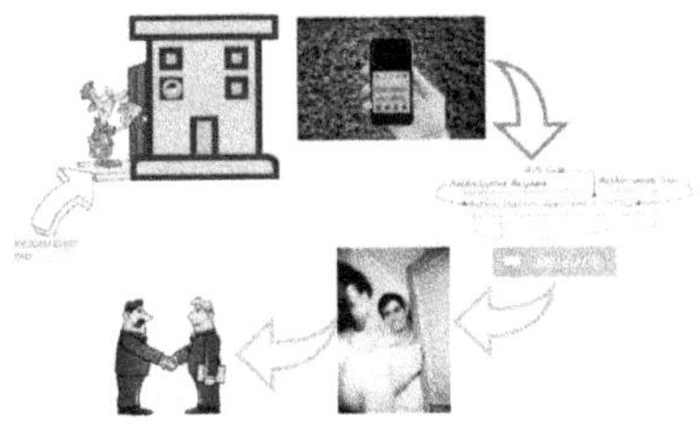

FUTURE RESEARCH DIRECTIONS

From the time immemorial it has been our prime rationale to improve the manners in which that can facilitate human endeavors, and make security a genuine cakewalk. Out of numerous methods for giving continuous security through the sensor, control, and Cyber-Physical Systems used around us, we conveyed the most appropriate Intrusion Detection System and the most ideal method for Cyber-Physical security system in a few viewpoints.

Research work utilizes two of the spearheading advances in their fields and guarantees an enormous prospect of IDS. Consolidating RFID with the current Piezo-based IDS ended up being another involvement in the field of Cyber-Physical Security Systems. The module planned by us was another idea in itself as the extent of this kind of framework can be interminable. The future viewpoints which can be seen rising through this module are excesses of extending from

1. RFID controlled GSM
2. Security for deaf and blind
3. Watt-less Security system or self-powered IDS
4. Automatic car parking system
5. Child safety and parental control systems.
6. Piezo-triggered security camera and other features.
7. Security for small children and women living at home alone.

In the entire procedure of making such a security framework we have found out about RFID, Piezo sensors, LCD, Zero PCB associations, Arduino, and its programming which helped us to comprehend the domain of IDS and its segments. The working of Intrusion Detection System (IDS) which is also an example of Cyber-Physical System is made according to the human interactive product as it would be easy to interact with this Cyber-Physical System and can complete all the necessities required by the home alone people by using hardware and software in such a way that it will become an example for another product in the coming future. Henceforth it tends to be in all respects gladly finished up by us that the undertaking on Cyber-Physical System (CPS) worked precisely according to the decision and in this way denoting the heavenly end of this venture session.

CONCLUSION

The Advantage of this Cyber-Physical System over other systems is that it requires low or almost negligible external supply, précised, Clean System of Intrusion Detection. It can be made more convenient using a GSM module for remote Alarming. It is very easy to install and most important it is cost-efficient also.

Because of business plans, our security system serves at an affordable price with better security features as compared to other expensive Cyber-Physical systems. It has two checks at the doorstep. It targets women and children living in flats and apartments. Our key competitor is GODREJ whose security systems are quite expensive but are the most reliable ones in Cyber-Physical Systems. This system also features access control services and Theft and fire detection alarm. Also being cost-efficient will over a huge market as people always buy that costs less.

ACKNOWLEDGMENT

Our Sincere thanks to all direct and indirect supporters and well-wishers. Esteem gratitude to Management of ABESEC, Ghaziabad India to provide excellent infrastructure to complete this work. The sincere guidance of Director sir, Prof. Shailesh Tiwari, and HoD-ECE, Prof. Sanjay Singh is also recognized. ABESEC provided the infrastructure, staff support, and students' feedback, and fund to complete this. We especially thank for their time contribution to be part of this study. We also convey gratitude to Prof. D.K. Chaturvedi Sir, professor of DayalBagh Educational Institute, Agra for his mentorship and timely valuable suggestions.

REFERENCES

Agarwal, A., Goel, D., Tyagi, A., Aggarwal, A., & Rastogi, R. (2018). A Smarter Approach for Better Lifestyle in Indian Societies. In Progress in Advanced Computing and Intelligent Engineering. Advances in Intelligent Systems and Computing (vol 563, pp. 355-362). Springer. DOI doi:10.1007/978-981-10-6872-0_33

Aggarwal, K., Rastogi, R., & Mondal, P. (2015). An exhaustive review for infix to postfix conversion with applications and benefits. *Proceedings of 9th INDIACom;2nd International Conference on Computing for Sustainable Global Development*, 95-100. https://ieeexplore.ieee.org/document/7100228

Agarwal, S., Sharma, P., Rastogi, R., Kaul, U., & Jain, S. (2014) Unsupervised Classification of Mixed Data Type of Attributes Using Genetic Algorithm (Numeric, Categorical, Ordinal, Binary, Ratio-Scaled). In *Proceedings of the Third International Conference on Soft Computing for Problem Solving. Advances in Intelligent Systems and Computing* (vol. 258, pp. 121-131). Springer. DOI 10.1007/978-81-322-1771-8_11

Aggarwal, S., Rastogi, R., & Mittal, S. (2015). A Novel approch for Communication Among Blind Deaf and Dumb People. *Proceedings of 09th INDIACom; 2015 2nd International Conference on Computing for Sustainable Global Development*. https://ieeexplore.ieee.org/abstract/document/7100321

Arora, S. R., Goel, P., Maini, J., & Mallick, P. (2016). Increasing Efficiency in Online Studies through Web Socket. *Proceedings of 3rd International Conference on Computing for Sustainable Global Development*. https://ieeexplore.ieee.org/document/7724316

Asensio, J. A., Criado, J., Padilla, N., & Iribarne, L. (2019). *Emulating home automation installations through component-based web technology*. Communication Magazine. doi:10.1016/j.future.2017.09.062

Gluhak & Krco. (2011). A survey on facilities for experimental internet of things research. *IEEE Communications Magazine, 49*(11), 58-67.

Goel, D., Agarwal, A., & Rastogi, R. (2016). A Novel Approach for Residential Society Maintenance Problem for Better Human Life. In R. Rajesh & B. Mathivanan (Eds.), *Communication and Power Engineering* (pp. 177–185). doi:10.1515/9783110469608-017

Goel, D., Agarwal, A., & Rastogi, R. (2016). A Novel Approach for Residential Society Maintenance Problem for Better Human Life. *International Journal of Urban Design for Ubiquitous Computing*, *4*, 1–8. doi:10.21742/ijuduc.2016.4.2.01

Gupta, R., Rastogi, R., & Singh, A. (2015). A novel approach for vehicle tracking system for traffic jam problem. *Proceedings of 09th INDIACom; 2015 2nd International Conference on Computing for Sustainable Global Development*. https://ieeexplore.ieee.org/document/7100240

Gupta, R., Rastogi, R., Mondal, P., & Aggarwal, K. (2015). GA Based Clustering of Mixed Data Type of Attributes (Numeric, Categorical, Ordinal, Binary, Ratio-Scaled). *BIJIT*, *7*(2). http://bvicam.ac.in/bjit/issues.asp?issue=14

Hofmanna, P., Walchb, A., Dinkelmanna, A., Selvarayan, S. K., & Gressera, G. T. (2019). Woven piezoelectric sensors as part of the textile reinforcement of fiber-reinforced plastic. *Composites. Part A, Applied Science and Manufacturing*, *116*, 79–86. doi:10.1016/j.compositesa.2018.10.019

Karpinskya, N. D., Chanceyb, E. T., Palmera, D. B., & Yamania, Y. (2018). *Automation trust and attention allocation in multitasking workspace*. Communication Magazine. doi:10.1016/j.apergo.2018.03.008

Kaul, U., Rastogi, R., Agarwal, S., Sharma, P., & Jain, S. (2015). A Novel D&C Approach for Efficient Fuzzy Unsupervised Classification for Mixed Variety of Data. In *Emerging ICT for Bridging the Future - Proceedings of the 49th Annual Convention of the Computer Society of India, CSI Volume 2. Advances in Intelligent Systems and Computing*, (vol. 338, pp. 553-563). Springer. https://link.springer.com/chapter/10.1007/978-3-319-13731-5_60

Li, H., Zhang, X., & Tzou, H. (2017). Diagonal piezoelectric sensors on cylindrical shells. *Journal of Sound and Vibration*, *400*, 201–212. doi:10.1016/j.jsv.2017.03.039

Marie, A., Celebre, D., Benedict, I., Medina, A., Zandrae, A., Dubouzet, D., Neil, A., Surposa, M., & Gustilo, E. R. C. (2015). Home automation using Raspberry Pi through Siri enabled mobile devices. *Proc. 8th IEEE International Conference Humanoid, Nanotechnology, Information Technology Communication and Control, Environment and Management (HNICEM) The Institute of Electrical and Electronics Engineers Inc. (IEEE) – Philippine.*

Nigam, A., Rastogi, R., Mishra, R., Arya, P., & Sharma, S. (2015).S ecurity of Data Transmission Using Logic Gates and Crypt Analysis. *CSI Communication*. http://www.csi-india.org/communications/CSI%20June15%20Combine.pdf

Nigam, A., Rastogi, R., Mishra, R., Arya, P., & Sharma, S. (2015).Security of data transmission using logic gates and crypt analysis. *Proceedings of 9th INDIACom; 2015 2nd International Conference on Computing for Sustainable Global Development*, 101-105. https://ieeexplore.ieee.org/document/7100229

Niua, J., Ganga, H., Zhangb, Y., & Duc, X. (2018). Relationship between automation trust and operator performance for the novice and expert in spacecraft rendezvous and docking (RVD). *Applied Ergonomics, 35*(4), 1–8. doi:10.1016/j.apergo.2018.03.014 PMID:29764609

Qun, C., Mei, L., & Xinyu, L. (2018). Bike Fleet Allocation Models for Repositioning in Bike-Sharing Systems. *IEEE Intelligent Transportation Systems Magazine, 10*(1), 19–29. doi:10.1109/MITS.2017.2776129

Rungta, S., Srivastava, S., Yadav, U. S., & Rastogi, R. (2014) A Comparative Analysis of New Approach with an Existing Algorithm to Detect Cycles in a Directed Graph. *Proceedings of the 48th Annual Convention of Computer Society of India*. Springer. 10.1007/978-3-319-03095-1_5

Saquib, S. T., & Hameed, S. (2013). Wireless Control of Miniaturized Mobile Vehicle for Indoor Surveillance. *IOP Conference Series Material Science and Engineering, 51*(3), 452-465. 10.1088/1757-899X/51/1/012025

Sharma, R., Jain, A., & Rastogi, R. (2013). A New Face To Photo Security Of Facebook. *Proceedings of the Sixth International Conference on Contemporary Computing (IC3-2013).* https://ieeexplore.ieee.org/document/6612231

Sharma, S., Rastogi, R., & Kumar, S. (2015). A revolutionary technology to help the differently abled person. *Proceedings of 9th INDIACom; 2015 2nd International Conference on Computing for Sustainable Global Development*. https://ieeexplore.ieee.org/document/7100324

Sharma, P., Rastogi, R., Aggarwal, S., Kaul, U., & Jain, S. (2014). Business Analysis and Decision Making Through Unsupervised Classification of Mixed Data Type of Attributes Through Genetic Algorithm. *BIJIT*. http://bvicam.ac.in/bjit/issues.asp?issue=11

Shekhar, S., Rastogi, R., & Mittal, S. (2015). Linear algorithm for Imbricate Cryptography using Pseudo Random Number Generator. *Proceedings of 9th INDIACom, pp. 89-94; 2015 2nd International Conference on Computing for Sustainable Global Development*. https://ieeexplore.ieee.org/document/7100227

Srivatava, S., Rastogi, R., Rungta, S., & Yadav, U. (2014).A Methodology to Find the Cycle in a Directed Graph Using Linked List. *BIJIT*. http://bvicam.ac.in/bjit/issues.asp?issue=12

Suresh, S., Yuthika, S., & Adityavardhini, G. (2016). *Home Based Fire Monitoring and Warning System*. Academic Press.

Zhang, L., & Wang, G. (2019). Design and Implementation of Automatic Fire Alarm System Based on Wireless Sensor Networks. *Proceedings of the International Symposium on Information Processing (ISIP'09)*, 410-413.

Zheng, S. R. (2011). The Internet of Things. *IEEE Communications Magazine*, *11*, 30–31. doi:10.1109/MCOM.2011.6069706

ADDITIONAL READING

Applied Cyber-Physical Systems. (2014). In U. J. Tanik, J. N. Carbone, & A. Eroglu, Eds.). Editors. https://www.springer.com/gp/book/9781461473350

KEY TERMS AND DEFINITIONS

Artificial Intelligence: In computer science, artificial intelligence (AI), sometimes called machine intelligence, is intelligence demonstrated by machines, in contrast to the natural intelligence displayed by humans and animals. Leading AI textbooks define the field as the study of "intelligent agents": any device that perceives its environment and takes actions that maximize its chance of successfully achieving its goals.

Automated Security Systems: The smart home promises a futuristic home experience where everything you need is just a simple voice command away, even when you're far from home. There are some very cool gadgets that let you do a lot with home automation, many of which are incorporated into home security systems. These include smart doorbells, smart locks, smart cameras, smart thermostats, smart lights, and smart smoke alarms.

Big Data Analysis: Extremely large data sets that may be analysed computationally to reveal patterns, trends, and associations, especially relating to human behaviour and interactions.

Cyber-Physical System: It is a system in which a mechanism is controlled or monitored by computer-based algorithms. It includes autonomous automobile systems, robotics, automatic pilot avionics, and many more.

GSM Applications: The global system for mobile communications (*GSM*) is a standard developed by the European Telecommunications Standards Institute (ETSI) to describe the protocols for second-generation (2G) digital cellular networks used by mobile devices such as mobile phones and tablets.

Home Automation: Home automation or domestics is building automation for a home, called a smart home or smart house. A home automation system will control lighting, climate, entertainment systems, and appliances. It may also include home security such as access control and alarm systems.

IoT-Automation: The internet of things (IoT) is a system of interrelated computing devices, mechanical and digital machines provided with unique identifiers (UIDs) and the ability to transfer data over a network without requiring human-to-human or human-to-computer interaction.

Mobile Automation: Mobile automation, as the name suggests, refers to 'automation' that is done on mobile devices. Automation is the process whereby one automates testing of an application - in this case a mobile application - which can be a WAP site or an app. This can be done by using tools and helps in reducing the testing time cycle.

Piezoelectric Sensor: A piezoelectric sensor, also known as a piezoelectric transducer, is a device that uses the piezoelectric effect to measure changes in pressure, acceleration, temperature, strain, or force by converting these into an electrical charge. The prefix piezo is Greek for press or squeeze.

Smart Automation: Automation has never been more important than in today's business environment. The wave of digital transformation and evolution has created customer interactions that can happen at any time and in huge and unpredictable volumes. The rapid pace of market change means that agility is becoming an organizational survival skill, yet many organizations are suffering under a heavy load of legacy applications and associated manual processes.

Weighted Automata: Weighted finite automata (WFA) are finite automata whose transitions and states are augmented with some weights, elements of a smearing.

Compilation of References

Aaen, P., Plá, J. A., & Wood, J. (2007). *Modeling and characterization of RF and microwave power FETs*. Cambridge University Press. doi:10.1017/CBO9780511541124

Agarwal, A., Goel, D., Tyagi, A., Aggarwal, A., & Rastogi, R. (2018). A Smarter Approach for Better Lifestyle in Indian Societies. In Progress in Advanced Computing and Intelligent Engineering. Advances in Intelligent Systems and Computing (vol 563, pp. 355-362). Springer. DOI doi:10.1007/978-981-10-6872-0_33

Agarwal, S., Sharma, P., Rastogi, R., Kaul, U., & Jain, S. (2014) Unsupervised Classification of Mixed Data Type of Attributes Using Genetic Algorithm (Numeric, Categorical, Ordinal, Binary, Ratio-Scaled). In *Proceedings of the Third International Conference on Soft Computing for Problem Solving. Advances in Intelligent Systems and Computing* (vol. 258, pp. 121-131). Springer. DOI 10.1007/978-81-322-1771-8_11

Aggarwal, K., Rastogi, R., & Mondal, P. (2015). An exhaustive review for infix to postfix conversion with applications and benefits. *Proceedings of 9th INDIACom;2nd International Conference on Computing for Sustainable Global Development*, 95-100. https://ieeexplore.ieee.org/document/7100228

Aggarwal, S., Rastogi, R., & Mittal, S. (2015). A Novel approch for Communication Among Blind Deaf and Dumb People. *Proceedings of 09th INDIACom; 2015 2nd International Conference on Computing for Sustainable Global Development*. https://ieeexplore.ieee.org/abstract/document/7100321

Ahmad, H. M. N., Arifuzzman, A. K. M., Nowshin, N., & Akram, C. (2011). Effects of Series resistance, Effective Mobility and Output Conductance on Si-NWFET Based on Y-function Technique. *International Journal of Engineering Science and Technology, 3*(1).

Arora, S. R., Goel, P., Maini, J., & Mallick, P. (2016). Increasing Efficiency in Online Studies through Web Socket. *Proceedings of 3rd International Conference on Computing for Sustainable Global Development*. https://ieeexplore.ieee.org/document/7724316

Asensio, J. A., Criado, J., Padilla, N., & Iribarne, L. (2019). *Emulating home automation installations through component-based web technology*. Communication Magazine. doi:10.1016/j.future.2017.09.062

Atashbar, M. Z., Banerji, D., Singamaneni, S., & Bliznyuk, V. (2004). Deposition of parallel arrays of palladium nanowires and electrical characterization using microelectrode contacts. *Nanotechnology, 15*(3), 374–378. doi:10.1088/0957-4484/15/3/025

Atias, L., Teman, A., Giterman, R., Meinerzhagen, P., & Fish, A. (2016). *A Low-Voltage Radiation-Hardened 13T SRAM Bitcell for Ultralow Power Space Applications. IEEE Transactions on Very Large Scale Integration (VLSI) Systems*. doi:10.1109/TVLSI.2016.2518220

ATLAS User's Manual 3-D Device Simulator. (2010). SILVACO International, Version 5.14.0.R.

Atlas, R. M. (2003). Bioterrorism and biodefence research: Changing the focus of microbiology. *Nature Reviews. Microbiology, 1*(1), 70–74. doi:10.1038/nrmicro728

Autran, J. L., Nehari, K., & Munteanu, D. (2005). Compact modeling of the threshold voltage in silicon nanowire MOSFET including 2D-quantum confinement effects. *Molecular Simulation, 31*(12), 839–843. doi:10.1080/08927020500314027

Ávila-Herrera, F., Paz, B. C., Cerdeira, A., Estrada, M., & Pavanello, M. A. (2016). Charge-based compact analytical model for triple-gate junctionless nanowire transistors. *Solid-State Electronics, 122*, 23–31. doi:10.1016/j.sse.2016.04.013

Babinec, T. M., Hausmann, B. J., Khan, M., Zhang, Y., Maze, J. R., Hemmer, P. R., & Lončar, M. (2010). A diamond nanowire single-photon source. *Nature Nanotechnology, 5*(3), 195–199. doi:10.1038/nnano.2010.6

Baidya, A., Baishya, S., & Lenka, T. R. (2017). Impact of thin high-k dielectrics and gate metals on RF characteristics of 3D double gate junctionless transistor. *Materials Science in Semiconductor Processing, 71*(April), 413–420. doi:10.1016/j.mssp.2017.08.031

Baidya, A., Lenka, T. R., & Baishya, S. (2016). Mixed-mode simulation and analysis of 3D double gate junctionless nanowire transistor for CMOS circuit applications. *Superlattices and Microstructures, 100*, 14–23. doi:10.1016/j.spmi.2016.08.028

Bala, J., & Singh, D. (2020, June 4). Elastic and ultrasonic properties of fermium monopnictides. *Engineering and Applied Science Research, 47*(2), 182–189.

Bala, J., Singh, D., Pandey, D., & Yadav, C. (2020, February 26). Mechanical and Thermophysical Properties of ScM (M: Ru, Rh, Pd, Ag) Intermetallics. *International Journal of Thermophysics, 41*(4), 1–13. doi:10.100710765-020-02624-9

Bala, S., & Khosla, M. (2018). Design and simulation of nanoscale double-gate TFET/tunnel CNTFET. *Journal of Semiconductors, 39*(4), 044001. Advance online publication. doi:10.1088/1674-4926/39/4/044001

Bao, H., Cui, X., Li, C. M., Gan, Y., Zhang, J., & Guo, J. (2007). Photoswitchable semiconductor bismuth sulfide (Bi2S3) nanowires and their self-supported nanowire arrays. *The Journal of Physical Chemistry C, 111*(33), 12279–12283. doi:10.1021/jp073504t

Baral, K., Singh, P. K., Kumar, S., Chander, S., & Jit, S. (2020). Ultrathin body nanowire hetero-dielectric stacked asymmetric halo doped junctionless accumulation mode MOSFET for enhanced electrical characteristics and negative bias stability. *Superlattices and Microstructures*, *138*, 106364. doi:10.1016/j.spmi.2019.106364

Baumann, R. C. (2005). Radiation-induced soft errors in advanced semiconductor technologies. *IEEE Transactions on Device and Materials Reliability*, *5*(3), 305–316. Advance online publication. doi:10.1109/TDMR.2005.853449

Bellaouar, A., Elmasry, M. I., Bellaouar, A., & Elmasry, M. I. (1995). Low-Power CMOS Random Access Memory Circuits. In Low-Power Digital VLSI Design (pp. 313–407). doi:10.1007/978-1-4615-2355-0_6

Bergveld, P. (1970). Development of an ion-sensitive solid-state device for neurophysiological measurements. *IEEE Transactions on Biomedical Engineering*, *BME-17*(1), 70–71. doi:10.1109/TBME.1970.4502688 PMID:5441220

Beyer, R. T., & Letcher, S. V. (1969). *Physical Ultrasonics*. Academic Press.

Bhalla, V., Singh, D., Jain, S., & Kumar, R. (2016, March 17). Ultrasonic attenuation in rare-earth monoarsenides. *Pramana -. Journal of Physics*, *86*(6), 1355–1367. doi:10.100712043-015-1183-5

Bhattacharrya, S., Saha, S. K., & Chakravorty, D. (2000). Nanowire formation in a polymeric film. *Applied Physics Letters*, *76*(26), 3896–3898. doi:10.1063/1.126813

Björk, M. T., Ohlsson, B. J., Sass, T., Persson, A. I., Thelander, C., Magnusson, M. H., Deppert, K., Wallenberg, L. R., & Samuelson, L. (2002). One-dimensional steeplechase for electrons realized. *Nano Letters*, *2*(2), 87–89. doi:10.1021/nl010099n

Bland, H. A., Thomas, E. L., Klemencic, G. M., Mandal, S., Morgan, D. J., Papageorgiou, A., ... Williams, O. A. (2019). superconducting Diamond on silicon Nitride for Device Applications. *Scientific Reports*, *9*(1), 1–9. doi:10.103841598-019-39707-z PMID:30626917

Bohr, M. T. (2002). Nanotechnology goals and challenges for electronic applications. *IEEE Transactions on Nanotechnology*, *1*(1), 56–62. doi:10.1109/TNANO.2002.1005426

Bohr, M. T., & Young, I. A. (2017). CMOS Scaling Trends and beyond. *IEEE Micro*, *37*(6), 20–29. Advance online publication. doi:10.1109/MM.2017.4241347

Bömmel, H. E. (1954, October 1). Ultrasonic attenuation in superconducting lead. *Physical Review*, *96*(1), 220–221. doi:10.1103/PhysRev.96.220

Bondyopadhyay, P. K. (1998). Moore's law governs the silicon revolution. *Proceedings of the IEEE*, *86*(1), 78–81. doi:10.1109/5.658761

Borgström, M. T., Wallentin, J., Heurlin, M., Fält, S., Wickert, P., Leene, J., Magnusson, M. H., Deppert, K., & Samuelson, L. (2010). Nanowires with promise for photovoltaics. *IEEE Journal of Selected Topics in Quantum Electronics*, *17*(4), 1050–1061. doi:10.1109/JSTQE.2010.2073681

Borgström, M. T., Wallentin, J., Trägårdh, J., Ramvall, P., Ek, M., Wallenberg, L. R., Samuelson, L., & Deppert, K. (2010). In situ etching for total control over axial and radial nanowire growth. *Nano Research*, *3*(4), 264–270. doi:10.100712274-010-1029-x

Borgström, M. T., Zwiller, V., Müller, E., & Imamoglu, A. (2005). Optically bright quantum dots in single nanowires. *Nano Letters*, *5*(7), 1439–1443. doi:10.1021/nl050802y

Born, M., & Huang, K. (1954). *Dynamical theory of crystal lattices. U.K.* Clarendon press.

Boukai, A. I., Bunimovich, Y., Tahir-Kheli, J., Yu, J.-K., Goddard, W. A. III, & Heath, J. R. (2008). Silicon nanowires as efficient thermoelectric materials. *Nature*, *451*(7175), 168–171. doi:10.1038/nature06458 PMID:18185583

Brabec, C. J., & Durrant, J. R. (2008). Solution-processed organic solar cells. *MRS Bulletin*, *33*(7), 670–675. doi:10.1557/mrs2008.138

Bradley, K., Gabriel, J. C. P., Star, A., & Grüner, G. (2003). Short-channel effects in contact-passivated nanotube chemical sensors. *Applied Physics Letters*, *83*(18), 3821–3823. doi:10.1063/1.1619222

Bricout, P. H., & Dubois, E. (1996). Short-channel effect immunity and current capability of sub-0.1-micron MOSFET's using a recessed channel. *IEEE Transactions on Electron Devices*, *43*(8), 1251–1255. doi:10.1109/16.506776

Briseno, A. L., Mannsfeld, S. C., Lu, X., Xiong, Y., Jenekhe, S. A., Bao, Z., & Xia, Y. (2007). Fabrication of field-effect transistors from hexathiapentacene single-crystal nanowires. *Nano Letters*, *7*(3), 668–675. doi:10.1021/nl0627036 PMID:17288487

Briseno, A. L., Mannsfeld, S. C., Reese, C., Hancock, J. M., Xiong, Y., Jenekhe, S. A., Bao, Z., & Xia, Y. (2007). Perylenediimide nanowires and their use in fabricating field-effect transistors and complementary inverters. *Nano Letters*, *7*(9), 2847–2853. doi:10.1021/nl071495u PMID:17696562

Brubaker, M. D., Blanchard, P. T., Schlager, J. B., Sanders, A. W., Roshko, A., Duff, S. M., Gray, J. M., Bright, V. M., Sanford, N. A., & Bertness, K. A. (2013). On-chip optical interconnects made with gallium nitride nanowires. *Nano Letters*, *13*(2), 374–377. doi:10.1021/nl303510h PMID:23324057

Bryllert, T., Wernersson, L. E., Froberg, L. E., & Samuelson, L. (2006a, May). Vertical high-mobility wrap-gated InAs nanowire transistor. *IEEE Electron Device Letters*, *27*(5), 323–325. doi:10.1109/LED.2006.873371

Bryllert, T., Wernersson, L.-E., Löwgren, T., & Samuelson, L. (2006b). Vertical wrap-gated nanowire transistors. *Nanotechnology*, *17*(11), S227–S230. doi:10.1088/0957-4484/17/11/S01

Bulgarini, G., Reimer, M. E., Hocevar, M., Bakkers, E. P., Kouwenhoven, L. P., & Zwiller, V. (2012). Avalanche amplification of a single exciton in a semiconductor nanowire. *Nature Photonics*, *6*(7), 455–458. doi:10.1038/nphoton.2012.110

C., L., Y, Z., S, D., & Han, R. (2013). Subthreshold behavior models for nanoscale short-channel junctionless cylindrical surrounding-gate MOSFETs. *IEEE Transactions on Electron Devices, 60*(11), 3655–3662.

Cairns, D. R., Witte, R. P. II, Sparacin, D. K., Sachsman, S. M., Paine, D. C., Crawford, G. P., & Newton, R. R. (2000). Strain-dependent electrical resistance of tin-doped indium oxide on polymer substrates. *Applied Physics Letters, 76*(11), 1425–1427. doi:10.1063/1.126052

Calin, T., Nicolaidis, M., & Velazco, R. (1996). Upset hardened memory design for submicron CMOS technology. *IEEE Transactions on Nuclear Science, 43*(6), 2874–2878. doi:10.1109/23.556880

Canham, L. T. (1990). Silicon quantum wire array fabrication by electrochemical and chemical dissolution of wafers. *Applied Physics Letters, 57*(10), 1046–1048. doi:10.1063/1.103561

Cao, G. (2004). *Nanostructures and nanomaterials: synthesis, properties and applications*. World scientific. doi:10.1142/p305

Cellini, F., Gao, Y., & Riedo, E. (2019, March 11). Å-Indentation for non-destructive elastic moduli measurements of supported ultra-hard ultra-thin films and nanostructures. *Scientific Reports, 9*(1), 1–16. doi:10.103841598-019-40636-0 PMID:30858472

Ch, N. R., Manohar Reddy, D., Kishore Kumar, P., & Kaushal, G. (2019). Robust SRAM Cell Development for Single-Event Multiple Effects. In Communications in Computer and Information Science (Vol. 892). doi:10.1007/978-981-13-5950-7_29

Chan, C. K., Peng, H., Liu, G., McIlwrath, K., Zhang, X. F., Huggins, R. A., & Cui, Y. (2008). High-performance lithium battery anodes using silicon nanowires. *Nature Nanotechnology, 3*(1), 31–35. doi:10.1038/nnano.2007.411 PMID:18654447

Chang, M.-H., Chiu, Y.-T., Lai, S.-L., & Hwang, W. (2011). A 1kb 9T subthreshold SRAM with bit-interleaving scheme in 65nm CMOS. *IEEE/ACM International Symposium on Low Power Electronics and Design*, 291–296. 10.1109/ISLPED.2011.5993652

Chatterjee, I., Narasimham, B., Mahatme, N. N., Bhuva, B. L., Reed, R. A., & Schrimpf, R. D. … Monzel, C. (2014). Impact of technology scaling on SRAM soft error rates. *IEEE Transactions on Nuclear Science*. Advance online publication. doi:10.1109/TNS.2014.2365546

Chaturvedi, P., & Kumar, M. J. (2014). Impact of gate leakage considerations in tunnel field effect transistor design. *Japanese Journal of Applied Physics, 53*(7), 074201. doi:10.7567/JJAP.53.074201

Chaudhry, A., & Kumar, M. J. (2004). Controlling Short-Channel Effects in Deep-Submicron SOI MOSFETs for Improved Reliability: A Review. *IEEE Transactions on Device and Materials Reliability, 4*(1), 99–109. doi:10.1109/TDMR.2004.824359

Chen, X.-Q., Niu, H., Li, D., & Li, Y. (2011, September). Modeling hardness of polycrystalline materials and bulk metallic glasses. *Intermetallics, 19*(9), 1275–1281. doi:10.1016/j.intermet.2011.03.026

Che, Y., Datar, A., Balakrishnan, K., & Zang, L. (2007). Ultralong nanobelts self-assembled from an asymmetric perylene tetracarboxylic diimide. *Journal of the American Chemical Society, 129*(23), 7234–7235. doi:10.1021/ja071903w PMID:17506565

Chiang, T. K. (2012). A new quasi-2-D threshold voltage model for short-channel junctionless cylindrical surrounding gate (JLCSG) MOSFETs. *IEEE Transactions on Electron Devices, 59*(11), 3127–3129. doi:10.1109/TED.2012.2212904

Cho, H., Kapur, P., & Saraswat, K. C. (2004). Power comparison between high-speed electrical and optical interconnects for interchip communication. *Journal of Lightwave Technology, 22*(9), 2021–2033. doi:10.1109/JLT.2004.833531

Choi, Y. K., Asano, K., Lindert, N., Subramanian, V., King, T. J., Bokor, J., & Hu, C. (1999, December). Ultra-thin body SOI MOSFET for deep-sub-tenth micron era. In *International Electron Devices Meeting 1999. Technical Digest* (Cat. No. 99CH36318) (pp. 919-921). IEEE.

Choi, Y. K., Ha, D., King, T. J., & Hu, C. (2001, June). Ultra-thin body PMOSFETs with selectively deposited Ge source/drain. In *2001 Symposium on VLSI Technology. Digest of Technical Papers* (IEEE Cat. No. 01 CH37184) (pp. 19-20). IEEE.

Chueh, Y. L., Ko, M. T., Chou, L. J., Chen, L. J., Wu, C. S., & Chen, C. D. (2006). TaSi2 nanowires: A potential field emitter and interconnect. *Nano Letters, 6*(8), 1637–1644. doi:10.1021/nl060614n

Chung, D., & Buessem, W. (1967). The elastic anisotropy of crystals. *Journal of Applied Physics, 38*(5), 2010–2012. doi:10.1063/1.1709819

Clarke, D. R. (2003, January 30). Materials selection guidelines for low thermal conductivity thermal barrier coatings. *Surface and Coatings Technology, 163-164*, 67–74. doi:10.1016/S0257-8972(02)00593-5

Coffinier, Y., Szunerits, S., Drobecq, H., Melnyk, O., & Boukherroub, R. (2012). Diamond nanowires for highly sensitive matrix-free mass spectrometry analysis of small molecules. *Nanoscale, 4*(1), 231–238. doi:10.1039/C1NR11274K

Cohen, G. M., Rooks, M. J., Chu, J. O., Laux, S. E., Solomon, P. M., Ott, J. A., Miller, R. J., & Haensch, W. (2007). Nanowire metal-oxide-semiconductor field effect transistor with doped epitaxial contacts for source and drain. *Applied Physics Letters, 90*(23), 233110. doi:10.1063/1.2746946

Cohen-Karni, T., Casanova, D., Cahoon, J. F., Qing, Q., Bell, D. C., & Lieber, C. M. (2012). Synthetically encoded ultrashort-channel nanowire transistors for fast, pointlike cellular signal detection. *Nano Letters, 12*(5), 2639–2644. doi:10.1021/nl3011337

Colinge, J. P., Lee, C. W., Afzalian, A., Akhavan, N. D., Yan, R., Ferain, I., ... Kelleher, A. M. (2010). Nanowire transistors without junctions. *Nature Nanotechnology, 5*(3), 225-229.

Colinge, J. P. (2008). The new generation of soi mosfets. *Romanian Journal of Information Science and Technology, 11*(1), 3–15.

Colinge, J. P. (2008). The new generation of SOI MOSFETs. *Romanian Journal of Information Science and Technology, 11*(1), 3–15.

Colinge, J. P., Kranti, A., Yan, R., Lee, C. W., Ferain, I., Yu, R., Akhavan, N. D., & Razavi, P. (2011). Junctionless Nanowire Transistor (JNT): Properties and design guidelines. *Solid-State Electronics, 65–66*(1), 33–37. doi:10.1016/j.sse.2011.06.004

Colinge, J. P., Lee, C.-W., Afzalian, A., Akhavan, N. D., Yan, R., Ferain, I., Razavi, P., O'Neill, B., Blake, A., White, M., Kelleher, A.-M., McCarthy, B., & Murphy, R. (2010a). Nanowire transistors without junctions. *Nature Nanotechnology, 5*(3), 225–229. doi:10.1038/nnano.2010.15 PMID:20173755

Colinge, J. P., Lee, C.-W., Ferain, I., Akhavan, N. D., Yan, R., Razavi, P., Yu, R., Nazarov, A. N., & Doria, R. T. (2010). Reduced electric field in junctionless transistors. *Applied Physics Letters, 96*(7), 23–26. doi:10.1063/1.3299014

Crupi, F., Kaczer, B., Degraeve, R., Subramanian, V., Srinivasan, P., Simoen, E., Dixit, A., Jurczak, M., & Groeseneken, G. (2006). Reliability comparison of triple-gate versus planar SOI FETs. *IEEE Transactions on Electron Devices, 53*(9), 2351–2357. doi:10.1109/TED.2006.880824

Cui, Y., Zhong, Z. H., Wang, D. L., Wang, W. U., & Lieber, C. M. (2003, January). High performance silicon nanowire field effect transistors. *Nano Letters, 3*(2), 149–152. doi:10.1021/nl0258751

Dang, L. D. T., Kim, J. S., & Chang, I. J. (2017). We-Quatro: Radiation-Hardened SRAM Cell With Parametric Process Variation Tolerance. *IEEE Transactions on Nuclear Science*. Advance online publication. doi:10.1109/TNS.2017.2728180

Daniels, J. S., & Pourmand, N. (2007). Label-free impedance biosensors: Opportunities and challenges. Electroanalysis. *An International Journal Devoted to Fundamental and Practical Aspects of Electroanalysis, 19*(12), 1239–1257. PMID:18176631

Das, S. R. (2019). *Advances in InSb and InAs Nanowire Based Nanoelectronic Field Effect Transistors*. Academic Press.

Dasgupta, N. P., Sun, J., Liu, C., Brittman, S., Andrews, S. C., Lim, J., Gao, H., Yan, R., & Yang, P. (2014). 25th anniversary article: Semiconductor nanowires–synthesis, characterization, and applications. *Advanced Materials, 26*(14), 2137–2184. doi:10.1002/adma.201305929 PMID:24604701

Das, K., Chakraborty, A. K., NandaGoswami, M. L., Shingha, R. K., Dhar, A., Coleman, K. S., & Ray, S. K. (2007). Temperature dependent shape transformation of Ge nanostructures by the vapor-liquid-solid method. *Journal of Applied Physics, 101*(7), 074307. doi:10.1063/1.2718282

Davies, P. R., & Roberts, M. W. (2007). *Atom resolved surface reactions: nanocatalysis*. Royal Society of Chemistry.

De Luca, G., Liscio, A., Maccagnani, P., Nolde, F., Palermo, V., Muellen, K., & Samori, P. (2007). Nucleation-Governed Reversible Self-Assembly of an Organic Semiconductor at Surfaces: Long-Range Mass Transport Forming Giant Functional Fibers. *Advanced Functional Materials*, *17*(18), 3791–3798. doi:10.1002/adfm.200700549

Dehdashti Akhavan, N., Ferain, I., Razavi, P., Yu, R., & Colinge, J. P. (2011). Improvement of carrier ballisticity in junctionless nanowire transistors. *Applied Physics Letters*, *98*(10), 2011–2014. doi:10.1063/1.3559625

De, S., Higgins, T. M., Lyons, P. E., Doherty, E. M., Nirmalraj, P. N., Blau, W. J., Boland, J. J., & Coleman, J. N. (2009). Silver nanowire networks as flexible, transparent, conducting films: Extremely high DC to optical conductivity ratios. *ACS Nano*, *3*(7), 1767–1774. doi:10.1021/nn900348c

Dhawan, P. K., Wan, M., Verma, S., Pandey, D., & Yadav, R. (2015, February 20). Effect of diameter and surface roughness on ultrasonic properties of GaAs nanowires. *Journal of Applied Physics*, *117*(7), 074307. doi:10.1063/1.4913289

Dick, K. A., Kodambaka, S., Reuter, M. C., Deppert, K., Samuelson, L., Seifert, W., Wallenberg, L. R., & Ross, F. M. (2007). The morphology of axial and branched nanowire heterostructures. *Nano Letters*, *7*(6), 1817–1822. doi:10.1021/nl0705900 PMID:17489640

Ding, X., Briggs, G., Zhou, W., Chen, Q., & Peng, L. M. (2006). In situ growth and characterization of Ag and Cu nanowires. *Nanotechnology*, *17*(11), S376–S380. doi:10.1088/0957-4484/17/11/S24

Division, C. T., Company, M., & Mc, T. S. M. (2014). *Mu ultigate transistors*. Academic Press.

Dodd, P. E., & Massengill, L. W. (2003). Basic mechanisms and modeling of single-event upset in digital microelectronics. *IEEE Transactions on Nuclear Science*, *50*(3), 583–602. doi:10.1109/TNS.2003.813129

Dodd, P. E., Sexton, F. W., Hash, G. L., Shaneyfelt, M. R., Draper, B. L., Farino, A. J., & Flores, R. S. (1996). Impact of technology trends on SEJJ in CMOS SRAMs. *IEEE Transactions on Nuclear Science*, *43*(6 PART 1), 2797–2804. doi:10.1109/23.556869

Dorvel, B. R., Reddy Jr, B., Go, J., Duarte Guevara, C., Salm, E., Alam, M. A., & Bashir, R. (2012). Silicon nanowires with high-k hafnium oxide dielectrics for sensitive detection of small nucleic acid oligomers. *ACS Nano*, *6*(7), 6150–6164. doi:10.1021/nn301495k

Dowdy, R., Walko, D. A., Fortuna, S. A., & Li, X. (2012). Realization of unidirectional planar GaAs. *IEEE Electron Device Letters*, *33*(4), 522–524. doi:10.1109/LED.2012.2186115

Duan, X., Gao, R., Xie, P., Cohen-Karni, T., Qing, Q., Choe, H. S., Tian, B., Jiang, X., & Lieber, C. M. (2012). Intracellular recordings of action potentials by an extracellular nanoscale field-effect transistor. *Nature Nanotechnology*, *7*(3), 174–179. doi:10.1038/nnano.2011.223

Duan, X., Huang, Y., Cui, Y., Wang, J., & Lieber, C. M. (2001). Indium phosphide nanowires as building blocks for nanoscale electronic and optoelectronic devices. *Nature*, *409*(6816), 66–69. doi:10.1038/35051047

Duan, X., Wang, J., & Lieber, C. M. (2000). Synthesis and optical properties of gallium arsenide nanowires. *Applied Physics Letters*, *76*(9), 1116–1118. doi:10.1063/1.125956

Dura, J., Martinie, S., Munteanu, D., Triozon, F., Barraud, S., Niquet, Y. M., & Autran, J. L. (2011, September). Analytical model of drain current in nanowire MOSFETs including quantum confinement, band structure effects and quasi-ballistic transport: device to circuit performances analysis. In *2011 International Conference on Simulation of Semiconductor Processes and Devices* (pp. 43-46). IEEE. 10.1109/SISPAD.2011.6035045

Elfström, N., Juhasz, R., Sychugov, I., Engfeldt, T., Karlström, A. E., & Linnros, J. (2007). Surface charge sensitivity of silicon nanowires: Size dependence. *Nano Letters*, *7*(9), 2608–2612. doi:10.1021/nl0709017 PMID:17691849

Fagas, G., & Greer, J. C. (2009). Ballistic conductance in oxidized Si. *Nanoletters, 9*(5), 1856-1860. Available at: http://apps.isiknowledge.com

Fahad, H. M., & Hussain, M. M. (2012). Are nanotube architectures more advantageous than nanowire architectures for field effect transistors? *Scientific Reports*, *2*(1), 1–7. doi:10.1038rep00475 PMID:22741059

Fahad, H. M., Smith, C. E., Rojas, J. P., & Hussain, M. M. (2011). Silicon nanotube field effect transistor with core–shell gate stacks for enhanced high-performance operation and area scaling benefits. *Nano Letters*, *11*(10), 4393–4399. doi:10.1021/nl202563s PMID:21923117

Ferlet-Cavrois, V., Massengill, L. W., & Gouker, P. (2013). Single event transients in digital CMOS - A review. *IEEE Transactions on Nuclear Science*, *60*(3), 1767–1790. Advance online publication. doi:10.1109/TNS.2013.2255624

Gan, F., Hou, L., Wang, G., Liu, H., & Li, J. (2000). Optical and recording properties of short wavelength optical storage materials. *Materials Science and Engineering B*, *76*(1), 63–68. doi:10.1016/S0921-5107(00)00400-1

Gao, A., Lu, N., Dai, P., Li, T., Pei, H., Gao, X., Gong, Y., Wang, Y., & Fan, C. (2011). Silicon-nanowire-based CMOS-compatible field-effect transistor nanosensors for ultrasensitive electrical detection of nucleic acids. *Nano Letters*, *11*(9), 3974–3978. doi:10.1021/nl202303y PMID:21848308

Gao, T., Li, Q. H., & Wang, T. H. (2005). CdS nanobelts as photoconductors. *Applied Physics Letters*, *86*(17), 173105. doi:10.1063/1.1915514

Gates, B., Wu, Y., Yin, Y., Yang, P., & Xia, Y. (2001). Single-crystalline nanowires of Ag2Se can be synthesized by templating against nanowires of trigonal Se. *Journal of the American Chemical Society*, *123*(46), 11500–11501. doi:10.1021/ja0166895 PMID:11707137

Gautam, Saxena, Gupta, & Gupta. (2013). Analytical Model of Double Gate MOSFET for High Sensitivity Low Power Photosensor. *Journal of Semiconductor Technology and Science*, *13*(5).

Gautam, R., Saxena, M., Gupta, R. S., & Gupta, M. (2013). Gate all around MOSFET with vacuum gate dielectric for improved hot carrier reliability and RF performance. *IEEE Transactions on Electron Devices*, *60*(6), 1820–1827. doi:10.1109/TED.2013.2256912

Ge, J. P., Wang, J., Zhang, H. X., Wang, X., Peng, Q., & Li, Y. D. (2005). Orthogonal PbS nanowire arrays and networks and their Raman scattering behavior. *Chemistry (Weinheim an der Bergstrasse, Germany)*, *11*(6), 1889–1894. doi:10.1002/chem.200400633

Ghate, P. (1965, August 30). Third-order elastic constants of alkali halide crystals. *Physical Review*, *139*(5A), A1666–A1674. doi:10.1103/PhysRev.139.A1666

Ghosh, P., Haldar, S., Gupta, R. S., & Gupta, M. (2012). An analytical drain current model for dual material engineered cylindrical/surrounded gate MOSFET. *Microelectronics Journal*, *43*(1), 17–24. doi:10.1016/j.mejo.2011.10.001

Gluhak & Krco. (2011). A survey on facilities for experimental internet of things research. *IEEE Communications Magazine, 49*(11), 58-67.

Gnani, E., Gnudi, A., Reggiani, S., & Baccarani, G. (2008). Quasi-ballistic transport in nanowire field-effect transistors. *IEEE Transactions on Electron Devices*, *55*(11), 2918–2930. doi:10.1109/TED.2008.2005178

Goel, D., Agarwal, A., & Rastogi, R. (2016). A Novel Approach for Residential Society Maintenance Problem for Better Human Life. In R. Rajesh & B. Mathivanan (Eds.), *Communication and Power Engineering* (pp. 177–185). doi:10.1515/9783110469608-017

Goel, D., Agarwal, A., & Rastogi, R. (2016). A Novel Approach for Residential Society Maintenance Problem for Better Human Life. *International Journal of Urban Design for Ubiquitous Computing*, *4*, 1–8. doi:10.21742/ijuduc.2016.4.2.01

Goel, E., Kumar, S., Singh, B., Singh, K., & Jit, S. (2017). Two-dimensional model for subthreshold current and subthreshold swing of graded-channel dual-material double-gate (GCDMDG) MOSFETs. *Superlattices and Microstructures*, *106*, 147–155. doi:10.1016/j.spmi.2017.03.047

Gol'tsman, G. N., Okunev, O., Chulkova, G., Lipatov, A., Semenov, A., Smirnov, K., Voronov, B., Dzardanov, A., Williams, C., & Sobolewski, R. (2001). Picosecond superconducting single-photon optical detector. *Applied Physics Letters*, *79*(6), 705–707. doi:10.1063/1.1388868

Gong, X., Tong, M. H., Xia, Y. J., Cai, W. Z., Moon, J. S., Cao, Y., Yu, G., Shieh, C. L., Nilsson, B., & Heeger, A. J. (2009). High-Detectivity Polymer Photodetectors with Spectral Response from 300 nm to 1450 nm. *Science*, *325*(5948), 1665–1667. doi:10.1126cience.1176706 PMID:19679770

Gracia-Moran, J., Saiz-Adalid, L. J., Gil-Tomas, D., & Gil-Vicente, P. J. (2018). *Improving error correction codes for multiple-cell upsets in space applications. IEEE Transactions on Very Large Scale Integration (VLSI) Systems*. doi:10.1109/TVLSI.2018.2837220

Granato, A., & Lücke, K. (1956). Theory of mechanical damping due to dislocations. *Journal of Applied Physics*, *27*(6), 583–593. doi:10.1063/1.1722436

Gray, D. E. (1972). *American Institute of Physics handbook*. McGraw-Hill.

Greaves, G. N., Greer, A., Lakes, R. S., & Rouxel, T. (2011, October 24). Poisson's ratio and modern materials. *Nature Materials*, *10*(11), 823–837. doi:10.1038/nmat3134 PMID:22020006

Guang-Xi, H., Ling-Li, W., Ran, L., Ting-Ao, T., & Zhi-Jun, Q. (2010). Quantum-mechanical study on surrounding-gate metal-oxide-semiconductor field-effect transistors. *Communications in Theoretical Physics*, *54*(4), 763–767. doi:10.1088/0253-6102/54/4/33

Gudiksen, M. S., Lauhon, L. J., Wang, J., Smith, D. C., & Lieber, C. M. (2002). Growth of nanowire superlattice structures for nanoscale photonics and electronics. *Nature*, *415*(6872), 617–620. doi:10.1038/415617a PMID:11832939

Gu, J. J., Wang, X. W., Wu, H., Shao, J., Neal, A. T., Manfra, M. J., ... Ye, P. D. (2012, December). 20–80nm Channel length InGaAs gate-all-around nanowire MOSFETs with EOT= 1.2 nm and lowest SS= 63mV/dec. In *2012 International Electron Devices Meeting* (pp. 27-6). IEEE.

Guo, J., Zhu, L., Liu, W., Huang, H., Liu, S., Wang, T., … Mao, Z. (2017). Novel Radiation-Hardened-by-Design (RHBD) 12T Memory Cell for Aerospace Applications in Nanoscale CMOS Technology. *IEEE Transactions on Very Large Scale Integration (VLSI) Systems*. doi:10.1109/TVLSI.2016.2645282

Guo, J., Zhu, L., Sun, Y., Cao, H., Huang, H., Wang, T., … Mao, Z. (2018). Design of Area-Efficient and Highly Reliable RHBD 10T Memory Cell for Aerospace Applications. *IEEE Transactions on Very Large Scale Integration (VLSI) Systems*. doi:10.1109/TVLSI.2017.2788439

Guo, J., Xiao, L., & Mao, Z. (2014). Novel Low-Power and Highly Reliable Radiation Hardened Memory Cell for 65 nm CMOS Technology. *IEEE Transactions on Circuits and Systems. I, Regular Papers*, *61*(7), 1994–2001. doi:10.1109/TCSI.2014.2304658

Gupta, J. B. (2018). *Electronic Devices and Circuits*. Pearson Education.

Gupta, R., Rastogi, R., & Singh, A. (2015). A novel approach for vehicle tracking system for traffic jam problem. *Proceedings of 09th INDIACom; 2015 2nd International Conference on Computing for Sustainable Global Development*. https://ieeexplore.ieee.org/document/7100240

Gupta, R., Rastogi, R., Mondal, P., & Aggarwal, K. (2015). GA Based Clustering of Mixed Data Type of Attributes (Numeric, Categorical, Ordinal, Binary, Ratio-Scaled). *BIJIT*, *7*(2). http://bvicam.ac.in/bjit/issues.asp?issue=14

Gutsche, C., Lysov, A., Braam, D., Regolin, I., Keller, G., Li, Z. A., Geller, M., Spasova, M., Prost, W., & Tegude, F. J. (2012). n-GaAs/InGaP/p-GaAs Core-Multishell Nanowire Diodes for Efficient Light-to-Current Conversion. *Advanced Functional Materials*, *22*(5), 929–936. doi:10.1002/adfm.201101759

Hadfield, R. H., Miller, A. J., Nam, S. W., Kautz, R. L., & Schwall, R. E. (2005). Low-frequency phase locking in high-inductance superconducting nanowires. *Applied Physics (Berlin)*.

Hahm, J. I., & Lieber, C. M. (2004). Direct ultrasensitive electrical detection of DNA and DNA sequence variations using nanowire nanosensors. *Nano Letters*, *4*(1), 51–54. doi:10.1021/nl034853b

Hashim, A. A. (Ed.). (2011). *Nanowires: Implementations and Applications*. BoD–Books on Demand. doi:10.5772/1025

He, J., & Lilley, C. M. (2008, May 30). Surface effect on the elastic behavior of static bending nanowires. *Nano Letters*, *8*(7), 1798–1802. doi:10.1021/nl0733233 PMID:18510370

He, L., Musick, M. D., Nicewarner, S. R., Salinas, F. G., Benkovic, S. J., Natan, M. J., & Keating, C. D. (2000). Colloidal Au-enhanced surface plasmon resonance for ultrasensitive detection of DNA hybridization. *Journal of the American Chemical Society*, *122*(38), 9071–9077. doi:10.1021/ja001215b

Hill, R. (1963, September). Elastic properties of reinforced solids: Some theoretical principles. *Journal of the Mechanics and Physics of Solids*, *11*(5), 357–372. doi:10.1016/0022-5096(63)90036-X

Hochbaum, A. I., Chen, R., Delgado, R. D., Liang, W., Garnett, E. C., Najarian, M., Majumdar, A., & Yang, P. (2008). Enhanced thermoelectric performance of rough silicon nanowires. *Nature*, *451*(7175), 163–167. doi:10.1038/nature06381 PMID:18185582

Hoeben, F. J., Jonkheijm, P., Meijer, E. W., & Schenning, A. P. (2005). About supramolecular assemblies of π-conjugated systems. *Chemical Reviews*, *105*(4), 1491–1546. doi:10.1021/cr030070z PMID:15826018

Hofmanna, P., Walchb, A., Dinkelmanna, A., Selvarayan, S. K., & Gressera, G. T. (2019). Woven piezoelectric sensors as part of the textile reinforcement of fiber-reinforced plastic. *Composites. Part A, Applied Science and Manufacturing*, *116*, 79–86. doi:10.1016/j.compositesa.2018.10.019

Holtij, T., Kloes, A., & Iñíguez, B. (2015). 3-D compact model for nanoscale junctionless triple-gate nanowire MOSFETs, including simple treatment of quantization effects. *Solid-State Electronics*, *112*, 85–98. doi:10.1016/j.sse.2015.02.002

Hong, S., Lee, H., Lee, J., Kwon, J., Han, S., Suh, Y. D., Cho, H., Shin, J., Yeo, J., & Ko, S. H. (2015). Highly stretchable and transparent metal nanowire heater for wearable electronics applications. *Advanced Materials*, *27*(32), 4744–4751. doi:10.1002/adma.201500917

Höök, F., Ray, A., Nordén, B., & Kasemo, B. (2001). Characterization of PNA and DNA immobilization and subsequent hybridization with DNA using acoustic-shear-wave attenuation measurements. *Langmuir*, *17*(26), 8305–8312. doi:10.1021/la0107704

Huang, B. H., Shen, S. S., Wei, N., Guo, X. F., & Wang, H. (2020). Fluorescence biosensor based on silicon quantum dots and 5, 5′-dithiobis-(2-nitrobenzoic acid) for thiols in living cells. *Spectrochimica Acta. Part A: Molecular and Biomolecular Spectroscopy*, *229*, 117972. doi:10.1016/j.saa.2019.117972

Huang, Y., Duan, X., Cui, Y., Lauhon, L. J., Kim, K. H., & Lieber, C. M. (2001). Logic gates and computation from assembled nanowire building blocks. *Science*, *294*(5545), 1313–1317. doi:10.1126cience.1066192 PMID:11701922

Huang, Y., Duan, X., Wei, Q., & Lieber, C. M. (2001). Directed assembly of one-dimensional nanostructures into functional networks. *Science*, *291*(5504), 630–633. doi:10.1126cience.291.5504.630 PMID:11158671

Huczko, A. (2002). Synthesis of aligned carbon nanotubes. *Applied Physics. A, Materials Science & Processing*, *74*(5), 617–638. doi:10.1007003390100929

Hu, L., & Chen, G. (2007). Analysis of optical absorption in silicon nanowire arrays for photovoltaic applications. *Nano Letters*, *7*(11), 3249–3252. doi:10.1021/nl071018b PMID:17927257

Hu, L., Kim, H. S., Lee, J. Y., Peumans, P., & Cui, Y. (2012). ACS Nano 4, 2955 (2010). *P. Lee, J. Lee, H. Lee, J. Yeo, S. Hong, KH Nam, D. Lee, SS Lee and SH Ko. Advanced Materials*, *24*, 3326.

Huo, H. B., Dai, L., Liu, C., You, L. P., Yang, W. Q., Ma, R. M., Ran, G. Z., & Qin, G. G. (2006). Electrical properties of Cu doped p-ZnTe nanowires. *Nanotechnology*, *17*(24), 5912–5915. doi:10.1088/0957-4484/17/24/002

Huo, H. B., Dai, L., Xia, D. Y., Ran, G. Z., You, L. P., Zhang, B. R., & Qin, G. G. (2006). Synthesis and optical properties of ZnTe single-crystalline nanowires. *Journal of Nanoscience and Nanotechnology*, *6*(4), 1182–1184. doi:10.1166/jnn.2006.179

Huo, J., Solanki, R., Freeouf, J. L., & Carruthers, J. R. (2004). Electroluminescence from silicon nanowires. *Nanotechnology*, *15*(12), 1848–1850. doi:10.1088/0957-4484/15/12/027

Hutchby, J. A., Bourianoff, G. I., Zhirnov, V. V., & Brewer, J. E. (2002). Extending the road beyond CMOS. *IEEE Circuits and Devices Magazine*, *18*(2), 28–41. doi:10.1109/101.994856

Ihn, S. G., Song, J. I., Kim, T. W., Leem, D. S., Lee, T., Lee, S. G., Koh, E. K., & Song, K. (2007). Morphology-and orientation-controlled gallium arsenide nanowires on silicon substrates. *Nano Letters*, *7*(1), 39–44. doi:10.1021/nl0618795

Inci, F., Tokel, O., Wang, S., Gurkan, U. A., Tasoglu, S., Kuritzkes, D. R., & Demirci, U. (2013). Nanoplasmonic quantitative detection of intact viruses from unprocessed whole blood. *ACS Nano*, *7*(6), 4733–4745. doi:10.1021/nn3036232

Iñiguez, B., Fjeldly, T. A., Lázaro, A., Danneville, F., & Deen, M. J. (2006). Compact-modeling solutions for nanoscale double-gate and gate-all-around MOSFETs. *IEEE Transactions on Electron Devices*, *53*(9), 2128–2142. doi:10.1109/TED.2006.881007

Isberg, J., Hammersberg, J., Johansson, E., Wikström, T., Twitchen, D. J., Whitehead, A. J., ... Scarsbrook, G. A. (2002). High carrier mobility in single-crystal plasma-deposited diamond. *Science*, *297*(5587), 1670–1672. doi:10.1126cience.1074374

Ishikawa, F., & Buyanova, I. (Eds.). (2017). *Novel compound semiconductor nanowires: materials, devices, and applications*. CRC Press. doi:10.1201/9781315364407

Ito, T., Nishimura, M., Yokoyama, M., Irie, M., & Wang, C. (2000). Highly efficient electron emitting diode fabricated with single-crystalline diamond. *Diamond and Related Materials*, *9*(9-10), 1561–1568. doi:10.1016/S0925-9635(00)00293-4

Iwai, H. (2009). Roadmap for 22 nm and beyond. *Microelectronic Engineering*, *86*(7-9), 1520–1528. doi:10.1016/j.mee.2009.03.129

Jain, A., Sharma, S. K., & Raj, B. (2016). Design and analysis of high sensitivity photosensor using Cylindrical Surrounding Gate MOSFET for low power applications. *Engineering Science and Technology, an International Journal, 19*(4), 1864-1870.

Jain, N., & Raj, B. (2018). Analysis and performance exploration of high performance (HfO2) SOI FinFETs over the conventional (Si3N4) SOI FinFET towards analog/RF design. *Journal of Semiconductors*, *39*(12), 124002. doi:10.1088/1674-4926/39/12/124002

Jassby, K. M., & Vreeland, T. Jr. (1970). An experimental study of the mobility of edge dislocations in pure copper single crystals. *Philosophical Magazine*, *21*(174), 1147–1168. doi:10.1080/14786437008238500

Javey, A., Nam, S., Friedman, R. S., Yan, H., & Lieber, C. M. (2007). Layer-by-layer assembly of nanowires for three-dimensional, multifunctional electronics. *Nano Letters*, *7*(3), 773–777. doi:10.1021/nl063056l PMID:17266383

Jazaeri, F., Barbut, L., & Sallese, J. M. (2014). Trans-capacitance modeling in junctionless gate-all-around nanowire FETs. *Solid-State Electronics*, *96*, 34–37. doi:10.1016/j.sse.2014.04.022

Jensen, K. K., Ørum, H., Nielsen, P. E., & Nordén, B. (1997). Kinetics for hybridization of peptide nucleic acids (PNA) with DNA and RNA studied with the BIAcore technique. *Biochemistry*, *36*(16), 5072–5077. doi:10.1021/bi9627525

Jiang, C., Liang, R., Wang, J., & Xu, J. (2015). A two-dimensional analytical model for short channel junctionless double-gate MOSFETs. *AIP Advances*, *5*(5), 057122. Advance online publication. doi:10.1063/1.4921086

Jin, C., Zhang, G., Qian, T., Li, X., & Yao, Z. (2005). Large-area Sb2Te3 nanowire arrays. *The Journal of Physical Chemistry B*, *109*(4), 1430–1432. doi:10.1021/jp046100z

Jin, S., Whang, D., McAlpine, M. C., Friedman, R. S., Wu, Y., & Lieber, C. M. (2004). Scalable interconnection and integration of nanowire devices without registration. *Nano Letters*, *4*(5), 915–919. doi:10.1021/nl049659j

Jin, X., Liu, X., Lee, J. H., & Lee, J. H. (2014). Modeling of subthreshold characteristics of short channel junctionless cylindrical surrounding-gate nanowire metal-oxide-silicon field effect transistors. *Physica Scripta*, *89*(1), 015804. Advance online publication. doi:10.1088/0031-8949/89/01/015804

Jones, S., & Menon, C. S. (2014, April 04). Non-linear elastic behavior of hexagonal silicon carbide. *Physica Status Solidi. B, Basic Research*, *251*(6), 1186–1191. doi:10.1002/pssb.201451024

Joyce, H. J., Gao, Q., Tan, H. H., Jagadish, C., Kim, Y., Zou, J., ... Johnston, M. B. (2011). III–V semiconductor nanowires for optoelectronic device applications. *Progress in Quantum Electronics*, *35*(2-3), 23–75. doi:10.1016/j.pquantelec.2011.03.002

Jung, I. S., & Kim, Y. B., & Lombardi, F. (2012). A novel sort error hardened 10T SRAM cells for low voltage operation. *Midwest Symposium on Circuits and Systems*. 10.1109/MWSCAS.2012.6292120

Jung, S., Lee, J., Hyeon, T., Lee, M., & Kim, D. H. (2014). Fabric-based integrated energy devices for wearable activity monitors. *Advanced Materials*, *26*(36), 6329–6334. doi:10.1002/adma.201402439

Kamitani, K., Grimsditch, M., Nipko, J., Loong, C.-K., Okada, M., & Kimura, I. (1997, June). The elastic constants of silicon carbide: A Brillouin-scattering study of 4H and 6H SiC single crystals. *Journal of Applied Physics*, *82*(6), 3152–3154. doi:10.1063/1.366100

Kang, C. C., Lai, C. W., Peng, H. C., Shyue, J. J., & Chou, P. T. (2007). Surfactant-and Temperature-Controlled CdS Nanowire Formation. *Small*, *3*(11), 1882–1885. doi:10.1002mll.200700390

Kao, L. T. H., Shankar, L., Kang, T. G., Zhang, G., Tay, G. K. I., Rafei, S. R. M., & Lee, C. W. H. (2011). Multiplexed detection and differentiation of the DNA strains for influenza A (H1N1 2009) using a silicon-based microfluidic system. *Biosensors & Bioelectronics*, *26*(5), 2006–2011. doi:10.1016/j.bios.2010.08.076

Karpinskya, N. D., Chanceyb, E. T., Palmera, D. B., & Yamania, Y. (2018). *Automation trust and attention allocation in multitasking workspace*. Communication Magazine. doi:10.1016/j.apergo.2018.03.008

Kar, S., Ghoshal, T., & Chaudhuri, S. (2006). Simple thermal evaporation route to synthesize Zn and Cd metal nanowires. *Chemical Physics Letters*, *419*(1-3), 174–178. doi:10.1016/j.cplett.2005.11.054

Kaul, U., Rastogi, R., Agarwal, S., Sharma, P., & Jain, S. (2015). A Novel D&C Approach for Efficient Fuzzy Unsupervised Classification for Mixed Variety of Data. In *Emerging ICT for Bridging the Future - Proceedings of the 49th Annual Convention of the Computer Society of India, CSI Volume 2. Advances in Intelligent Systems and Computing,* (vol. 338, pp. 553-563). Springer. https://link.springer.com/chapter/10.1007/978-3-319-13731-5_60

Kaur, H., Kabra, S., Bindra, S., Haldar, S., & Gupta, R. S. (2007). Impact of graded channel (GC) design in fully depleted cylindrical/surrounding gate MOSFET (FD CGT/SGT) for improved short channel immunity and hot carrier reliability. *Solid-State Electronics*, *51*(3), 398–404. doi:10.1016/j.sse.2007.01.025

Kaushal, G., Rathod, S. S., Raghuram, C. N., & Dasgupta, S. (2019). Radiation hard circuit design: flip-flop and SRAM. In VLSI and Post-CMOS Electronics. Volume 2: Devices, circuits and interconnects (Vol. 2, pp. 249–278). doi:10.1049/pbcs073g_ch12

Kawamura, T., Hori, D., Kangawa, Y., Kakimoto, K., Yoshimura, M., & Mori, Y. (2008, December 19). Thermal conductivity of SiC calculated by molecular dynamics. *Japanese Journal of Applied Physics*, *47*(12, 12R), 8898–8901. doi:10.1143/JJAP.47.8898

Kayes, B. M., Atwater, H. A., & Lewis, N. S. (2005). Comparison of the device physics principles of planar and radial p-n junction nanorod solar cells. *Journal of Applied Physics*, *97*(11), 114302. doi:10.1063/1.1901835

Kerman, A. J., Dauler, E. A., Keicher, W. E., Yang, J. K., Berggren, K. K., Gol'Tsman, G., & Voronov, B. (2006). Kinetic-inductance-limited reset time of superconducting nanowire photon counters. *Applied Physics Letters*, *88*(11), 111116. doi:10.1063/1.2183810

Khandelwal, S., Raj, B., & Gupta, R. D. (2015). FinFET Based 6T SRAM cell design: Analysis of performance metric, process variation and temperature effect. *Journal of Computational and Theoretical Nanoscience*, *12*(9), 2500–2506. Advance online publication. doi:10.1166/jctn.2015.4055

Kim, D. H., Lee, D. Y., Lee, H. S., Lee, W. H., Kim, Y. H., Han, J. I., & Cho, K. (2007). High-mobility organic transistors based on single-crystalline microribbons of triisopropylsilylethynyl pentacene via solution-phase self-assembly. *Advanced Materials*, *19*(5), 678–682. doi:10.1002/adma.200601259

Kim, F., Kwan, S., Akana, J., & Yang, P. (2001). Langmuir– Blodgett nanorod assembly. *Journal of the American Chemical Society*, *123*(18), 4360–4361. doi:10.1021/ja0059138 PMID:11457213

Kim, J., & Anderson, W. A. (2006). Direct electrical measurement of the self-assembled nickel silicide nanowire. *Nano Letters*, *6*(7), 1356–1359. doi:10.1021/nl0602894

Kim, J., Shin, D. H., Lee, E. S., Han, C. S., & Park, Y. C. (2007). Electrical characteristics of single and doubly connected Ni silicide nanowire grown by plasma-enhanced chemical vapor deposition. *Applied Physics Letters*, *90*(25), 253103. doi:10.1063/1.2749430

Kim, Y. B. (2010). Challenges for nanoscale MOSFETs and emerging nanoelectronics. *Transactions on Electrical and Electronic Materials*, *11*(3), 93–105. doi:10.4313/TEEM.2010.11.3.093

Kittel. (2007). *Introduction to solid state physics* (7th ed.). Wiley India Pvt. Limited.

Koblmüller, G., Hertenberger, S., Vizbaras, K., Bichler, M., Bao, F., Zhang, J. P., & Abstreiter, G. (2010). Self-induced growth of vertical free-standing InAs nanowires on Si (111) by molecular beam epitaxy. *Nanotechnology*, *21*(36), 365602. doi:10.1088/0957-4484/21/36/365602

Koizumi, S., Watanabe, K., Hasegawa, M., & Kanda, H. (2001). Ultraviolet emission from a diamond pn junction. *Science*, *292*(5523), 1899–1901. doi:10.1126cience.1060258

Kranti, A., Yan, R., Lee, C. W., Ferain, I., Yu, R., Akhavan, N. D., . . . Colinge, J. P. (2010, September). Junctionless nanowire transistor (JNT): Properties and design guidelines. In *2010 Proceedings of the European Solid State Device Research Conference* (pp. 357-360). IEEE.

Kranti, A., Haldar, S., & Gupta, R. S. (2001). An accurate 2D analytical model for short channel thin film fully depleted cylindrical/surrounding gate (CGT/SGT) MOSFET. *Microelectronics Journal, 32*(4), 305–313. doi:10.1016/S0026-2692(01)00008-8

Krogstrup, P., Jørgensen, H. I., Heiss, M., Demichel, O., Holm, J. V., Aagesen, M., Nygard, J., & Fontcuberta i Morral, A. (2013). Single-nanowire solar cells beyond the Shockley–Queisser limit. *Nature Photonics*, *7*(4), 306–310. doi:10.1038/nphoton.2013.32

Kumar, A., Bhushan, S., & Tiwari, P. K. (2017). A threshold voltage model of silicon-nanotube-based ultrathin double gate-all-around (DGAA) MOSFETs incorporating quantum confinement effects. *IEEE Transactions on Nanotechnology*, *16*(5), 868–875. doi:10.1109/TNANO.2017.2717841

Kumar, A., Bhushan, S., & Tiwari, P. K. (2017). Analytical modeling of subthreshold characteristics of ultra-thin double gate-all-around (DGAA) MOSFETs incorporating quantum confinement effects. *Superlattices and Microstructures*, *109*, 567–578. doi:10.1016/j.spmi.2017.05.038

Kumar, S., Kumari, A., & Das, M. K. (2017). Modeling gate-all-around Si/SiGe MOSFETs and circuits for digital applications. *Journal of Computational Electronics*, *16*(1), 47–60. doi:10.100710825-016-0941-z

Kumar, S., & Raj, B. (2016). Analysis of I ON and Ambipolar current for dual-material gate-drain overlapped DG-TFET. *Journal of Nanoelectronics and Optoelectronics*, *11*(3), 323–333. doi:10.1166/jno.2016.1902

Kung, S. C., van der Veer, W. E., Yang, F., Donavan, K. C., & Penner, R. M. (2010). 20 μs photocurrent response from lithographically patterned nanocrystalline cadmium selenide nanowires. *Nano Letters*, *10*(4), 1481–1485. doi:10.1021/nl100483v PMID:20334354

Kurkina, T., & Balasubramanian, K. (2012). Towards in vitro molecular diagnostics using nanostructures. *Cellular and Molecular Life Sciences*, *69*(3), 373–388. doi:10.100700018-011-0855-7

Lai, S. K. (2008). Flash memories: Successes and challenges. *IBM Journal of Research and Development, 52*(4.5), 529-535.

Larson, J. M., & Snyder, J. P. (2006). Overview and status of metal S/D Schottky-barrier MOSFET technology. *IEEE Transactions on Electron Devices*, *53*(5), 1048–1058. doi:10.1109/TED.2006.871842

Lauhon, L. J., Gudiksen, M. S., & Lieber, C. M. (2004). Semiconductor nanowire heterostructures. *Philosophical Transactions of the Royal Society of London. Series A: Mathematical, Physical and Engineering Sciences, 362*(1819), 1247-1260.

Lauhon, L. J., Gudiksen, M. S., Wang, D., & Lieber, C. M. (2002). Epitaxial core–shell and core–multishell nanowire heterostructures. *Nature*, *420*(6911), 57–61. doi:10.1038/nature01141 PMID:12422212

Lau, K., & McCurdy, A. (1998, October 1). Elastic anisotropy factors for orthorhombic, tetragonal, and hexagonal crystals. *Physical Review B: Condensed Matter and Materials Physics*, *58*(14), 8980–8984. doi:10.1103/PhysRevB.58.8980

Lee, C. S., Kim, S. K., & Kim, M. (2009). Ion-sensitive field-effect transistor for biological sensing. *Sensors (Basel)*, *9*(9), 7111–7131. doi:10.339090907111 PMID:22423205

Lee, W. H., Kim, D. H., Jang, Y., Cho, J. H., Hwang, M., Park, Y. D., Kim, Y. H., Han, J. I., & Cho, K. (2007). Solution-processable pentacene microcrystal arrays for high performance organic field-effect transistors. *Applied Physics Letters*, *90*(13), 132106. doi:10.1063/1.2717087

Łepkowski, S., Majewski, J., & Jurczak, G. (2005, December 1). Nonlinear elasticity in III-N compounds: Ab initio calculations. *Physical Review B: Condensed Matter and Materials Physics*, *72*(24), 245201. doi:10.1103/PhysRevB.72.245201

Leung, G., & Chui, C. O. (2011). Variability of inversion-mode and junctionless finfets due to line edge roughness. *IEEE Electron Device Letters*, *32*(11), 1489–1491. doi:10.1109/LED.2011.2164233

Levitt, A. P. (1970). *Whisker technology*. John Wiley & Sons.

Liang, H., Upmanyu, M., & Huang, H. (2005, June 8). Size-dependent elasticity of nanowires: Nonlinear effects. *Physical Review B: Condensed Matter and Materials Physics*, *71*(24), 241403. doi:10.1103/PhysRevB.71.241403

Liang, S., Islam, R., Smith, D. J., Bennett, P. A., O'brien, J. R., & Taylor, B. (2006). Magnetic iron silicide nanowires on Si (110). *Applied Physics Letters*, *88*(11), 113111. doi:10.1063/1.2185610

Liao, M., Hishita, S., Watanabe, E., Koizumi, S., & Koide, Y. (2010). Suspended Single-Crystal Diamond Nanowires for High-Performance Nanoelectromechanical Switches. *Advanced Materials*, *22*(47), 5393–5397. doi:10.1002/adma.201003074

Li, G., Zhu, R., & Yang, Y. (2012). Polymer Solar Cells. *Nature Photonics*, *6*(3), 153–161. doi:10.1038/nphoton.2012.11

Li, H., Zhang, X., & Tzou, H. (2017). Diagonal piezoelectric sensors on cylindrical shells. *Journal of Sound and Vibration*, *400*, 201–212. doi:10.1016/j.jsv.2017.03.039

Li, L., Zhang, Y., Li, G., & Zhang, L. (2003). A route to fabricate single crystalline bismuth nanowire arrays with different diameters. *Chemical Physics Letters*, *378*(3-4), 244–249. doi:10.1016/S0009-2614(03)01264-8

Lin, C., & Povinelli, M. L. (2009). Optical absorption enhancement in silicon nanowire arrays with a large lattice constant for photovoltaic applications. *Optics Express*, *17*(22), 19371–19381. doi:10.1364/OE.17.019371 PMID:19997158

Lin, H. W., Ku, S. Y., Su, H. C., Huang, C. W., Lin, Y. T., Wong, K. T., & Wu, C. C. (2005). Highly Efficient Visible-Blind Organic Ultraviolet Photodetectors. *Advanced Materials*, *17*(20), 2489–2493. doi:10.1002/adma.200401622

Lin, J. L., Petrovykh, D. Y., Kirakosian, A., Rauscher, H., Himpsel, F. J., & Dowben, P. A. (2001). Self-assembled Fe nanowires using organometallic chemical vapor deposition and CaF 2 masks on stepped Si (111). *Applied Physics Letters*, *78*(6), 829–831. doi:10.1063/1.1345830

Li, Q., Kumar, V., Li, Y., Zhang, H., Marks, T. J., & Chang, R. P. (2005). Fabrication of ZnO nanorods and nanotubes in aqueous solutions. *Chemistry of Materials*, *17*(5), 1001–1006. doi:10.1021/cm048144q

Li, Q., & Penner, R. M. (2005). Photoconductive cadmium sulfide hemicylindrical shell nanowire ensembles. *Nano Letters*, *5*(9), 1720–1725. doi:10.1021/nl050994x PMID:16159212

Liu, B., Wang, Y., Dilts, S., Mayer, T. S., & Mohney, S. E. (2007). Silicidation of silicon nanowires by platinum. *Nano Letters*, *7*(3), 818–824. doi:10.1021/nl062393r

Liu, C. W., & Lin, C. H. (2010). Metal-Insulator-Semiconductor Photodetectors. *Sensors (Basel)*, *10*(10), 8797–8826. doi:10.3390101008797 PMID:22163382

Liu, J., Wu, D., & Zeng, S. (2009). Da Wu, and Shengnan Zeng. "Influence of temperature and layers on the characterization of ITO films. *Journal of Materials Processing Technology*, *209*(8), 3943–3948. doi:10.1016/j.jmatprotec.2008.09.016

Liu, Y., Dong, L., Lu, J., Huo, W., Du, Y., Zhang, W., & Zhang, Y. (2020, April 5). Microstructure and mechanical properties of SiC nanowires reinforced titanium matrix composites. *Journal of Alloys and Compounds*, *819*, 152953. doi:10.1016/j.jallcom.2019.152953

Liu, Z., Bando, Y., & Tang, C. (2003). Synthesis of tungsten oxide nanowires. *Chemical Physics Letters*, *372*(1-2), 179–182. doi:10.1016/S0009-2614(03)00397-X

Li, Y.-J., Li, Y.-L., Li, S.-L., Gong, P., & Fang, X.-Y. (2017). Structural, electronic, and optical properties of hexagonal and triangular SiC NWs with different diameters. *Chinese Physics B*, *26*(4), 047309. doi:10.1088/1674-1056/26/4/047309

Li, Z., Chen, Y., Li, X., Kamins, T. I., Nauka, K., & Williams, R. S. (2004). Sequence-specific label-free DNA sensors based on silicon nanowires. *Nano Letters*, *4*(2), 245–247. doi:10.1021/nl034958e

Lu, W., Xie, P., & Lieber, C. M. (2008). Nanowire Transistor Performance Limits and Applications. *IEEE Transactions on Electron Devices, 55*(11), 2859-2876. Available at: http://apps.isiknowledge.com

Lubow, A., Ismail-Beigi, S., & Ma, T. P. (2010). Comparison of drive currents in metal-oxide-semiconductor field-effect transistors made of Si, Ge, GaAs, InGaAs, and InAs channels. *Applied Physics Letters*, *96*(12), 122105. doi:10.1063/1.3367708

Lu, S., Sivakumar, K., & Panchapakesan, B. (2007). Sonochemical Synthesis of Platinum Nanowires and its Application as Electro-Chemical Actuators. *Journal of Nanoscience and Nanotechnology*, *7*(7), 1. doi:10.1166/jnn.2007.427

Lu, W., & Lieber, C. M. (2006). Semiconductor nanowires. *Journal of Physics. D, Applied Physics*, *39*(21), R387–R406. doi:10.1088/0022-3727/39/21/R01

Ma, D. D. D., Lee, S. T., Mueller, P., & Alvarado, S. F. (2006). Scanning tunneling microscope excited cathodoluminescence from ZnS nanowires. *Nano Letters*, *6*(5), 926–929. doi:10.1021/nl052507j

Ma, L., Hao, T., & Hossain, Z. M. (2019, October 22). Size-dependent toughness and strength in defective 3C-SiC nanowires. *Journal of Applied Physics*, *126*(16), 164303. doi:10.1063/1.5120034

Mandal, K. C., Muzykov, P. G., Krishna, R., Hayes, T., & Sudarshan, T. S. (2011, April). Thermally stimulated current and high temperature resistivity measurements of 4H semi-insulating silicon carbide. *Solid State Communications*, *151*(7), 532–535. doi:10.1016/j.ssc.2011.01.019

Mandl, B., Stangl, J., Mårtensson, T., Mikkelsen, A., Eriksson, J., Karlsson, L. S., Bauer, G., Samuelson, L., & Seifert, W. (2006). Au-free epitaxial growth of InAs nanowires. *Nano Letters*, *6*(8), 1817–1821. doi:10.1021/nl060452v

Marie, A., Celebre, D., Benedict, I., Medina, A., Zandrae, A., Dubouzet, D., Neil, A., Surposa, M., & Gustilo, E. R. C. (2015). Home automation using Raspberry Pi through Siri enabled mobile devices. *Proc. 8th IEEE International Conference Humanoid, Nanotechnology, Information Technology Communication and Control, Environment and Management (HNICEM) The Institute of Electrical and Electronics Engineers Inc. (IEEE) – Philippine.*

Marin, E. G., Ruiz, F. G., Schmidt, V., Godoy, A., Riel, H., & Gámiz, F. (2015). Analytic drain current model for III–V cylindrical nanowire transistors. *Journal of Applied Physics*, *118*(4), 044502. doi:10.1063/1.4927041

Mårtensson, T., Wagner, J. B., Hilner, E., Mikkelsen, A., Thelander, C., Stangl, J., Ohlsson, B. J., Gustafsson, A., Lundgren, E., Samuelson, L., & Seifert, W. (2007). Epitaxial growth of indium arsenide nanowires on silicon using nucleation templates formed by self-assembled organic coatings. *Advanced Materials*, *19*(14), 1801–1806. doi:10.1002/adma.200700285

Mason, W. (1955b, January 15). Ultrasonic attenuation due to lattice-electron interaction in normal conducting metals. *Physical Review*, *97*(2), 557–558. doi:10.1103/PhysRev.97.557

Mason, W. P. (1965a). Effect of impurities and phonon processes on the ultrasonic attenuation of germanium, crystal quartz, and silicon. In *Physical Acoustics* (Vol. 3, pp. 235–286). Elsevier. doi:10.1016/B978-0-12-395669-9.50013-8

Mason, W. P., & Baerwald, H. (1951). Piezoelectric crystals and their applications to ultrasonics. *Physics Today*, *4*(5), 23–24. doi:10.1063/1.3067231

Mason, W. P., & Bateman, T. (1964). Ultrasonic-wave propagation in pure silicon and germanium. *The Journal of the Acoustical Society of America*, *36*(4), 644–652. doi:10.1121/1.1919031

Mattila, M., Hakkarainen, T., Lipsanen, H., Jiang, H., & Kauppinen, E. I. (2007). Enhanced luminescence from catalyst-free grown InP nanowires. *Applied Physics Letters*, *90*(3), 033101. doi:10.1063/1.2431711

McLain, M. L., Barnaby, H. J., Esqueda, I. S., Oder, J., & Vermeire, B. (2009a). Reliability of high performance standard two-edge and radiation hardened by design enclosed geometry transistors. *IEEE International Reliability Physics Symposium Proceedings.* 10.1109/IRPS.2009.5173247

Mertens, H., Ritzenthaler, R., Chasin, A., Schram, T., Kunnen, E., Hikavyy, A., & Devriendt, K. (2016, December). Vertically stacked gate-all-around Si nanowire CMOS transistors with dual work function metal gates. In *2016 IEEE International Electron Devices Meeting (IEDM)* (pp. 19-7). IEEE. 10.1109/IEDM.2016.7838456

Miranda, A., & Pérez, L. A. (2016, January). Electronic properties of fluorinated silicon carbide nanowires. *Computational Materials Science*, *111*, 294–300. doi:10.1016/j.commatsci.2015.09.028

Molares, M. T., Buschmann, V., Dobrev, D., Neumann, R., Scholz, R., Schuchert, I. U., & Vetter, J. (2001). Single-crystalline copper nanowires produced by electrochemical deposition in polymeric ion track membranes. *Advanced Materials*, *13*(1), 62–65. doi:10.1002/1521-4095(200101)13:1<62::AID-ADMA62>3.0.CO;2-7

Molnar, W., Lugstein, A., Pongratz, P., Auner, N., Bauch, C., & Bertagnolli, E. (2010). Subeutectic synthesis of epitaxial Si-NWs with diverse catalysts using a novel Si precursor. *Nano Letters*, *10*(10), 3957–3961. doi:10.1021/nl101744q PMID:20843058

Mondal, S. P., Das, K., Dhar, A., & Ray, S. K. (2007). Characteristics of CdS nanowires grown in a porous alumina template using a two-cell method. *Nanotechnology*, *18*(9), 095606. doi:10.1088/0957-4484/18/9/095606

Monroy, E., Omnès, F., & Calle, F. (2003). Wide-bandgap semiconductor ultraviolet photodetectors. *Semiconductor Science and Technology*, *18*(4), R33–R51. doi:10.1088/0268-1242/18/4/201

Monson, C. F., & Woolley, A. T. (2003). DNA-templated construction of copper nanowires. *Nano Letters*, *3*(3), 359–363. doi:10.1021/nl034016+

Moon, D., Choi, S.-J., Duarte, J. P., & Choi, Y.-K. (2013). Investigation of silicon nanowire gate-all-around junctionless transistors built on a bulk substrate. *IEEE Transactions on Electron Devices*, *60*(4), 1355–1360. doi:10.1109/TED.2013.2247763

Morales, A. M., & Lieber, C. M. (1998). A laser ablation method for the synthesis of crystalline semiconductor nanowires. *Science*, *279*(5348), 208–211. doi:10.1126cience.279.5348.208 PMID:9422689

Mori, S., & Hiki, Y. (1978, May). Calculation of the third-and fourth-order elastic constants of alkali halide crystalsCalculation of the third-and fourth-order elastic constants of alkali halide crystals. *Journal of the Physical Society of Japan*, *45*(5), 1449–1456. doi:10.1143/JPSJ.45.1449

Morse, R., Myers, A., & Walker, C. (1960, June 15). Fermi surfaces of gold and silver from ultrasonic attenuation. *Physical Review Letters*, *4*(12), 605–607. doi:10.1103/PhysRevLett.4.605

Muskins, O. L. (2008). Article. *Nano Letters*, *8*, 2638. doi:10.1021/nl0808076 PMID:18700806

Najam, S. F. (2008). *Silicon Nanowire: Synthesis and Applications*. Available at: http://swot.ecs.soton.ac.uk/137/3/IRP_edited.pdf

Nandanpawar, M., & Rajagopalan, S. (1982). Grüneisen numbers in hexagonal crystals. *The Journal of the Acoustical Society of America, 71*(6), 1469–1472. doi:10.1121/1.387844

Narasimham, B., Gupta, S., Reed, D., Wang, J. K., Hendrickson, N., & Taufique, H. (2018). Scaling trends and bias dependence of the soft error rate of 16 nm and 7 nm FinFET SRAMs. *IEEE International Reliability Physics Symposium Proceedings*. 10.1109/IRPS.2018.8353583

Natarajan, C. M., Tanner, M. G., & Hadfield, R. H. (2012). Superconducting nanowire single-photon detectors: Physics and applications. *Superconductor Science and Technology, 25*(6), 063001. doi:10.1088/0953-2048/25/6/063001

Neamen, D. A. (2019). *Semiconductor physics and devices: basic principles*. doi:10.1017/CBO9781107415324.004

Neretina, S., Hughes, R. A., Britten, J. F., Sochinskii, N. V., Preston, J. S., & Mascher, P. (2007). Vertically aligned wurtzite CdTe nanowires derived from a catalytically driven growth mode. *Nanotechnology, 18*(27), 275301. doi:10.1088/0957-4484/18/27/275301

Neubrech, F., Kolb, T., Lovrincic, R., Fahsold, G., Pucci, A., Aizpurua, J., Cornelius, T. W., Toimil-Molares, M. E., Neumann, R., & Karim, S. (2006). Resonances of individual metal nanowires in the infrared. *Applied Physics Letters, 89*(25), 253104. doi:10.1063/1.2405873

Nigam, A., Rastogi, R., Mishra, R., Arya, P., & Sharma, S. (2015).S ecurity of Data Transmission Using Logic Gates and Crypt Analysis. *CSI Communication*. http://www.csi-india.org/communications/CSI%20June15%20Combine.pdf

Nigam, A., Rastogi, R., Mishra, R., Arya, P., & Sharma, S. (2015).Security of data transmission using logic gates and crypt analysis. *Proceedings of 9th INDIACom; 2015 2nd International Conference on Computing for Sustainable Global Development*, 101-105. https://ieeexplore.ieee.org/document/7100229

Nirmal, D., & Kumar, P. V. J. (2013). *Nanoscale tri gate MOSFET for ultra low power applications using high-k dielectrics*. Academic Press.

Niua, J., Ganga, H., Zhangb, Y., & Duc, X. (2018). Relationship between automation trust and operator performance for the novice and expert in spacecraft rendezvous and docking (RVD). *Applied Ergonomics, 35*(4), 1–8. doi:10.1016/j.apergo.2018.03.014 PMID:29764609

Noborisaka, J., Motohisa, J., & Fukui, T. (2005). Catalyst-free growth of GaAs nanowires by selective-area metalorganic vapor-phase epitaxy. *Applied Physics Letters, 86*(21), 213102. doi:10.1063/1.1935038

Ohno, T., Shiraishi, K., & Ogawa, T. (1992). Intrinsic origin of visible light emission from silicon quantum wires: Electronic structure and geometrically restricted exciton. *Physical Review Letters, 69*(16), 2400–2403. doi:10.1103/PhysRevLett.69.2400 PMID:10046475

Oh, S. S., Kim, D. H., Moon, M. W., Vaziri, A., Kim, M., Yoon, E., Oh, K. H., & Hutchinson, J. W. (2008). Indium nanowires synthesized at an ultrafast rate. *Advanced Materials*, *20*(6), 1093–1098. doi:10.1002/adma.200702134

Okano, K., Koizumi, S., Silva, S. R. P., & Amaratunga, G. A. (1996). Low-threshold cold cathodes made of nitrogen-doped chemical-vapour-deposited diamond. *Nature*, *381*(6578), 140–141. doi:10.1038/381140a0

Okino, H., Matsuda, I., Hobara, R., Hosomura, Y., Hasegawa, S., & Bennett, P. A. (2005). In situ resistance measurements of epitaxial cobalt silicide nanowires on Si (110). *Applied Physics Letters*, *86*(23), 233108. doi:10.1063/1.1948519

Oktyabrsky, S., & Ye, P. D. (2010). Fundamentals of III-V semiconductor MOSFETs. *Fundamentals of III-V Semiconductor MOSFETs*. doi:10.1007/978-1-4419-1547-4

Oligschleger, C., Jones, R., Reimann, S., & Schober, H. (1996, March 1). Model interatomic potential for simulations in selenium. *Physical Review B: Condensed Matter and Materials Physics*, *53*(10), 6165–6173. doi:10.1103/PhysRevB.53.6165 PMID:9982015

Oproglidis, T. A., Karatsori, T. A., Barraud, S., Ghibaudo, G., & Dimitriadis, C. A. (2017). Leakage current conduction in metal gate junctionless nanowire transistors. *Solid-State Electronics*, *131*, 20–23. doi:10.1016/j.sse.2017.02.003

Ouyang, H., Striemer, C. C., & Fauchet, P. M. (2006). Quantitative analysis of the sensitivity of porous silicon optical biosensors. *Applied Physics Letters*, *88*(16), 163108. doi:10.1063/1.2196069

Paladugu, M., Zou, J., Guo, Y. N., Auchterlonie, G. J., Joyce, H. J., Gao, Q., Hoe Tan, H., Jagadish, C., & Kim, Y. (2007). Novel growth phenomena observed in axial InAs/GaAs nanowire heterostructures. *Small*, *3*(11), 1873–1877. doi:10.1002mll.200700222

Pandey, D. K., & Pandey, S. (2010). Ultrasonics: A technique of material characterization. In D. Dissanayake (Ed.), Acoustic Waves (pp. 397–430). Academic Press.

Pandey, D., Singh, D., & Yadav, R. (2007, July). Ultrasonic wave propagation in IIIrd group nitrides. *Applied Acoustics*, *68*(7), 766–777. doi:10.1016/j.apacoust.2006.04.004

Pan, Z. W., Dai, Z. R., Xu, L., Lee, S. T., & Wang, Z. L. (2001). Temperature-controlled growth of silicon-based nanostructures by thermal evaporation of SiO powders. *The Journal of Physical Chemistry B*, *105*(13), 2507–2514. doi:10.1021/jp004253q

Parkinson, P., Lloyd-Hughes, J., Gao, Q., Tan, H. H., Jagadish, C., Johnston, M. B., & Herz, L. M. (2007). Transient terahertz conductivity of GaAs nanowires. *Nano Letters*, *7*(7), 2162–2165. doi:10.1021/nl071162x

Park, J. T., Colinge, J. P., & Diaz, C. H. (2001). Pi-gate soi mosfet. *IEEE Electron Device Letters*, *22*(8), 405–406. doi:10.1109/55.936358

Park, T. S., Yoon, E., & Lee, J. H. (2003). A 40nm body-tied FinFET (OMEGA MOSFET) using bulk Si wafer. *Physica E, Low-Dimensional Systems and Nanostructures*, *19*(1-2), 6–12. doi:10.1016/S1386-9477(03)00285-6

Park, W. I., & Yi, G. C. (2004, January). Electroluminescence in n-ZnOnanorod arrays vertically grown on p-GaN. *Advanced Materials*, *16*(1), 87–90. doi:10.1002/adma.200305729

Pavan, P., Bez, R., Olivo, P., & Zanoni, E. (1997). Flash memory cells-an overview. *Proceedings of the IEEE*, *85*(8), 1248–1271. doi:10.1109/5.622505

Peng, C., Hu, Z., En, Y., Chen, Y., Lei, Z., Zhang, Z., … Li, B. (2018). Radiation hardening by the modification of shallow trench isolation process in partially depleted SOI MOSFETs. IEEE Transactions on Nuclear Science. doi:10.1109/TNS.2018.2798295

Peng, C., Huang, J., Liu, C., Zhao, Q., Xiao, S., Wu, X., … Zeng, X. (2019). Radiation-hardened 14T SRAM bitcell with speed and power optimized for space application. *IEEE Transactions on Very Large Scale Integration (VLSI) Systems*. doi:10.1109/TVLSI.2018.2879341

Peng, M. T. L. (2007). *Modeling the Effect of Velocity Saturation in Nanoscale Mosfet* (Doctoral dissertation). Universiti Teknologi Malaysia.

Peng, F., Su, Y., Ji, X., Zhong, Y., Wei, X., & He, Y. (2014). Doxorubicin-loaded silicon nanowires for the treatment of drug-resistant cancer cells. *Biomaterials*, *35*(19), 5188–5195. doi:10.1016/j.biomaterials.2014.03.032

Peng, X. S., Meng, G. W., Zhang, J., Zhao, L. X., Wang, X. F., Wang, Y. W., & Zhang, L. D. (2001). Electrochemical fabrication of ordered Bi2S3 nanowire arrays. *Journal of Physics. D, Applied Physics*, *34*(22), 3224–3228. doi:10.1088/0022-3727/34/22/304

Pettersson, H., Trägårdh, J., Persson, A. I., Landin, L., Hessman, D., & Samuelson, L. (2006). Infrared photodetectors in heterostructure nanowires. *Nano Letters*, *6*(2), 229–232. doi:10.1021/nl052170l

Pettifor, D. (1992). Theoretical predictions of structure and related properties of intermetallics. *Materials Science and Technology*, *8*(4), 345–349. doi:10.1179/mst.1992.8.4.345

Pillai, S. O. (2010). *Solid State Physics* (7th ed.). New Age Science.

Ponraj, J. S., Dhanabalan, S. C., Attolini, G., & Salviati, G. (2016, May 17). SiC nanostructures toward biomedical applications and its future challenges. *Critical Reviews in Solid State and Material Sciences*, *41*(5), 430–446. doi:10.1080/10408436.2016.1150806

Poole, P. J., Lefebvre, J., & Fraser, J. (2003). Spatially controlled, nanoparticle-free growth of InP nanowires. *Applied Physics Letters*, *83*(10), 2055–2057. doi:10.1063/1.1608486

Prasad, V., & Sandya, S. (2014). Single event transient tolerant high speed phase frequency detector for PLL based frequency synthesizer. *Proceedings of International Conference on Circuits, Communication, Control and Computing, I4C 2014*. 10.1109/CIMCA.2014.7057761

Pratap, Y., Haldar, S., Gupta, R. S., & Gupta, M. (2016). Gate-Material-Engineered Junctionless Nanowire Transistor (JNT) with Vacuum Gate Dielectric for Enhanced Hot-Carrier Reliability. *IEEE Transactions on Device and Materials Reliability*, *16*(3), 360–369. doi:10.1109/TDMR.2016.2583262

Prieto, A. L., Sander, M. S., Martín-González, M. S., Gronsky, R., Sands, T., & Stacy, A. M. (2001). Electrodeposition of ordered Bi2Te3 nanowire arrays. *Journal of the American Chemical Society*, *123*(29), 7160–7161. doi:10.1021/ja015989j

Prokhorov, A. S., Tyurin, S. F., & Chudinov, M. A. (2018). A radiation resistant voting circuit. *Proceedings of the 2018 IEEE Conference of Russian Young Researchers in Electrical and Electronic Engineering, ElConRus 2018*. 10.1109/EIConRus.2018.8317068

Pugh, S. (1954). XCII. Relations between the elastic moduli and the plastic properties of polycrystalline pure metals. *The London, Edinburgh and Dublin Philosophical Magazine and Journal of Science*, *45*(367), 823–843. doi:10.1080/14786440808520496

Pushpakaran, B. N., Hinojosa, M., Bayne, S. B., Veliadis, V., Urciuoli, D., El-Hinnawy, N., Borodulin, P., Gupta, S., & Scozzie, C. (2013). High temperature unclamped inductive switching mode evaluation of SiC JFET. *IEEE Electron Device Letters*, *34*(4), 526–528. doi:10.1109/LED.2013.2247020

Qi, C., Goncher, G., Solanki, R., & Jordan, J. (2007). SiGe nanowire growth and characterization. *Nanotechnology*, *18*(7), 075302. doi:10.1088/0957-4484/18/7/075302

Qi, C., Musetti, S., Fu, L.-H., Zhu, Y.-J., & Huang, L. (2019, May 13). Biomolecule-assisted green synthesis of nanostructured calcium phosphates and their biomedical applications. *Chemical Society Reviews*, *48*(10), 2698–2737. doi:10.1039/C8CS00489G PMID:31080987

Qing, Q., Jiang, Z., Xu, L., Gao, R., Mai, L., & Lieber, C. M. (2014). Free-standing kinked nanowire transistor probes for targeted intracellular recording in three dimensions. *Nature Nanotechnology*, *9*(2), 142–147. doi:10.1038/nnano.2013.273

Qing, Q., Pal, S. K., Tian, B., Duan, X., Timko, B. P., Cohen-Karni, T., Murthy, V. N., & Lieber, C. M. (2010). Nanowire transistor arrays for mapping neural circuits in acute brain slices. *Proceedings of the National Academy of Sciences of the United States of America*, *107*(5), 1882–1887. doi:10.1073/pnas.0914737107

Qun, C., Mei, L., & Xinyu, L. (2018). Bike Fleet Allocation Models for Repositioning in Bike-Sharing Systems. *IEEE Intelligent Transportation Systems Magazine*, *10*(1), 19–29. doi:10.1109/MITS.2017.2776129

Raghuram, C. H. N., & Member, S. (2020). *Double Node Upset Tolerant RHBD15T SRAM Cell Design for Space Applications, 4388*(c). doi:10.1109/TDMR.2020.2970089

Rahman, A., & Lundstrom, M. S. (2002). A compact scattering model for the nanoscale double-gate MOSFET. *IEEE Transactions on Electron Devices*, *49*(3), 481–489. doi:10.1109/16.987120

Rai-Choudhury, P. (Ed.). (1997). *Handbook of microlithography, micromachining, and microfabrication. 2. Micromachining and microfabrication.* SPIE Optical Engineering Press. doi:10.1117/3.2265071

Rajaei, R., Asgari, B., Tabandeh, M., & Fazeli, M. (2015). Design of Robust SRAM Cells Against Single-Event Multiple Effects for Nanometer Technologies. *IEEE Transactions on Device and Materials Reliability, 15*(3), 429–436. doi:10.1109/TDMR.2015.2456832

Rajaei, R., Asgari, B., Tabandeh, M., & Fazeli, M.RAJAEI. (2017). Single event multiple upset-tolerant SRAM cell designs for nano-scale CMOS technology. *Turkish Journal of Electrical Engineering and Computer Sciences, 25*, 1035–1047. doi:10.3906/elk-1502-124

Raj, B., Saxena, A. K., & Dasgupta, S. (2009). Analytical modeling for the estimation of leakage current and subthreshold swing factor of nanoscale double gate FinFET device. *Microelectronics International, 26*(1), 53–63. doi:10.1108/13565360910923188

Raj, S. K. (2007). Analytical Modeling of Nanoscale Double Gate FinFET Device. *International Journal of Intelligent Electronics System, 1*, 633–637.

Rakovich, Y. P., Volkov, Y., Sapra, S., Susha, A. S., Döblinger, M., Donegan, J. F., & Rogach, A. L. (2007). CdTe nanowire networks: Fast self-assembly in solution, internal structure, and optical properties. *The Journal of Physical Chemistry C, 111*(51), 18927–18931. doi:10.1021/jp076622p

Ranganathan, S. I., & Ostoja-Starzewski, M. (2008, August 1). Universal elastic anisotropy index. *Physical Review Letters, 101*(5), 055504. doi:10.1103/PhysRevLett.101.055504 PMID:18764407

Razavi, P., Fagas, G., Ferain, I., Akhavan, N. D., Yu, R., & Colinge, J. P. (2011). Performance investigation of short-channel junctionless multigate transistors. *2011 12th International Conference on Ultimate Integration on Silicon, ULIS 2011*, 122–125. 10.1109/ULIS.2011.5758005

Razavi, P., & Fagas, G. (2013). Electrical performance of III-V gate-all-around nanowire transistors. *Applied Physics Letters, 103*(6), 1–4. doi:10.1063/1.4817997

Reddy, G. V., Kumar, M. J., & Member, S. (2005). *A New Dual-Material Double-Gate (DMDG).* Academic Press.

Reuss, A. (1929). Calculation of the flow limits of mixed crystals on the basis of the plasticity of monocrystals. *Zeitschrift für Angewandte Mathematik und Mechanik, 9*, 49–58. doi:10.1002/zamm.19290090104

Rich, R. M., Stelmakh, S., Patyk, J., Wieligor, M., Zerda, T., & Guo, Q. (2009, June 1). Bulk modulus of silicon carbide nanowires and nanosize grains. *Journal of Materials Science, 44*(11), 3010–3013. doi:10.100710853-009-3431-x

Richter, C. A., Xiong, H. D., Zhu, X., Wang, W., Stanford, V. M., Hong, W. K., Lee, T., Ioannou, D. E., & Li, Q. (2008). Metrology for the electrical characterization of semiconductor nanowires. *IEEE Transactions on Electron Devices, 55*(11), 3086–3095. doi:10.1109/TED.2008.2005394

Roy, K., Mahmoodi, H., Mukhopadhyay, S., Ananthan, H., Bansal, A., & Cakici, T. (2005, November). Double-gate SOI devices for low-power and high-performance applications. In *ICCAD-2005. IEEE/ACM International Conference on Computer- Computer Aided Design, 2005*, 217–224.

Rungta, S., Srivastava, S., Yadav, U. S., & Rastogi, R. (2014) A Comparative Analysis of New Approach with an Existing Algorithm to Detect Cycles in a Directed Graph. *Proceedings of the 48th Annual Convention of Computer Society of India*. Springer. 10.1007/978-3-319-03095-1_5

Sahay, S., & Kumar, M. J. (2016). Insight into Lateral Band-to-Band-Tunneling in Nanowire Junctionless FETs. *IEEE Transactions on Electron Devices*, *63*(10), 4138–4142. doi:10.1109/TED.2016.2601239

Sahay, S., & Kumar, M. J. (2017). Spacer Design Guidelines for Nanowire FETs from Gate-Induced Drain Leakage Perspective. *IEEE Transactions on Electron Devices*, *64*(7), 3007–3015. doi:10.1109/TED.2017.2702067

Saib, S. S., Yadav, S., Rahul, J., Srivastava, A., & Raj, B. (2015). Impact of HfO2 in Graded Channel Dual Insulator Double Gate MOSFET. *Journal of Computational and Theoretical Nanoscience*, *12*(6), 950–953. doi:10.1166/jctn.2015.3833

Sakaki, H. (1980). Scattering suppression and high-mobility effect of size-quantized electrons in ultrafine semiconductor wire structures. *Japanese Journal of Applied Physics*, *19*(12), L735–L738. doi:10.1143/JJAP.19.L735

Salfi, J., Philipose, U., De Sousa, C. F., Aouba, S., & Ruda, H. E. (2006). Electrical properties of Ohmic contacts to ZnSe nanowires and their application to nanowire-based photodetection. *Applied Physics Letters*, *89*(26), 261112. doi:10.1063/1.2424653

Sang, S., Wang, Y., Feng, Q., Wei, Y., Ji, J., & Zhang, W. (2016). Progress of new label-free techniques for biosensors: A review. *Critical Reviews in Biotechnology*, *36*(3), 465–481.

Saquib, S. T., & Hameed, S. (2013). Wireless Control of Miniaturized Mobile Vehicle for Indoor Surveillance. *IOP Conference Series Material Science and Engineering*, *51*(3), 452-465. 10.1088/1757-899X/51/1/012025

Sarasamak, K., Limpijumnong, S., & Lambrecht, W. R. (2010, July 1). Pressure-dependent elastic constants and sound velocities of wurtzite SiC, GaN, InN, ZnO, and CdSe, and their relation to the high-pressure phase transition: A first-principles study. *Physical Review B: Condensed Matter and Materials Physics*, *82*(3), 035201. doi:10.1103/PhysRevB.82.035201

Sarasqueta, G., Choudhury, K. R., Subbiah, J., & So, F. (2011). Organic and Inorganic Blocking Layers for Solution-Processed Colloidal PbSe Nanocrystal Infrared Photodetectors. *Advanced Functional Materials*, *21*(1), 167–171. doi:10.1002/adfm.201001328

Scarlet, S. P., Ambika, R., & Srinivasan, R. (2017). Effect of eccentricity on junction and junctionless based silicon nanowire and silicon nanotube FETs. *Superlattices and Microstructures*, *107*, 178–188. doi:10.1016/j.spmi.2017.04.015

Schaefer, H. E. (2010). *Nanoscience: the science of the small in physics, engineering, chemistry, biology and medicine*. Springer Science & Business Media. doi:10.1007/978-3-642-10559-3

Schöning, M. J., & Poghossian, A. (2002). Recent advances in biologically sensitive field-effect transistors (BioFETs). *Analyst (London)*, *127*(9), 1137–1151. doi:10.1039/B204444G PMID:12375833

Schwarz, K. W., & Tersoff, J. (2009). From droplets to nanowires: Dynamics of vapor-liquid-solid growth. *Physical Review Letters*, *102*(20), 206101. doi:10.1103/PhysRevLett.102.206101 PMID:19519043

Schwarz, K. W., & Tersoff, J. (2011). Elementary processes in nanowire growth. *Nano Letters*, *11*(2), 316–320. doi:10.1021/nl1027815 PMID:21188963

Seker, F., Meeker, K., Kuech, T. F., & Ellis, A. B. (2000). Surface chemistry of prototypical bulk II– VI and III– V semiconductors and implications for chemical sensing. *Chemical Reviews*, *100*(7), 2505–2536. doi:10.1021/cr980093r

Shao, M., Ma, D. D. D., & Lee, S. T. (2010). Silicon nanowires–synthesis, properties, and applications. *European Journal of Inorganic Chemistry*, *2010*(27), 4264–4278. doi:10.1002/ejic.201000634

Sharma, P., Rastogi, R., Aggarwal, S., Kaul, U., & Jain, S. (2014). Business Analysis and Decision Making Through Unsupervised Classification of Mixed Data Type of Attributes Through Genetic Algorithm. *BIJIT*. http://bvicam.ac.in/bjit/issues.asp?issue=11

Sharma, R., Jain, A., & Rastogi, R. (2013). A New Face To Photo Security Of Facebook. *Proceedings of the Sixth International Conference on Contemporary Computing (IC3-2013)*. https://ieeexplore.ieee.org/document/6612231

Sharma, Raj, & Khosla. (2016a). Comparative Analysis of MOSFET, CNTFET and NWFET for High Performance VLSI Circuit Design: A Review. *Journal of VLSI Design Tools & Technology*, *6*(2), 19-32.

Sharma, Raj, & Khosla. (2016b). A Gaussian Approach for Analytical Subthreshold Current Model of Cylindrical Nanowire FET with Quantum Mechanical Effects. *Microelectronics Journal*, *53*, 65-72.

Sharma, Raj, & Khosla. (2017). Subthreshold Performance of In1-xGaxAs based Dual Metal with Gate Stack Cylindrical/Surrounding Gate Nanowire MOSFET for Low Power Analog Applications. *Journal of Nanoelectronics and Optoelectronics*, *12*, 171–176.

Sharma, S. K., Khosla, M., & Raj, B. (2016). Performance Enhancement of Junctionless Nanowire FET with Laterally Graded Channel Doping and High-κ Spacers. *IEEE Global Conference on Consumer Electronics*.

Sharma, S. K., Raj, B., & Khosla, M. (2015, October). Performance enhancement of junctionless nanowire FET with laterally graded channel doping and high-K spacers. In *2015 IEEE 4th Global Conference on Consumer Electronics (GCCE)* (pp. 556-559). IEEE. 10.1109/GCCE.2015.7398608

Sharma, S., Rastogi, R., & Kumar, S. (2015). A revolutionary technology to help the differently abled person. *Proceedings of 9th INDIACom; 2015 2nd International Conference on Computing for Sustainable Global Development*. https://ieeexplore.ieee.org/document/7100324

Sharma, S. (2017). An Extensive Evaluation of Futuristic Gate All Around Junctionless Nanowire MOSFET Using Numerical Simulation. *International Journal for Research in Applied Science and Engineering Technology*, *5*(8), 1974–1979. doi:10.22214/ijraset.2017.8279

Sharma, S. K., Jain, A., & Raj, B. (2016). *Design and Analysis of High Sensitivity Photosensor Using Cylindrical Surrounding Gate MOSFET. In Engineering Science and Technology, an International Journal*. Elsevier.

Sharma, S. K., Raj, B., & Khosla, M. (2016). A Gaussian approach for analytical subthreshold current model of cylindrical nanowire FET with quantum mechanical effects. *Microelectronics Journal*, *53*, 65–72. doi:10.1016/j.mejo.2016.04.002

Sharma, S. K., Raj, B., & Khosla, M. (2017). Subthreshold Performance of In 1–x Ga x As Based Dual Metal with Gate Stack Cylindrical/Surrounding Gate Nanowire MOSFET for Low Power Analog Applications. *Journal of Nanoelectronics and Optoelectronics*, *12*(2), 171–176. doi:10.1166/jno.2017.1961

Sharma, S. K., Singh, J., Raj, B., & Khosla, M. (2018). Analysis of Barrier Layer Thickness on Performance of In1–x Ga x As Based Gate Stack Cylindrical Gate Nanowire MOSFET. *Journal of Nanoelectronics and Optoelectronics*, *13*(10), 1473–1477. doi:10.1166/jno.2018.2374

Sharma, S., Jain, A., & Raj, B. (2018, May). Analysis of Triple Metal Surrounding Gate (TM-SG) III-V Nanowire MOSFET for Photosensing Application. *Opto-electronics Journal, Elsevier*, *26*(2), 141–148. doi:10.1016/j.opelre.2018.03.001

Sharma, S., Raj, B., & Khosla, M. (2016). Comparative Analysis of MOSFET CNTFET and NWFET for High Performance VLSI Circuit Design. A Review. *Journal of VLSI Design Tools & Technology*, *6*(2), 19–32.

Sharma, S., Raj, B., & Khosla, M. (2016). Comparative Analysis of MOSFET CNTFET and NWFET for High Performance VLSI Circuit Design. *Journal of VLSI Design Tools & Technology*, *6*(2), 19–32.

Sharma, V. K., Pattanaik, M., & Raj, B. (2014a). PVT variations aware low leakage INDEP approach for nanoscale CMOS Circuits. *Microelectronics Reliability, Elsevier*, *54*(1), 90–99. doi:10.1016/j.microrel.2013.09.018

Sharma, V. K., Pattanaik, M., & Raj, B. (2014b). ONOFIC approach: Low power high speed nanoscale VLSI circuits design. *International Journal of Electronics, Taylor & Francis*, *101*(1), 61–73. doi:10.1080/00207217.2013.769186

Shaygan, M., Davami, K., Shaygan, M., Davami, K., Kheirabi, N., Baek, C. K., Cuniberti, G., Meyyappan, M., & Lee, J. S. (2014). Single-crystalline CdTe nanowire field effect transistors as nanowire-based photodetector. *Physical Chemistry Chemical Physics, 16*(41), 22687–22693. doi:10.1039/C4CP03322A PMID:25230829

Shehada, N., Brönstrup, G., Funka, K., Christiansen, S., Leja, M., & Haick, H. (2015). Ultrasensitive silicon nanowire for real-world gas sensing: Noninvasive diagnosis of cancer from breath volatolome. *Nano Letters, 15*(2), 1288–1295. doi:10.1021/nl504482t

Shehata, N., Gaber, A.-R., Naguib, A., Selmy, A. E., Hassan, H., Shoeer, I., Ahmadien, O., & Nabeel, R. (2015). 3D Multi-gate Transistors: Concept, Operation, and Fabrication. *J. of Electrical Engineering, 3*(1), 1–14. doi:10.17265/2328-2223/2015.01.001

Shekhar, S., Rastogi, R., & Mittal, S. (2015). Linear algorithm for Imbricate Cryptography using Pseudo Random Number Generator. *Proceedings of 9th INDIACom, pp. 89-94; 2015 2nd International Conference on Computing for Sustainable Global Development.* https://ieeexplore.ieee.org/document/7100227

Shen, F., Wang, J., Xu, Z., Wu, Y., Chen, Q., Li, X., Jie, X., Li, L., Yao, M., Guo, X., & Zhu, T. (2012). Rapid flu diagnosis using silicon nanowire sensor. *Nano Letters, 12*(7), 3722–3730. doi:10.1021/nl301516z

Shen, Z., Chen, J., Li, B., Li, G., Zhang, Z., & Hou, X. (2020, January 30). Recent progress in SiC nanowires as electromagnetic microwaves absorbing materials. *Journal of Alloys and Compounds, 815*, 152388. doi:10.1016/j.jallcom.2019.152388

Simpson, M. L., Ericson, M. N., Jellison, G. E., Dress, W. B., Wintenberg, A. L., & Bobrek, M. (1999). Application specific spectral response with CMOS compatible photodiodes. *IEEE Transactions on Electron Devices, 46*(5), 905–913. doi:10.1109/16.760396

Singh, A., Khosla, M., & Raj, B. (2016). Circuit compatible model for electrostatic doped Schottky barrier CNTFET. *Journal of Electronic Materials, 45*(10), 5381–5390. doi:10.100711664-016-4743-7

Singh, J., & Raj, B. (2018). Modeling of mean barrier height levying various image forces of metal–insulator–metal structure to enhance the performance of conductive filament based memristor model. *IEEE Transactions on Nanotechnology, 17*(2), 268–275. doi:10.1109/TNANO.2018.2793953

Singh, J., & Raj, B. (2019). An accurate and generic window function for nonlinear memristor models. *Journal of Computational Electronics, 18*(2), 640–647. doi:10.100710825-019-01306-6

Singh, J., & Raj, B. (2019). Design and Investigation of 7T2M-NVSRAM With Enhanced Stability and Temperature Impact on Store/Restore Energy. *IEEE Transactions on Very Large Scale Integration (VLSI) Systems, 27*(6), 1322–1328.

Singh, J., & Raj, B. (2019). Enhanced Nonlinear Memristor Model Encapsulating Stochastic Dopant Drift. *Journal of Nanoelectronics and Optoelectronics*, *14*(7), 958–963. doi:10.1166/jno.2019.2570

Singh, J., & Raj, B. (2019). Tunnel current model of asymmetric MIM structure levying various image forces to analyze the characteristics of filamentary memristor. *Applied Physics. A, Materials Science & Processing*, *125*(3), 203. doi:10.100700339-019-2482-3

Singh, J., Raj, B., & Khosla, M. (2018). Design and performance analysis of nano-scale memristor-based nonvolatile static random access memory. *Sensor Letters*, *16*(10), 798–805. doi:10.1166l.2018.4021

Skocpol, W. J., Beasley, M. R., & Tinkham, M. (1974). Self-heating hotspots in superconducting thin-film microbridges. *Journal of Applied Physics*, *45*(9), 4054–4066. doi:10.1063/1.1663912

Sköld, N., Karlsson, L. S., Larsson, M. W., Pistol, M. E., Seifert, W., Trägårdh, J., & Samuelson, L. (2005). Growth and Optical Properties of Strained GaAs– Ga x In1-x P Core– Shell Nanowires. *Nano Letters*, *5*(10), 1943–1947. doi:10.1021/nl051304s PMID:16218714

Skotnicki, T., Hutchby, J. A., & King, T. (2005). *Toward the Introduction of New Materials and Structural Changes to Improve MOSFET Performance*. Academic Press.

Skotnicki, T., Hutchby, J. A., King, T. J., Wong, H. S., & Boeuf, F. (2005). The end of CMOS scaling: Toward the introduction of new materials and structural changes to improve MOSFET performance. *IEEE Circuits and Devices Magazine*, *21*(1), 16–26. doi:10.1109/MCD.2005.1388765

Soci, C., Zhang, A., Bao, X. Y., Kim, H., Lo, Y., & Wang, D. (2010). Nanowire photodetectors. *Journal of Nanoscience and Nanotechnology*, *10*(3), 1430–1449. doi:10.1166/jnn.2010.2157 PMID:20355534

Sohn, Y. S., Park, J., Yoon, G., Song, J., Jee, S. W., Lee, J. H., Na, S., Kwon, T., & Eom, K. (2009). Mechanical Properties of Silicon Nanowires. *Nanoscale Research Letters*, 1-6. Available at: www.scopus.com

Song, J. Y., Choi, W. Y., Park, J. H., Lee, J. D., & Park, B. G. (2006). Design optimization of gate-all-around (GAA) MOSFETs. *IEEE Transactions on Nanotechnology*, *5*(3), 186–191. doi:10.1109/TNANO.2006.869952

Song, Y., Zhang, C., Dowdy, R., Chabak, K., Mohseni, P. K., Choi, W., & Li, X. (2014). III-V junctionless gate-all-around nanowire MOSFETs for high linearity low power applications. *IEEE Electron Device Letters*, *35*(3), 324–326. doi:10.1109/LED.2013.2296556

SOPRA infobase. (n.d.). http://refractiveindex.info

SOPRA Infobase. (n.d.). http://refractiveindex.info

Spanier, J. E. (2006). One-dimensional semiconductor and oxide nanostructures. *Nanotubes and Nanofibers*, 199-232.

Spaziani, L., & Lu, L. (2018). Silicon, GaN and SiC: There's room for all: An application space overview of device considerations. In *2018 IEEE 30th International Symposium on Power Semiconductor Devices and ICs (ISPSD)* (pp. 8-11). IEEE. doi:10.1109/ISPSD.2018.8393590

Srivatava, S., Rastogi, R., Rungta, S., & Yadav, U. (2014).A Methodology to Find the Cycle in a Directed Graph Using Linked List. *BIJIT*. http://bvicam.ac.in/bjit/issues.asp?issue=12

Stadler, K., Masignani, V., Eickmann, M., Becker, S., Abrignani, S., Klenk, H. D., & Rappuoli, R. (2003). SARS—Beginning to understand a new virus. *Nature Reviews. Microbiology*, *1*(3), 209–218. doi:10.1038/nrmicro775

Steinhögl, W., Schindler, G., Steinlesberger, G., & Engelhardt, M. (2002). Size-dependent resistivity of metallic wires in the mesoscopic range. *Physical Review B: Condensed Matter and Materials Physics*, *66*(7), 075414. doi:10.1103/PhysRevB.66.075414

Suk, S. D. (2005). *Tech.* Dig. IEDM.

Suresh, S., Yuthika, S., & Adityavardhini, G. (2016). *Home Based Fire Monitoring and Warning System*. Academic Press.

Sze, S. M., & Lee, M. K. (2008). *Semiconductor devices: physics and technology*. Academic Press.

Sze, S. M. (1981). *Physics of Semiconductor Devices* (2nd ed.). John Wiley and Sons.

Sze, S. M. (2008). *Semiconductor devices: physics and technology*. John Wiley & Sons.

Tai, C. H., Lin, J. T., Eng, Y. C., & Lin, P. H. (2010). A novel high-performance junctionless vertical MOSFET produced on bulk-si wafer. *ICSICT-2010 - 2010 10th IEEE International Conference on Solid-State and Integrated Circuit Technology Proceedings*, *3*, 108–110. doi:10.1109/ICSICT.2010.5667835

Takato, H., Sunouchi, K., Okabe, N., Nitayama, A., Hieda, K., Horiguchi, F., & Masuoka, F. (1991). Impact of surrounding gate transistor (SGT) for ultra-high-density LSI's. *IEEE Transactions on Electron Devices*, *38*(3), 573–578. doi:10.1109/16.75168

Tanaka, J., Toyabe, T., Ihara, S., Kimura, S. I., Noda, H., & Itoh, K. (1993). Simulation of sub-0.1-mu m MOSFETs with completely suppressed short-channel effect. *IEEE Electron Device Letters*, *14*(8), 396–399. doi:10.1109/55.225591

Tang, C., Bando, Y., Liu, Z., & Golberg, D. (2003). Synthesis and structure of InP nanowires and nanotubes. *Chemical Physics Letters*, *376*(5-6), 676–682. doi:10.1016/S0009-2614(03)01077-7

Tan, L. K., Chong, A. M., Tang, X. E., & Gao, H. (2007). Combining atomic layer deposition with a template-assisted approach to fabricate size-reduced nanowire arrays on substrates and their electrochemical characterization. *The Journal of Physical Chemistry C*, *111*(13), 4964–4968. doi:10.1021/jp066841v

Tao, Y., & Degen, C. L. (2015). Single-crystal diamond nanowire tips for ultrasensitive force microscopy. *Nano Letters*, *15*(12), 7893–7897. doi:10.1021/acs.nanolett.5b02885

Taylor, R. F., & Schultz, J. S. (1996). Handbook of Chemical and Biological Sensors. Institute of Physics Publishing.

Tekleab, D., Tran, H. H., Sleight, J. W., & Chidambarrao, D. (2014). U.*S. Patent No. 8,871,576.* Washington, DC: U.S. Patent and Trademark Office.

Tekleab, D. (2014). Device performance of silicon nanotube field effect transistor. *IEEE Electron Device Letters*, *35*(5), 506–508. doi:10.1109/LED.2014.2310175

Thelander, C., Mårtensson, T., Björk, M. T., Ohlsson, B. J., Larsson, M. W., Wallenberg, L. R., & Samuelson, L. (2003). Single-electron transistors in heterostructure nanowires. *Applied Physics Letters*, *83*(10), 2052–2054. doi:10.1063/1.1606889

Thelander, C., Nilsson, H. A., Jensen, L. E., & Samuelson, L. (2005). Nanowire single-electron memory. *Nano Letters*, *5*(4), 635–638. doi:10.1021/nl050006s PMID:15826100

Tian, B., Kempa, T. J., & Lieber, C. M. (2009). Single nanowire photovoltaics. *Chemical Society Reviews*, *38*(1), 16–24. doi:10.1039/B718703N PMID:19088961

Tian, F., Zhu, J., & Wei, D. (2007). Phase transition and magnetism of Ni nanowire arrays. *The Journal of Physical Chemistry C*, *111*(19), 6994–6997. doi:10.1021/jp070344c

Tian, M., Wang, J., Kumar, N., Han, T., Kobayashi, Y., Liu, Y., Mallouk, T. E., & Chan, M. H. (2006). Observation of superconductivity in granular Bi nanowires fabricated by electrodeposition. *Nano Letters*, *6*(12), 2773–2780. doi:10.1021/nl0618041

Tiwari, P. K., Samoju, V. R., Sunkara, T., Dubey, S., & Jit, S. (2016). Analytical modeling of threshold voltage for symmetrical silicon nano-tube field-effect-transistors (Si-NT FETs). *Journal of Computational Electronics*, *15*(2), 516–524. doi:10.100710825-016-0819-0

Tosi, M. P. (1964). Cohesion of ionic solids in the Born model. In F. S. Turnbull (Ed.), *Solid state physics* (Vol. 16, pp. 1–120). Academic Press., doi:10.1016/S0081-1947(08)60515-9

Trahey, L., Becker, C. R., & Stacy, A. M. (2007). Electrodeposited bismuth telluride nanowire arrays with uniform growth fronts. *Nano Letters*, *7*(8), 2535–2539. doi:10.1021/nl070711w

Trentler, T. J., Hickman, K. M., Goel, S. C., Viano, A. M., Gibbons, P. C., & Buhro, W. E. (1995). Solution-liquid-solid growth of crystalline III-V semiconductors: An analogy to vapor-liquid-solid growth. *Science*, *270*(5243), 1791–1794. doi:10.1126cience.270.5243.1791

Tripathi, S., Agarwal, R., & Singh, D. (2019a, July). Size dependent elastic and thermophysical properties of zinc oxide nanowires. *Johnson Matthey Technology Review*, *63*(3), 166–176. doi:10.1595/205651319X15514400132039

Tripathi, S., Agarwal, R., & Singh, D. (2020b, March 7). Size-Dependent Ultrasonic and Thermophysical Properties of Indium Phosphide Nanowires. *Zeitschrift fur Naturforschung. Section A. Physical Sciences*, *75*(4), 373–380. doi:10.1515/zna-2019-0351

Tse, G., Pal, J., Monteverde, U., Garg, R., Haxha, V., Migliorato, M., & Tomić, S. (2013, August). Non-linear piezoelectricity in zinc blende GaAs and InAs semiconductors. *Journal of Applied Physics, 114*(7), 073515. doi:10.1063/1.4818798

Tsormpatzoglou, A., Dimitriadis, C. A., Clerc, R., Pananakakis, G., & Ghibaudo, G. (2008). Threshold voltage model for short-channel undoped symmetrical double-gate MOSFETs. *IEEE Transactions on Electron Devices, 55*(9), 2512–2516. doi:10.1109/TED.2008.927394

Ulman, A. (2013). *An Introduction to Ultrathin Organic Films: From Langmuir—Blodgett to Self—Assembly*. Academic press.

Vaddiraju, S., Chandrasekaran, H., & Sunkara, M. K. (2003). Vapor phase synthesis of tungsten nanowires. *Journal of the American Chemical Society, 125*(36), 10792–10793. doi:10.1021/ja035868e

Vaddiraju, S., Sunkara, M. K., Chin, A. H., & Ning, C. Z. (2007). Dholakia GR and Meyyappan M. *The Journal of Physical Chemistry C, 2007*, 111.

Van Tilburg, J. W. W., Algra, R. E., Immink, W. G. G., Verheijen, M., Bakkers, E. P. A. M., & Kouwenhoven, L. P. (2010). Surface passivated InAs/InP core/shell nanowires. *Semiconductor Science and Technology, 25*(2), 024011. doi:10.1088/0268-1242/25/2/024011

Vega, R. A., & Liu, T. J. K. (2009). Three-dimensional FinFET source/drain and contact design optimization study. *IEEE Transactions on Electron Devices, 56*(7), 1483–1492. doi:10.1109/TED.2009.2021439

Vilouras, A., Heidari, H., Gupta, S., & Dahiya, R. (2017). Modeling of CMOS devices and circuits on flexible ultrathin chips. *IEEE Transactions on Electron Devices, 64*(5), 2038–2046. doi:10.1109/TED.2017.2668899

Vishvakarma, Agarwal, Raj, Saxena, & Dasgupta. (2007). Two Dimensional Analytical Potential Modeling of Nanoscale Symmetric Double Gate (SDG) MOSFET with Ultra Thin Body (UTB). *Journal of Computational and Theoretical Nanoscience, 4*(5), 1144–1148.

Vishvakarma, S. K., Agrawal, V., Raj, B., Dasgupta, S., & Saxena, A. K. (2007). Two dimensional analytical potential modeling of nanoscale symmetric double gate (SDG) MOSFET with ultra thin body (UTB). *Journal of Computational and Theoretical Nanoscience, 4*(6), 1144–1148. doi:10.1166/jctn.2007.2390

Voigt, W. (1928). *Lehrbuch der kristallphysik* (Vol. 962). Teubner Leipzig.

Wallentin, J., Persson, J. M., Wagner, J. B., Samuelson, L., Deppert, K., & Borgström, M. T. (2010). High-performance single nanowire tunnel diodes. *Nano Letters, 10*(3), 974–979. doi:10.1021/nl903941b

Wan, Dattoli, & Lu. (2007). Transparent metallic Sb-doped SnO2 nanowires. *Appl. Phys. Lett., 90*(22).

Wang, C. A. D. C. M. M. E., Garcia, A., Ingram, D. C., Lake, M., & Kordesch, M. E. (1991). Cold field emission from CVD diamond films observed in emission electron microscopy. *Electronics Letters*, *27*(16), 1459–1461. doi:10.1049/el:19910914

Wang, D. W., Wang, Q., Javey, A., Tu, R., Dai, H. J., Kim, H., McIntyre, P. C., Krishnamohan, T., & Saraswat, K. C. (2003, September). Germanium nanowire fieldeffect transistors with SiO2 and high-κ HfO2 gate dielectrics. *Applied Physics Letters*, *83*(12), 2432–2434. doi:10.1063/1.1611644

Wang, F., Gao, C., Kuklane, K., & Holmér, I. (2010). A review of technology of personal heating garments. *International Journal of Occupational Safety and Ergonomics*, *16*(3), 387–404. doi:10.1080/10803548.2010.11076854

Wang, R., Wang, S., & Wu, X. (2010, August). On third-order elastic constants for ductile rare-earth intermetallic compounds: A first-principles study. *Intermetallics*, *18*(8), 1653–1658. doi:10.1016/j.intermet.2010.04.021

Wang, S., Zeng, Q., Yang, L., Zhang, Z., Wang, Z., Pei, T., Ding, L., Liang, X., Gao, M., Li, Y., & Peng, L. M. (2011). High-performance carbon nanotube light-emitting diodes with asymmetric contacts. *Nano Letters*, *11*(1), 23–29. doi:10.1021/nl101513z

Wang, Y., Zhang, L., Liang, C., Wang, G., & Peng, X. (2002). Catalytic growth and photoluminescence properties of semiconductor single-crystal ZnS nanowires. *Chemical Physics Letters*, *357*(3-4), 314–318. doi:10.1016/S0009-2614(02)00530-4

Wan, Q., Dattoli, E. N., Fung, W. Y., Guo, W., Chen, Y., Pan, X., & Lu, W. (2006, December). High-performance transparent conducting oxide nanowires. *Nano Letters*, *6*(12), 2909–2915. doi:10.1021/nl062213d PMID:17163729

Wan, Q., Li, Q. H., Chen, Y. J., Wang, T. H., He, X. L., Li, J. P., & Lin, C. L. (2004, May). Fabrication and ethanol sensing characteristics of ZnO nanowire gas sensors. *Applied Physics Letters*, *84*(18), 3654–3656. doi:10.1063/1.1738932

Wan, Y., Sha, J., Chen, B., Fang, Y., Wang, Z., & Wang, Y. (2009). Nanodevices based on silicon nanowires. *Recent Patents on Nanotechnology*, *3*(1), 1–9. doi:10.2174/187221009787003348 PMID:19149750

Wei, W., Bao, X. Y., Soci, C., Ding, Y., Wang, Z. L., & Wang, D. (2009). Direct heteroepitaxy of vertical InAs nanowires on Si substrates for broad band photovoltaics and photodetection. *Nano Letters*, *9*(8), 2926–2934. doi:10.1021/nl901270n

Werner, P., Zakharov, N. D., Gerth, G., Schubert, L., & Gösele, U. (2006). On the formation of Si nanowires by molecular beam epitaxy: Dedicated to Professor Dr. Knut Urban on the occasion of his 65th birthday. *Zeitschrift fur Metallkunde*, *97*(7), 1008–1015.

Whang, D., Jin, S., Wu, Y., & Lieber, C. M. (2003). Large-scale hierarchical organization of nanowire arrays for integrated nanosystems. *Nano Letters*, *3*(9), 1255–1259. doi:10.1021/nl0345062

Wiederrecht, G. (2010). *Handbook of nanoscale optics and electronics*. Academic Press.

Wikipedia. (2009). *Nanowire*. Available at: https://en.wikipedia.org/wiki/Nanowire

Wolf, E. L. (2004). *Nanophysics and nanotechnology*. Weinaheim: WIEYaVCH.

Wong, H. S. P., Chan, K. K., & Taur, Y. (1997). Self-aligned (top and bottom) double-gate MOSFET with a 25 nm thick silicon channel. *Technical Digest - International Electron Devices Meeting, IEDM*, 427–430. 10.1109/iedm.1997.650416

Wu, X., Kulkarni, J. S., Collins, G., Petkov, N., Almecija, D., Boland, J. J., Erts, D., & Holmes, J. D. (2008). Synthesis and Electrical and Mechanical Properties of Silicon and Germanium Nanowires. *Progress in Chemistry, 20*(19), 5954-5967. Available at: http://apps.isiknowledge.com

Wu, K. H., Chien, H. C., Chan, C. C., Chen, T. S., & Kao, C. H. (2005). SONOS device with tapered bandgap nitride layer. *IEEE Transactions on Electron Devices*, *52*(5), 987–992. doi:10.1109/TED.2005.846347

Wu, M., Jin, X., Kwon, H., Chuai, R., Liu, X., & Lee, J. (2013). The optimal design of junctionless transistors with double-gate structure for reducing the effect of band-to-band tunneling. *JSTS: Journal of Semiconductor Technology and Science*, *13*(3), 245–251. doi:10.5573/JSTS.2013.13.3.245

Würthner, F. (2004). Perylene bisimide dyes as versatile building blocks for functional supramolecular architectures. *Chemical Communications*, (14), 1564–1579. doi:10.1039/B401630K PMID:15263926

Wu, Y., Fan, R., & Yang, P. (2002). Block-by-block growth of single-crystalline Si/SiGe superlattice nanowires. *Nano Letters*, *2*(2), 83–86. doi:10.1021/nl0156888

Wu, Y., Xiang, J., Yang, C., Lu, W., & Lieber, C. M. (2004). Single-crystal metallic nanowires and metal/semiconductor nanowire heterostructures. *Nature*, *430*(6995), 61–65. doi:10.1038/nature02674 PMID:15229596

Xiang, B., Wang, Q. X., Wang, Z., Zhang, X. Z., Liu, L. Q., Xu, J., & Yu, D. P. (2005). Synthesis and field emission properties of TiSi 2 nanowires. *Applied Physics Letters*, *86*(24), 243103. doi:10.1063/1.1948515

Xiang, J., Vidan, A., Tinkham, M., Westervelt, R. M., & Lieber, C. M. (2006, December). Ge/Si nanowire mesoscopic Josephson junctions. *Nature Nanotechnology*, *1*(3), 208–213. doi:10.1038/nnano.2006.140 PMID:18654188

Xiao, S., Tang, J., Beetz, T., Guo, X., Tremblay, N., Siegrist, T., Zhu, Y., Steigerwald, M., & Nuckolls, C. (2006). Transferring self-assembled, nanoscale cables into electrical devices. *Journal of the American Chemical Society*, *128*(33), 10700–10701. doi:10.1021/ja0642360 PMID:16910663

Xiao, Y., Zhang, B., Lou, H., Zhang, L., & Lin, X. (2016). A Compact Model of Subthreshold Current with Source/Drain Depletion Effect for the Short-Channel Junctionless Cylindrical Surrounding-Gate MOSFETs. *IEEE Transactions on Electron Devices*, *63*(5), 2176–2181. doi:10.1109/TED.2016.2535247

Xie, X., Kwok, S. Y., Lu, Z., Liu, Y., Cao, Y., Luo, L., Zapien, J. A., Bello, I., Lee, C.-S., Lee, S.-T., & Zhang, W. (2012). Visible–NIR photodetectors based on CdTe nanoribbons. *Nanoscale*, *4*(9), 2914–2919. doi:10.1039/c2nr30277b PMID:22499254

Yadav, A., Yadav, R., Pandey, D., & Singh, D. (2008, June 30). Ultrasonic study of fission products precipitated in the nuclear fuel. *Materials Letters*, *62*(17-18), 3258–3261. doi:10.1016/j.matlet.2008.02.036

Yadav, C. P., Pandey, D. K., & Singh, D. (2019, February 8). Ultrasonic study of Laves phase compounds ScOs 2 and YOs 2. *Indian Journal of Physics*, *93*(9), 1147–1153. doi:10.100712648-019-01389-8

Yadav, R. R., & Singh, D. (2001). Ultrasonic attenuation in lanthanum monochalcogenides. *Journal of the Physical Society of Japan*, *70*(6), 1825–1832. doi:10.1143/JPSJ.70.1825

Yamada, T., Hayamizu, Y., Yamamoto, Y., Yomogida, Y., Izadi-Najafabadi, A., Futaba, D. N., & Hata, K. (2011). A stretchable carbon nanotube strain sensor for human-motion detection. *Nature Nanotechnology*, *6*(5), 296–301. doi:10.1038/nnano.2011.36

Yang, N., Uetsuka, H., Williams, O. A., Osawa, E., Tokuda, N., & Nebel, C. E. (2009). Vertically aligned diamond nanowires: fabrication, characterization, and application for DNA sensing. *Physica Status Solidi (a), 206*(9), 2048-2056.

Yang, C., Zhong, Z., & Lieber, C. M. (2005). Encoding electronic properties by synthesis of axial modulation-doped silicon nanowires. *Science*, *310*(5752), 1304–1307. doi:10.1126cience.1118798 PMID:16311329

Yang, J. K., Kerman, A. J., Dauler, E. A., Anant, V., Rosfjord, K. M., & Berggren, K. K. (2007). Modeling the electrical and thermal response of superconducting nanowire single-photon detectors. *IEEE Transactions on Applied Superconductivity*, *17*(2), 581–585. doi:10.1109/TASC.2007.898660

Yang, W., Auciello, O., Butler, J. E., Cai, W., Carlisle, J. A., Gerbi, J. E., ... Smith, L. M. (2002). DNA-modified nanocrystalline diamond thin-films as stable, biologically active substrates. *Nature Materials*, *1*(4), 253–257. doi:10.1038/nmat779

Yang, Z., Bai, S., Yue, H., Li, X., Liu, D., Lin, S., Li, F., & He, D. (2014). Germanium anode with lithiated-copper-oxide nanorods as an electronic-conductor for high-performance lithium-ion batteries. *Materials Letters*, *136*, 107–110. doi:10.1016/j.matlet.2014.08.034

Yao, D., Zhang, G., & Li, B. (2008). A universal expression of band gap for silicon nanowires of different cross-section geometries. *Nano Letters*, *8*(12), 4557–4561. doi:10.1021/nl802807t PMID:19368006

Young, K. K. (1989). Short-channel effect in fully depleted SOI MOSFETs. *IEEE Transactions on Electron Devices*, *36*(2), 399–402. doi:10.1109/16.19942

You, P., Liu, Z., Tai, Q., Liu, S., & Yan, F. (2015). Efficient semitransparent perovskite solar cells with graphene electrodes. *Advanced Materials*, *27*(24), 3632–3638. doi:10.1002/adma.201501145

Yu, B., & Meyyappan, M. (2006). Nanotechnology-Role in emerging nanoelectronics. *Solid-State Electronics, 50*(4), 536-544. Available at: http://apps.isiknowledge.com

Yuan, X. Y., Wu, G. S., Xie, T., Lin, Y., & Zhang, L. D. (2003). Self-assembly synthesis and magnetic studies of Co–P alloy nanowire arrays. *Nanotechnology, 15*(1), 59–61. doi:10.1088/0957-4484/15/1/011

Yuan, Y., Yu, B., Song, J., & Taur, Y. (2009). An analytic model for threshold voltage shift due to quantum confinement in surrounding gate MOSFETs with anisotropic effective mass. *Solid-State Electronics, 53*(2), 140–144. doi:10.1016/j.sse.2008.10.010

Yu, D. P., Bai, Z. G., Wang, J. J., Zou, Y. H., Qian, W., Fu, J. S., ... Xu, J. (1999). Direct evidence of quantum confinement from the size dependence of the photoluminescence of silicon quantum wires. *Physical Review B: Condensed Matter and Materials Physics, 59*(4), R2498–R2501. doi:10.1103/PhysRevB.59.R2498

Yu, G., Li, X., Lieber, C. M., & Cao, A. (2008). Nanomaterial-incorporated blown bubble films for large-area, aligned nanostructures. *Journal of Materials Chemistry, 18*(7), 728–734. doi:10.1039/b713697h

Zach, M. P. (2000). KH Ng and RM Penner. *Science, 290*, 2120. doi:10.1126cience.290.5499.2120

Zagorac, D., Zagorac, J., Djukic, M., Jordanov, D., & Matović, B. (2019, October). Theoretical study of AlN mechanical behaviour under high pressure regime. *Theoretical and Applied Fracture Mechanics, 103*, 102289. doi:10.1016/j.tafmec.2019.102289

Zeng, D.-J., & Zheng, Q.-S. (2007, August 15). Resonant frequency-based method for measuring the Young's moduli of nanowires. *Physical Review B: Condensed Matter and Materials Physics, 76*(7), 075417. doi:10.1103/PhysRevB.76.075417

Zhang, G. J., Huang, M. J., Liu, E. T., & Desai, K. V. (2011). Self-assembled monolayer-assisted silicon nanowire biosensor for detection of protein–DNA interactions in nuclear extracts from breast cancer cell. *Biosensors & Bioelectronics, 26*(7), 3233–3239. doi:10.1016/j.bios.2010.12.032

Zhang, G. J., Zhang, L., Huang, M. J., Luo, Z. H. H., Tay, G. K. I., Lim, E. J. A., Kang, T. G., & Chen, Y. (2010). Silicon nanowire biosensor for highly sensitive and rapid detection of Dengue virus. *Sensors and Actuators. B, Chemical, 146*(1), 138–144. doi:10.1016/j.snb.2010.02.021

Zhang, H., Zuo, M., Tan, S., Li, G., & Zhang, S. (2006). Carbothermal reduction/sulfidation synthesis and structural characterization of PbS nanobelts and nanowires. *Nanotechnology, 17*(12), 2931–2936. doi:10.1088/0957-4484/17/12/018

Zhang, L., & Wang, G. (2019). Design and Implementation of Automatic Fire Alarm System Based on Wireless Sensor Networks. *Proceedings of the International Symposium on Information Processing (ISIP'09)*, 410-413.

Zhang, Z., Sun, X., Dresselhaus, M. S., Ying, J. Y., & Heremans, J. (2000). Electronic transport properties of single-crystal bismuth nanowire arrays. *Physical Review B: Condensed Matter and Materials Physics, 61*(7), 4850–4861. doi:10.1103/PhysRevB.61.4850

Zheng, G. F., Lu, W., Jin, S., & Lieber, C. M. (2004, September). Synthesis and fabrication of high-performance n-type silicon nanowire transistors. *Advanced Materials*, *16*(21), 1890–1893. doi:10.1002/adma.200400472

Zheng, J., Huang, J., & Xu, S. (2020, April). Multiscale micro-/nanostructures on single crystalline SiC fabricated by hybridly polarized femtosecond laser. *Optics and Lasers in Engineering*, *127*, 105940. doi:10.1016/j.optlaseng.2019.105940

Zheng, Q., Cui, J., Lu, W., Guo, H., Liu, J., & Yu, X. (2018). The Increased Single-Event Upset Sensitivity of 65-nm DICE SRAM Induced by Total Ionizing Dose. *IEEE Transactions on Nuclear Science*. Advance online publication. doi:10.1109/TNS.2018.2816583

Zheng, S. R. (2011). The Internet of Things. *IEEE Communications Magazine*, *11*, 30–31. doi:10.1109/MCOM.2011.6069706

Zhou, R., Zuo, R., Wang, L., Zhang, B., & Pan, B. (2011, April 20). Size-and surface-dependent electronic structures of crystalline SiC nanotubes. *Journal of Applied Physics*, *109*(8), 084318. doi:10.1063/1.3567114

Zhu, H., Bonevich, J. E., Li, H., Richter, C. A., Yuan, H., Kirillov, O., & Li, Q. (2014). Discrete charge states in nanowire flash memory with multiple Ta2O5 charge-trapping stacks. *Applied Physics Letters*, *104*(23), 233504. doi:10.1063/1.4883717

Zhu, J., Peng, H., Chan, C. K., Jarausch, K., Zhang, X. F., & Cui, Y. (2007). Hyperbranched lead selenide nanowire networks. *Nano Letters*, *7*(4), 1095–1099. doi:10.1021/nl0700393

Zhuo, S., Xun, M., Li, M., Kong, X., Shao, R., Zheng, T., & Li, Q. (2020). Rapid and Label-free Optical Assay of S-layer Protein with High Sensitivity Using TiO2-coated Porous Silicon-based Microfluidic Biosensor. *Sensors and Actuators. B, Chemical*, *321*, 128524. doi:10.1016/j.snb.2020.128524

Zhu, Y., Dou, X., Huang, X., Li, L., & Li, G. (2006). Thermal properties of Bi nanowire arrays with different orientations and diameters. *The Journal of Physical Chemistry B*, *110*(51), 26189–26193. doi:10.1021/jp0652493

Zou, Z. Q., Wang, H., Wang, D., Wang, Q. K., Mao, J. J., & Kong, X. Y. (2007). Epitaxial growth of manganese silicide nanowires on Si (111)-7× 7 surfaces. *Applied Physics Letters*, *90*(13), 133111. doi:10.1063/1.2717580

About the Contributors

Balwinder Raj (MIEEE'2006) was born in Pathankot, a city of Punjab, India, on 25 Nov. 1980. He did B. Tech, Electronics Engineering (PTU Jalandhar), M. Tech-Microelectronics (PU Chandigarh) and Ph.D-VLSI Design (IIT Roorkee), India in 2004, 2006 and 2010 respectively. For further research work, European Commission awarded him "Erasmus Mundus" Mobility of life research fellowship for postdoc research work at University of Rome, Tor Vergata, Italy in 2010-2011. Dr. Raj also received India4EU (India for European Commission) Fellowship and worked as visiting researcher at KTH University, Stockholm, Sweden, Oct-Nov 2013. Currently, he is working as Associate Professor at NITTTR Chandigarh since 2019. He worked as Assistant Professor at National Institute of Technology (NIT Jalandhar) (An Institute of national Importance, MHRD), Punjab, India from May 2012 to Dec 2019. Dr. Raj worked as Assistant Professor at ABV-IIITM Gwalior (An autonomous institute established by Ministry of HRD, Govt. of India) July 2011 to Apl 2012. Dr. Raj has authored/co-authored three books, five book chapters and more than 60 research papers in peer reviewed international/national journals and conferences. His areas of interest in research are Nanoscale Semiconductor Device Modeling.

* * *

Rekha Agarwal received her BEng (Electronics and communication) from Madhav Institute of Technology and Science, Jiwaji University, Gwalior, Madhya Pradesh, India; MEng (Communications) from Malviya National Institute of Technology,Jaipur Rajasthan, India and PhD from Guru Gobind Singh Indraprastha University, New Delhi, India. At present she is working as Professor, Department of Electronics and Communication Engineering at Amity School of Engineering and Technology, New Delhi, India. Her research interest lies in the area of wireless communications,coding techniques, and antenna arrays.

Puru Jain is currently Btech Third Year Student. He is Young and Dynamic. He is Involved in Hardware implementation on Smart Home applications.

Rishabh Jain is currently Btech Third Year Student. He is Young and Dynamic. Involved in RFID applications.

Rohit Rastogi received his B.E. degree in Computer Science and Engineering from C.C.S.Univ. Meerut in 2003, the M.E. degree in Computer Science from NITTTR-Chandigarh (National Institute of Technical Teachers Training and Research-affiliated to MHRD, Govt. of India), Punjab Univ. Chandigarh in 2010. Currently he is pursuing his Ph.D. In computer science from Dayalbagh Educational Institute, Agra under renowned professor of Electrical Engineering Dr. D.K. Chaturvedi in area of spiritual consciousness. Dr. Santosh Satya of IIT-Delhi and dr. Navneet Arora of IIT-Roorkee have happily consented him to co supervise. He is also working presently with Dr. Piyush Trivedi of DSVV Hardwar, India in center of Scientific spirituality. He is a Associate Professor of CSE Dept. in ABES Engineering. College, Ghaziabad (U.P.-India), affiliated to Dr. A.P. J. Abdul Kalam Technical Univ. Lucknow (earlier Uttar Pradesh Tech. University).Also, He is preparing some interesting algorithms on Swarm Intelligence approaches like PSO, ACO and BCO etc.Rohit Rastogi is involved actively with Vichaar Krnati Abhiyaan and strongly believe that transformation starts within self.

Devraj Singh is Professor & Head in the Department of Physics and Director of Professor Rajendra Singh (Rajju Bhaiya) Institute of Physical Sciences for Study & Research, Veer Bahadur Singh Purvanchal University, Uttar Pradesh, India. He obtained his D.Phil. in Science at the University of Allahabad, India, in 2002 on 'Study of Ultrasonic Attenuation in Condensed Materials'. His research and publication interests include the mechanical, thermophysical ultrasonic characterisation of condensed and advanced materials. He is a fellow of the Ultrasonics Society of India and the life member of a number of scientific societies: India Association of Physics Teachers, Acoustical Society of India, Materials Research Society of India, Metrology Society of India, Indian Physics Association, Ion Beam Society of India. He is the Associate Editor of MAPAN–Journal of Metrology Society of India (Springer) and member of the Editorial Boards of the Journal of Pure & Applied Ultrasonics, Journal of the Acoustical Society of India, ISST Journal of Applied Physics. With 18 years of teaching and research experience, he has to his credit 91 research articles in national and international journals of ultrasonics and materials science and 24 books including "Fundamentals of Optics", "Applied Optics", "Modern Physics for Scientist and Engineers" (PHI Learning, Delhi), "Thermal Physics", "Electronic Devices" (Alpha Science International Ltd, Oxford), "Engineering Physics" (Dhanpat Rai & Co, Delhi).

Jeetendra Singh received the B.Tech. degree in electronics and communication engineering from Uttar Pradesh Technical University, Lucknow, India, in 2009, and the M.Tech. degree in microwave electronics from the University of Delhi, Delhi, India, in 2012. He is doctorate in the VLSI domain from the National Institute of Technology Jalandhar, India. He is currently working as assistant professor in the Dept. of ECE, NIT Sikkim. His research interest includes modeling and simulation of semiconductor devices, especially memristor modeling for memory application, low power vlsi design, and nanowire FETs.

Sudhanshu Tripathi obtained his Master of Technology from Panjab University, Chandigarh, India. He has Joined Amity School of Engineering and Technology as Lecturer in 2006. Presently he is working at Amity School of Engineering and Technology as Assistant Professor. Currently he is pursuing his PhD at University School of Information, Communication and Technology, Guru Gobind Singh Indraprastha University, New Delhi, India. His research interest are nano-material characterization using ultrasonic techniques, nano electronics and nano devices. With 15 years of teaching and research experience, he has to his credit many research articles in various national and international journals and also he has authored two books namely "Circuit fundamentals and Basic Electronics"(I.K. International Publishing House Pvt. Ltd.) and "Electronic Devices" (Dhanpat Rai & Co, Delhi). He is a life member of Ultrasonic Society of India (USI)

Rashmi Vashisth has done M.Sc. (Electronics) and M.Tech. (Instrumentation and Control) in 2002 and 2006 respectively and PhD (Engg) in 2019. She has more than 14 years of teaching experience in academic institutions. She has guided various M.Tech and B.Tech projects and published research papers in various Journals and Conferences. Her current research interests include MEMS, Ultrasonic Transducers, Nano wires, Intelligent systems, Embedded Systems and Biomedical Instrumentation.

Index

Ensure Quality Research is Introduced to the Academic Community

Become an IGI Global Reviewer for Authored Book Projects

The overall success of an authored book project is dependent on quality and timely reviews.

In this competitive age of scholarly publishing, constructive and timely feedback significantly expedites the turnaround time of manuscripts from submission to acceptance, allowing the publication and discovery of forward-thinking research at a much more expeditious rate. Several IGI Global authored book projects are currently seeking highly-qualified experts in the field to fill vacancies on their respective editorial review boards:

Applications and Inquiries may be sent to:
development@igi-global.com

Applicants must have a doctorate (or an equivalent degree) as well as publishing and reviewing experience. Reviewers are asked to complete the open-ended evaluation questions with as much detail as possible in a timely, collegial, and constructive manner. All reviewers' tenures run for one-year terms on the editorial review boards and are expected to complete at least three reviews per term. Upon successful completion of this term, reviewers can be considered for an additional term.

If you have a colleague that may be interested in this opportunity, we encourage you to share this information with them.

www.ingramcontent.com/pod-product-compliance
Lightning Source LLC
LaVergne TN
LVHW061537290926
849162LV00004B/827